Basic Thermodynamics

Basic Thermodynamics

GERALD CARRINGTON

University of Otago

OXFORD NEW YORK TOKYO

OXFORD UNIVERSITY PRESS

1994

Oxford University Press, Walton Street, Oxford OX2 6DP

Oxford New York Toronto
Delhi Bombay Calcutta Madras Karachi
Kuala Lumpur Singapore Hong Kong Tokyo
Nairobi Dar es Salaam Cape Town
Melbourne Auckland Madrid
and associated companies in
Berlin Ibadan

Oxford is a trade mark of Oxford University Press

Published in the United States
by Oxford University Press Inc., New York

A catalogue record for this book is available from the British Library

Library of Congress Cataloging in Publication Data
Carrington, Gerald.
Basic thermodynamics / Gerald Carrington.
Includes bibliographical references and index.
1. Thermodynamics. I. Title.
QC311.C38   1994   536'.7—dc20   93–34952

ISBN 0 19 851748 3 (Hbk)
ISBN 0 19 851747 5 (Pbk)

Typeset by Colset Pte. Ltd., Singapore
Printed by St Edmundsbury Press,
Bury St Edmunds, Suffolk

Preface

This book is intended as a first text in classical thermodynamics. The material is suitable for undergraduates taking courses in pure or applied physics, or taking introductory courses in engineering. I hope the book will also be useful to others, including perhaps those who did not take thermodynamics as an undergraduate, but now find they must acquire a working knowledge of the subject.

It is well known that thermodynamics presents students with particular difficulties. They find the concepts evasive and the methods obscure. I believe these problems arise because it is traditional to emphasize how general thermodynamics is, at the outset. Unfortunately when ideas are introduced in an unspecific context, they fail to make contact with the experience of the student. Such ideas do not become part of the physical intuition of the student, as they should. For the lecturer with enough experience, a generalized introduction may make sense. For students, however, the subject is often so frustratingly abstruse that their interest and level of achievement are adversely affected.

This situation contrasts with introductory quantum mechanics where students are usually excited by the ideas, but it is not usual to introduce quantum mechanics in a wholly general context. Instead considerable time is devoted to the formation of basic ideas using simple illustrations. In this book I have attempted to introduce classical thermodynamics in this way. As Denbigh (1968) noted, 'thermodynamics is a subject which needs to be studied not once but several times over at advancing levels'. Thus, at the risk of being repetitive, I have developed thermodynamics at three levels, succeeding treatments providing increasing generality. It is rather like painting. While successive coats of paint may cover the same object, each has a specific purpose. In this book the same basic material is covered several times, but each layer has a different function and focus.

The first step in this process is devoted mainly to an ideal gas as an illustrative system, in order to introduce primitive concepts, such as *system, state*, and *process* in a non-abstract way. The need for linguistic precision in using the terms *heat* and *work* is illustrated, and the language of the subject is developed along with basic ideas on the energy and entropy functions.

In Part II students are ready to accept a more general expression of the principles of thermodynamics. Well-prepared undergraduates might begin at this point. Here, I have used the classical laws, since they are such an established part of the literature of physics. In writing the second law I have emphasized the notion of irreversibility, because this highlights the operational status of the law, and shows why it has so many equivalent forms. The second section also includes a detailed analysis of work processes due to electric and magnetic fields. These issues are non-trivial, and detailed treatments are difficult to find elsewhere.

The third section introduces the Gibbs maximum entropy approach, which in turn provides a systematic way to develop the thermodynamic potentials and their applications. I have also discussed the link between the classical laws and Gibbsian thermodynamics at some length. Finally, in Part IV, the context is enlarged by introducing a number of illustrative systems, including those in which matter undergoes a transformation. Among the latter are multi-phase systems, systems near a critical point, chemically reactive systems, and systems of particles at relativistic energies.

Throughout the book I have placed considerable emphasis on problem solving, since I believe that involved participation is the only way to acquire an intuitive understanding of the subject. On the whole, passive reading is ineffective. Can you imagine how you would learn to ride a bicycle from a book? To succeed you must just try to get on, accepting that you will probably fall off quite a lot at the start. So in thermodynamics you must try to use the concepts — and risk misusing them — by attempting the problems. This allows new ideas to be introduced in an implicit way, just as you first learned to speak or ride a bicycle.

As a guide, I suggest that all students should complete at least 30 per cent of the questions for homework or supervised tutorials. Most students who do not achieve this level of independent work have difficulty later. I must confess that my questions have more practical applications and numerical answers in the introductory chapters than is usual in a physics text. Nevertheless, I believe they have great teaching value at that level. I have also found that the use of numerical data presented in a graphical format is useful. Accordingly I have prepared a set of diagrams showing the thermodynamic properties of water and steam, for use in the questions. These are given in Appendix B.

In writing this book I have incurred a number of debts, the greatest being to the classes I have taught at Otago. Many of the exercises have been used by these classes and the evolution of the questions has been guided by the responses of the students. The ability of the students has spanned a wide range, and I have been guilty of confusing even the best. Yet all have contributed to this book and I am most grateful.

It is a pleasure to acknowledge the welcome accorded me by Keble College and the Department of Engineering Science, Oxford, during the formative stages of this book in 1987, and I am especially gateful to Alan Corney and Colin Bailey for their hospitality. I should also like to thank Peter Coveney, Mike Dadd, and Ewart Neal for their comments on early drafts. At Otago I have been generously supported by my colleagues and it is a pleasure to thank Rob Ballagh, Warren Kennedy, Wes Sandle, Zhifa Sun, and Eli Yasni for many helpful discussions. In addition I am grateful to Fred Ansbacher and Marlyn Jakub, both of whom read the manuscript, to Alan McCord, who provided invaluable assistance with graphing software, and to Helen Wreford, who typed much of the manuscript. Richard Docherty produced many of the figures. Finally, I am deeply grateful to Janet Carrington for her encouragement and support.

Otago   C. G. C.
November 1993

Contents

Contents

1
Introduction

1.1 What is thermodynamics?

Thermodynamics deals with the bulk properties of matter and the processes by which these properties are changed. We shall begin by considering briefly the two primary divisions of the subject, statistical mechanics and macroscopic thermodynamics.

In statistical mechanics, quantum theory is used to establish the properties of individual atoms. Large numbers of atoms together have many degrees of freedom and our information about the state of individual atoms is usually incomplete. Hence it is necessary to use statistical methods to determine the properties of a large system from basic quantum principles. For instance, given a quantity of gas in a certain volume we may obtain the allowed values for the energy of the individual atoms from quantum theory. The mean number of atoms in each energy state may then be calculated using statistical arguments. Finally the energy of the gas as a whole, and other properties, may be obtained.

In macroscopic thermodynamics the global properties of the system, such as the volume, the energy, and the pressure, are the primary features of interest. The aim is to find relationships between these macroscopic variables. These relationships may be established from statistical mechanics, but at some point the theory must focus on the global properties of the system. So macroscopic thermodynamics effectively begins when the statistical aspects move into the background. It is this formalism, known also as classical thermodynamics, which is the subject of this book. For brevity we shall call this topic thermodynamics.

Although many of the basic principles of thermodynamics were established by the end of the nineteenth century, the subject has not remained stationary. The development of statistical mechanics has greatly extended our understanding of the basic principles of thermodynamics, and we have a better appreciation of the limits of thermodynamics. In addition new systems have been discovered in which the methods of thermodynamics are applicable. Yet the basic principles have remained intact. Thus the foundations and methods of thermodynamics have proven to be surprisingly durable.

In the contemporary context, the special feature of thermodynamics is that it employs only global properties of the system of interest. Variables which express characteristics of the microscopic structure also appear, but they normally do so only in an indirect way, so the results obtained from thermodynamic arguments do not usually rely on our understanding of a physical system in detail. Thus thermodynamics is especially useful when the relevant microscopic theory is incomplete. It offers novel insights into problems which evade more detailed solutions and it provides an independent audit for the theory at the atomic level by raising questions such as, '...surely this violates the second law?'

For these reasons the methods of thermodynamics have a universal character. They are part of the language and thinking of almost all fields of contemporary science. The value of thermodynamics in this respect was recognized by Einstein in his autobiographical notes (Einstein 1979). 'A theory is the more impressive the greater the simplicity of its premises, the more different kinds of things it relates, and the more extended its area of applicability. Hence the deep impression that classical thermodynamics made upon me. It is the only physical theory of universal content concerning which I am convinced that, within the framework of the applicability of its basic concepts, it will never be overthrown (for the special attention of those who are skeptics on principle).'[†]

1.2 The laws of thermodynamics

Thermodynamics, like classical mechanics, has a rather limited observational basis. Both theories lend themselves to axiomatic formulations in which the concepts are developed as a crisp logical structure founded on a few basic laws. In this way classical mechanics is often regarded as a branch of mathematics in which Newton's laws are taken as the axioms. Similarly it is common to take an axiomatic approach to elementary thermodynamics, basing the development on the classical laws of thermodynamics.

The laws of thermodynamics, however, have simple expressions only because they have been abstracted from diverse empirical data. They represent an underlying structure derived from measurement and observation. But to use the laws in a particular situation it is necessary to reconstruct the flesh of the relevant physics around the dry bones of the laws. In this sense the laws of thermodynamics are incomplete. They are necessary for understanding thermodynamics, but they are insufficient by themselves. For those who are new to the subject, many new physical concepts are needed to give life to the laws in a given situation. Those concepts and

[†]Reprinted from *Autobiographical Notes* by Albert Einstein, translated and edited by Paul Arthur Schilpp, by permission of the publisher (La Salle, IL: Open Court Publishing Company, 1979). First published in The Library of Living Philosophers, Vol. VII, *Albert Einstein: Philosopher–Scientist*, edited by Paul Arthur Schilpp (La Salle, IL: Open Court Publishing Company, 1949).

principles must be drawn from elsewhere in physics, and the substance of thermodynamics lies as much in the development of those concepts as in the laws. Nevertheless the laws of thermodynamics remain important because they guide and focus our thinking. They provide an anchor for the thermodynamic paradigm.

1.3 Making a start

Thermodynamics is an inclusive discipline which touches all others in physics to some extent. This broad scope creates difficulties for those just starting out. Many new ideas and physical terms must be understood and the precise use of familiar terms must be refined. For instance, temperature, heat, and energy are words in everyday use, but this does not endow them with a consistent physical meaning, so these terms must be re-established. This is a difficult process because it is quite hard to abandon the familiar, yet vague, ideas on temperature and heat which one may have used for many years.

To address this problem we shall take a pluralistic approach to the introduction of the basic concepts. First the elements of thermodynamics are developed in the context of one simple thermodynamic system in Part I. Subsequently in Parts II and III the context is enlarged and in Part IV applications of the earlier methods are developed. Thus Part I is confined mainly to the thermodynamics of ideal gas systems, the physical properties of which are represented by the elementary equations of state, reviewed in Chapter 2. This stage allows important ideas and methods to be introduced without invoking the laws of thermodynamics in a general way. Some of the more abstract concepts are given a preliminary airing through examples and questions. The aim of Part I is to provide models which will help later to introduce new concepts through the use of analogies.

In Part II the earlier material is consolidated in a more general physical context. Here the primary physical principles are expressed using the classical laws of thermodynamics. Basic intuition established in Part I should be helpful and some notions, such as irreversibility, which could only be hinted at in Part I, are developed in an explicit way.

Part III encompasses all the principles developed in Parts I and II, but a more comprehensive formalism is employed. This approach, which is due to Gibbs, is based largely on the maximum entropy principle. The formalism enables us to deal in a natural way with processes in which the quantity of matter may change. Logically we could begin immediately with Gibbsian thermodynamics but many opportunities to establish a useful physical intuition for basic processes tend to get swept aside. Finally, in Part IV, the application of the theory is illustrated.

Part I
Basic Concepts: The Ideal Gas

2
The ideal gas temperature

We shall introduce first the properties of an ideal gas, starting with the ideal gas law and the elementary kinetic theory of gases. In effect we begin by taking a microscopic approach because most beginners feel comfortable with the basic concepts involved. Remember that the primary purpose of this chapter, and the two which follow, is not to develop the physical properties of ideal gases as such. Instead we shall use the ideal gas as an example to introduce some important thermodynamic ideas.

2.1 The ideal gas law

An ideal gas is one in which the long-range interactions between the atoms or molecules can be ignored. I assume that you know the atoms of the gas move about incessantly, colliding with the walls of the containing vessel and with each other. In an ideal gas the duration of these encounters is short compared with the time the atoms spend in free motion between collisions.

Clearly an ideal gas is an abstraction. Nevertheless it is a very useful one. For instance, it is quite sufficient to regard air as a mixture of ideal gases for many practical purposes. Even the water vapour in air may often be treated as an ideal gas.

Perhaps the best-known property of an ideal gas is represented by the equation of state

$$PV = NRT, \qquad (2.1.1)$$

which I assume is already familiar to you. Here we shall use the SI system in which the unit for pressure, P, is Pascal (Pa), the unit for volume, V, is m^3, and T, the temperature, is in kelvin (K). In addition R is the molar gas constant and N is the number of moles of gas. This is given by

$$N = \frac{\tilde{N}}{\tilde{N}_A} = \frac{m}{M_g}. \qquad (2.1.2)$$

Here $\tilde{N}$ is the number of atoms (or molecules) in the gas, $\tilde{N}_A$ is Avogadro's number (in atoms or molecules per mole), m is the mass of gas (in kg) and M_g is the molecular weight in kg mol^{-1}. Hence the ideal gas equation can also be written as

$$PV = \tilde{N}kT, \qquad (2.1.3)$$

where

$$k = \frac{R}{\tilde{N}_A} \qquad (2.1.4)$$

is Boltzmann's constant. Values for the constants R, $\tilde{N}_A$, and k are listed in Appendix A.

Note that (2.1.3) holds whether or not the atoms are identical and consequently it involves only the total number of atoms, $\tilde{N}$. This means the ideal gas equation applies equally to a single-component gas and to a mixture of different ideal gases.

The ideal gas equation illustrates two basic thermodynamic concepts. First, the quantities, P, V, N, and T are examples of *thermodynamic variables*. Such variables apply to the system as a whole. We may also say they are *macroscopic variables*. We shall not attempt to provide a general definition of a thermodynamic variable at this stage. For the present it is sufficient to accept this term as common usage for global variables of this kind. Second, we note that the ideal gas equation applies only when the gas is undisturbed. For instance, it must not be stirred so that it is subject to gross internal motions. To represent this restriction we say the equation applies to the gas in an *equilibrium state*. This state is identified by specifying the values of any three of the four variables P, V, N, and T. The fourth is then determined by the equation of state.

The equation of state for an ideal gas had already been established in a form similar to eqn (2.1.1) by 1820. In the most elementary sense it represents both an experimental law (Boyle's law) and a definition of the ideal gas temperature. For completeness, and later reference, it is useful to set out what this means.

For the present we shall adopt some pragmatic notions of temperature and thermal equilibrium. Later we shall express these ideas in a more formal way. In the meantime it is sufficient to have a practical method which provides a consistent way to measure the temperature of an object. In this spirit, we shall say that the temperature is the value obtained for T by using a constant volume ideal gas thermometer.

To illustrate, suppose we wish to determine the temperature of a large body of some fluid. We shall assume that it is in equilibrium with its surroundings. The constant volume gas thermometer, shown schematically in Fig. 2.1, would be used operationally as follows.

1.　Establish a reference temperature. By convention, a stable mixture of water, ice, and water vapour, called a triple-point mixture, provides the fundamental reference temperature for the ideal gas scale. For historical reasons the temperature of this mixture is defined to be 273.16 K.

2.　Immerse the bulb of the constant volume gas thermometer first in the

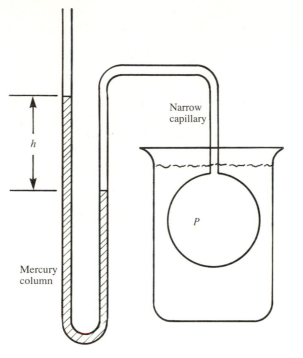

Fig. 2.1   Constant volume gas thermometer. The pressure, $P = \rho g h$ where ρ is the density of mercury and g is the acceleration due to gravity.

reference temperature bath and measure the pressure, P_1, when equilibrium is reached. Then, with the bulb immersed in the fluid whose temperature is to be determined, measure the equilibrium gas pressure, P_2.

3.   Calculate the ratio P_1/P_2.

It is found experimentally that this ratio is not strongly dependent on the amount of gas in the constant volume gas thermometer, nor does it depend on the particular gas selected for the thermometer, provided the density is low enough. If we extrapolate to the limit of zero density the ratio is found to be the same for all gases. Hence we can define the temperature in a consistent way without the need to select any particular gas. The temperature of the fluid, as measured by an ideal gas thermometer, is given by

$$T = 273.16 \left(\frac{P_2}{P_1} \right)_{\text{zero density limit}} \qquad \text{(kelvin)}. \qquad (2.1.5)$$

In everyday use we often express the temperature in degrees Celsius (°C).

The Celsius scale intervals are the same as the ideal gas scale, but the zero is shifted. For historical reasons this scale is defined by,

$$0°C = 273.15 \text{ K}. \tag{2.1.6}$$

Example

The molar composition of dry air is 78.08 per cent N_2, 20.95 per cent O_2, 0.93 per cent A, and 0.04 per cent CO_2 (ignoring other minor components). Given that the molecular weights of N_2, O_2, A, and CO_2 are 28.01, 32.00, 39.95, and 44.01 g mol^{-1} respectively, calculate the density of dry air at 20°C at a pressure of 1 standard atmosphere.

Solution

In any problem a good way to start is to re-express the information at your disposal in a way that is relevant to the question. As you do this, the path to the solution will often emerge. Here we first express the temperature in kelvin because there is no call for the temperature in °C.

$$T = 273.15 + 20 = 293.15 \text{ K}.$$

Second, we rewrite the ideal gas equation in a way which expresses the molar concentration:

$$\frac{N}{V} = \frac{P}{RT}.$$

From Appendix A, $P = 1.0133 \times 10^5$ Pa (standard atmosphere) and $R = 8.3145$ J K^{-1} mol^{-1} and hence we find $N/V = 41.573$ mol m^{-3}.

Now 1 mol of air comprises 0.7808 mol N_2, 0.2095 mol O_2, 0.0093 mol A, and 0.0004 mol CO_2. Thus for the mixture, the mass per mole is

$$M_g = 0.7808 \times 28.01 + 0.2095 \times 32.00 + 0.0093 \times 39.95 + 0.0004 \times 44.01$$
$$= 28.96 \text{ g} = 28.96 \times 10^{-3} \text{ kg}.$$

Hence the density of dry air at 20°C is $41.573 \times 28.96 \times 10^{-3} = 1.204$ kg m^{-3}.

Questions

2.1.1.   Assume that in deep space there is 1 hydrogen atom per cubic metre and the temperature is 3 K. What is the pressure (in Pa)?

2.1.2.   A hot air balloon, which is approximately spherical in shape, floats in a state of neutral buoyancy when the supported weight is 3000 N and the density of the surrounding air is 1.2 kg m^{-3}. Assume the gas within the balloon is air at 40°C having the same pressure and composition as the surrounding air. The temperature of the surrounding air is 20°C.

(a) Calculate the diameter of the balloon.

(b) Suppose the quantity of water vapour within the balloon envelope increases (without condensing) and the density of the surrounding air remains constant. Would the balloon rise or fall?

2.1.3. The temperature of a boiling water bath is measured with a constant volume gas thermometer using the triple point of water as the standard reference. The atmospheric pressure, measured using a mercury barometer, is 760.0 mm Hg. Seven measurements are made, each with a different amount of gas in the thermometer. Readings of the height of the mercury column are listed below. Here + means that the mercury level in the open tube is above the closed tube fixed point, − means below (see Fig. 2.1).

	Reading in triple-point bath. (mm Hg)	Reading in boiling water bath (mm Hg)
(1)	− 662.0	− 626.1
(2)	− 565.0	− 493.5
(3)	− 459.0	− 348.6
(4)	− 332.5	− 175.5
(5)	− 163.0	+ 56.6
(6)	− 26.0	+ 244.2
(7)	+ 61.0	+ 363.4

(a) For each measurement calculate the corresponding apparent temperature.

(b) Determine the temperature of the boiling water bath on the Celsius scale, correct to $\pm 0.02°C$.

2.1.4. A constant volume gas thermometer contains 0.1 mol of argon in a volume of $1 \times 10^{-3} \, m^3$. An experimenter measures the pressure, P_1, at the triple point of water and again at the steam point (100°C) obtaining P_2.

The equation of state for the gas is represented by the van der Waals equation rather better than by the ideal gas equation (2.1.1):

$$\left(P + a\left(\frac{N}{V}\right)^2\right)\left(\left(\frac{V}{N}\right) - b\right) = RT$$

where $a = 0.132 \, N \, m^4 \, mol^{-2}$ and $b = 30.2 \times 10^{-6} \, m^3 \, mol^{-1}$.

(a) Using the van der Waals equation, calculate the pressures P_1 and P_2 that the experimenter would have obtained.

(b) Use the pressures P_1 and P_2 obtained in (a) to find the

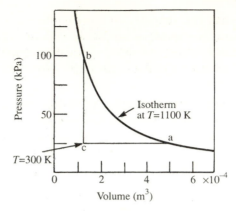

Fig. 2.2 P-V diagram for a process in helium.

apparent temperature of steam had (2.1.5) been used without extrapolation.

2.1.5. Suppose a gas cylinder contains 3 g of hydrogen gas at a pressure of 1.0×10^5 Pa. Determine the pressure in the cylinder when the hydrogen is replaced by 2 g of nitrogen at the same temperature. Regard both as ideal diatomic gases.

2.1.6. Some He, contained in a variable-volume vessel, is taken through the cyclic process a-b-c-a shown in Fig. 2.2. Here $V_a = 0.5 \times 10^{-3}\,\text{m}^3$ and the equilibrium pressure in state b, $P_b = 1.0 \times 10^5$ Pa. Given that the amount of helium in the vessel is constant, what is the mass of helium?

2.1.7. The free electrons responsible for conduction in a metal can be regarded as an exotic kind of ideal gas. In this case the equation of state is determined by T_f, a constant, called the Fermi temperature, given by

$$T_f = \left(\frac{3\tilde{N}}{\pi V}\right)^{2/3} \frac{h^2}{8mk},\tag{2.1.7}$$

where $\tilde{N}$ is the number of free electrons in a volume V, m is the electron mass, h is the Planck constant, and k is Boltzmann's constant. When the actual temperature $T \ll T_f$, the gas is said to be a degenerate Fermi gas. To a good approximation, the equation of state is then

$$PV = \tfrac{2}{5}\tilde{N}k\,T_f.\tag{2.1.8}$$

(a) Metallic sodium has one free electron per atom and the density of sodium is $970\,\text{kg mm}^{-3}$ at room temperature. The molar mass is 23 g mol^{-1}. Determine T_f.

(b)   Calculate the pressure of the free electron gas in metallic sodium, in Pascal, at room temperature. Comment on the magnitude of the result. For example, is the pressure large or small compared with an industrial cylinder of compressed gas at say 100 bar?

2.1.8.   Suppose the terrestrial atmosphere can be represented as a mixture of ideal gases having molar mass M_g kg mol^{-1}.

(a)   By considering the hydrostatic equilibrium of adjacent elements subject to gravity (Fig. 2.3), show that the vertical pressure gradient is given by

$$\frac{dP}{dh} = -\rho g, \qquad (2.1.9)$$

where ρ is the density (kg m^{-3}), g is the acceleration of gravity, and h is the vertical height measured from the earth's surface.

(b)   Assume that the atmosphere is isothermal. Use the ideal gas equation to show that

$$\left(\frac{\partial \rho}{\partial P}\right)_T = \frac{M_g}{RT}. \qquad (2.1.10)$$

(c)   Obtain $(\partial \rho / \partial h)_T$ from (2.1.9) and (2.1.10). Hence by taking g and T to be constants, show that,

$$\rho = \rho_0 \exp\left(-\frac{M_g gh}{RT}\right), \qquad (2.1.11)$$

where ρ_0 is the density at $h = 0$.

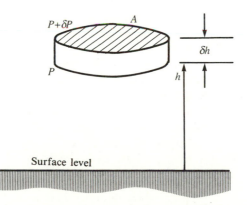

Fig. 2.3   Illustrating the conditions for hydrostatic equilibrium in the atmosphere. The downward gravitational force on the element shown is $\rho g A \delta h$, where A is the cross-sectional area of the element. The buoyancy force is $-A\delta P$.

(d) The quantity $h_0 = RT/(M_g g)$ is called the scale height for the isothermal atmosphere. Show that $h_0 = 8.6\,\text{km}$ for the earth's atmosphere, taking $T = 293\,\text{K}$. (Note $M_g = 28.96 \times 10^{-3}\,\text{kg}$.)

(e) Consider a fictitious isothermal gas of height h_0 having constant density ρ_0. Show that this would have the same mass as the atmosphere represented by (2.1.11).

(f) Given that the scale height of the atmosphere is small compared with the mean radius of the earth ($6.38 \times 10^6\,\text{m}$), show that the total mass, M_A of the atmosphere is given by,

$$M_A = A\rho_0 h_0, \qquad\qquad (2.1.12)$$
$$= AP_0/g, \qquad\qquad (2.1.13)$$

where P_0 is atmospheric pressure at surface-level and A is the surface area of the earth. Show that (2.1.13) holds whether or not the isothermal model for the atmosphere is applicable and confirm that $M_A \approx 5.3 \times 10^{18}\,\text{kg}$.

2.2 Kinetic theory of an ideal gas I

The purpose of this section is to illustrate how the internal energy of an ideal gas can be calculated using simple arguments. We shall consider an ideal monatomic gas of $\tilde{N}$ weakly interacting atoms, regarded as classical particles, and show that

$$PV = \tfrac{1}{3}\tilde{N}m\,\overline{v^2}. \qquad\qquad (2.2.1)$$

Here m is the mass of an atom (in kg) and $\overline{v^2}$ is the mean squared velocity for the atoms of the gas. From this equation and (2.1.1) it follows that the total kinetic energy for N moles of the gas, U, is given by,

$$U = \tfrac{3}{2}NRT. \qquad\qquad (2.2.2)$$

U is called the *internal energy* of the gas.

We first show this result using the following rather simple-minded argument. Consider an ideal gas of atoms in a box, taken to be a cube of side a, as in Fig. 2.4. If we ignore atom–atom collisions, then an atom, i, of mass m with x-component of velocity, $v_x(i)$, will make $v_x(i)/2a$ impacts per second with an end wall, assuming the collisions to be elastic. Each impact will impart an impulse of $2mv_x(i)$ on the wall.

The force on the wall due to all atoms in the box is obtained by the sum of all impulses per unit time:

$$\sum_i \frac{2mv_x(i)^2}{2a}.$$

Hence the pressure on the wall, having area a^2, is

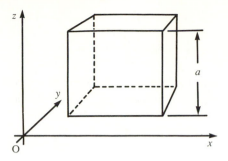

Fig. 2.4 Cube having sides of length a.

$$P = \sum_i \frac{m v_x(i)^2}{a^3}$$

or

$$PV = \sum_i m v_x(i)^2,$$

where $V = a^3$ is the volume. Now the mean squared velocity for the $\tilde{N}$ atoms is

$$\overline{v^2} = \sum_i \frac{(v_x(i)^2 + v_y(i)^2 + v_z(i)^2)}{\tilde{N}}.$$

We shall assume that the distribution of velocities is isotropic. In this situation,

$$\tilde{N}\overline{v_x^2} = \sum v_x^2(i) = \sum v_y^2(i) = \sum v_z(i)^2 = \tfrac{1}{3}\tilde{N}\overline{v^2}. \tag{2.2.3}$$

Hence

$$PV = \tfrac{1}{3}\tilde{N}m\overline{v^2} = \tfrac{2}{3}U \tag{2.2.4}$$

where

$$U = \tfrac{1}{2}\tilde{N}m\overline{v^2} = \tfrac{1}{2}M\overline{v^2}. \tag{2.2.5}$$

U is the total kinetic energy for all the atoms in the gas, and M is the total mass. Equation (2.2.2) follows from (2.2.4).

We also see that the mean kinetic energy per atom,

$$\frac{U}{\tilde{N}} = \frac{3}{2}\frac{NRT}{\tilde{N}} = \tfrac{3}{2}kT, \tag{2.2.6}$$

where

$$k = \frac{NR}{\tilde{N}} = R/\tilde{N}_A$$

is the Boltzmann constant and $\tilde{N}_A$ is Avogadro's number.

Thus the temperature is a measure of the kinetic energy of the atoms of an ideal gas due to their thermal motion. Equation (2.2.6) does not derive from the ideal gas equation of state (2.1.1) alone, so it represents a new thermodynamic relation for the ideal gas system. Note that the factor of 3 in the equation for U arises because the velocity vector has three independent components. These represent the *degrees of freedom* for translational motion of the atom.

Equation (2.2.6), which holds for monatomic gases, is a particular example of a more general result, called the *principle of equipartition*. This states that the mean energy of a system in thermal equilibrium is $\frac{1}{2}kT$ per degree of freedom. For instance, polyatomic molecules have vibrational and rotational modes in addition to the translational modes, and the mean thermal energy per molecule is then given by

$$\frac{U}{\tilde{N}} = \tfrac{1}{2}fkT, \tag{2.2.7}$$

where f is the number of degrees of freedom. In the case of diatomic molecules f may include vibrational and rotational modes in addition to the translational modes. However, equipartition is a good approximation only when the gas temperature is sufficiently high (see, for example, Riedi (1988)). This effect is related to the energy separation of the quantum states of the molecule compared with kT.

2.3   Kinetic theory of an ideal gas II

While the results obtained in Section 2.2 are correct, the physical argument ignores the collisions between the atoms. The argument also assumes the atoms are perfectly reflected in collisions at the wall. As we shall see here, these assumptions are not necessary.

Consider a gas of $\tilde{N}$ atoms, each of mass m. We write the velocity of a particular atom, i, as

$$\mathbf{v}(i) = v_x(i)\mathbf{i} + v_y(i)\mathbf{j} + v_z(i)\mathbf{k}$$

using orthogonal coordinates. Assume that in equilibrium the velocity distribution is isotropic, so that (2.2.3) holds.

Now suppose that, in each unit of volume, $d^3\tilde{N}(\mathbf{v})$ is the average number of atoms having velocity in the range

$$\mathbf{v} = (v_x, v_y, v_z)$$

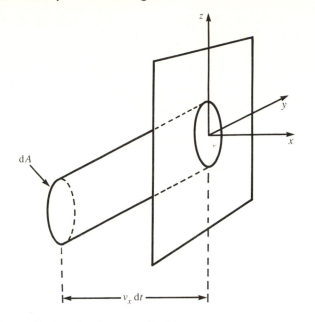

Fig. 2.5 Slant cylinder of volume $v_x \, \mathrm{d}t \, \mathrm{d}A$.

to

$$\mathbf{v} + \mathrm{d}\mathbf{v} = (v_x + \mathrm{d}v_x, \, v_y + \mathrm{d}v_y, \, v_z + \mathrm{d}v_z).$$

We assume that $\mathrm{d}^3\tilde{N}(\mathbf{v})$ is independent of time and position, which is intuitively reasonable given that the gas is in equilibrium. Near a wall perpendicular to the x-axis (Fig. 2.5), the number of atoms in this velocity class which strike an element of area, $\mathrm{d}A$, in a time, $\mathrm{d}t$, is the number of atoms contained in a slant cylinder of height $v_x \, \mathrm{d}t$, namely, $\mathrm{d}^3\tilde{N}(\mathbf{v}) \, v_x \, \mathrm{d}t \, \mathrm{d}A$. Note, we can ensure that all the atoms in the cylinder collide with the wall, rather than other atoms, by choosing $\mathrm{d}t$ to be sufficiently small. When v_x is positive this is the number of atoms in the selected velocity class which approach and collide with area $\mathrm{d}A$ of the wall in time $\mathrm{d}t$. The impulse delivered to the wall in time $\mathrm{d}t$, by atoms in this velocity class, is

$$\mathrm{d}^3\tilde{N}(\mathbf{v}) v_x \, \mathrm{d}t \, \mathrm{d}A \, v_x m.$$

Notice that we do not include the impulse delivered to the wall due to the rebound in this expression, because the velocity class after the impact is not known. Instead the rebound impulse is included later by allowing v_x to be negative as well as positive. When v_x is negative this expression gives the impulse due to atoms which rebound from the wall into the selected velocity class in time $\mathrm{d}t$.

The total impulse delivered to the wall in time $\mathrm{d}t$ is now obtained by integrating over all velocity classes. Hence, dividing by dt, the average force, dF, on dA is given by

$$dF = dA \, m \int_v v_x^2 d^3 \tilde{N}(\mathbf{v}).$$

Now if there are $\tilde{N}$ atoms in the volume V,

$$\int v_x^2 d^3 \tilde{N}(\mathbf{v}) = \frac{\tilde{N}}{V} \overline{v_x^2}.$$

Hence the pressure is given by

$$P = \frac{dF}{dA} = \frac{\tilde{N}}{V} m \overline{v_x^2},$$

which yields (2.2.1) again.

Notice, we have assumed that the distribution of velocities is isotropic, so that (2.2.3) holds. This highlights the significance of the term *equilibrium* in connection with the ideal gas equations, $PV = NRT$ and $U = \frac{3}{2}NRT$. The relations apply only to a gas in equilibrium, which we assume implies random motion of the atoms, and an isotropic homogeneous velocity distribution. The energy equation applies to monatomic ideal gases only.

The variables P and T, in particular, are equilibrium variables. That is, except for an equilibrium state, they do not have unique values for the gas as a whole. V and N and the total energy (U plus kinetic, potential energy, and so on) on the other hand, have well-defined values whether the gas is in equilibrium or not. This suggests that these parameters may be more useful than P and T when we want to represent processes in which a gas changes from one equilibrium state to another. We may think of them as thermodynamic constraints for such state changes.

Finally, note that the temperature is a uniquely thermodynamic property. Whereas the pressure, volume, mole number, and energy can be defined outside thermodynamics, the temperature cannot, because the notion of thermal equilibrium is required.

Questions

2.3.1. A cylinder, volume 10 litres ($1 \times 10^{-2} \, \text{m}^3$), contains 1 kg of an ideal gas, at a pressure of 2.0 MPa. Calculate the root mean square (r.m.s) velocity of the molecules.

2.3.2. (a) Calculate the r.m.s. velocity for a ^{4}He atom at 20°C. Determine the escape velocity for a projectile at the earth's surface. ($r_{\text{earth}} = 6.4 \times 10^6 \, \text{m}$, $m_{\text{earth}} = 6.0 \times 10^{24} \, \text{kg}$)

       (b) It is well known that helium released into the earth's atmosphere is ultimately lost into space. Do you expect ^{4}He,

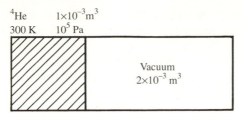

Fig. 2.6   Volume containing helium adjacent to an evacuated space prior to a free-expansion process.

having the r.m.s. velocity found in part (a), to remain in the atmosphere? If so, why is helium lost? You will need to make some assumptions; identify these explicitly.

2.3.3.   A cylinder (Fig. 2.6) contains a thin partition having $1 \times 10^{-3} \, m^3$ of helium, ^{4}He, on one side at a pressure of $10^5 \, Pa$ and a temperature of 300 K. (Assume the gas is ideal.) Initially the other side, having a volume of $2 \times 10^{-3} \, m^3$, is evacuated. The partition is suddenly removed, so that the gas is free to enter the evacuated space, without any energy being exchanged between the gas and the surroundings. This process is called free expansion because it is unimpeded. (It is also known as a Joule expansion process.) Because no energy is exchanged with the surroundings the energy of the gas is the same in the initial and final states.
   (a)   When equilibrium is re-established, what is the temperature of the gas?
   (b)   Calculate the new pressure.

2.3.4.   The partitioned volume shown in Fig. 2.7 initially contains helium on one side and argon on the other. Assume each can be treated as an ideal monatomic gas. The partition is suddenly removed and the two gases subsequently mix to form a homogeneous mixture. No energy is exchanged with the surroundings and, as in question 2.3.3, the energy of the total system remains constant. Calculate the temperature and pressure of the equilibrium mixture.

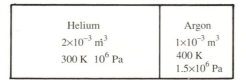

Fig. 2.7   Partitioned volume containing helium and argon before mixing.

2.3.5. An enclosed evacuated volume nevertheless contains radiation which is continuously being emitted and absorbed by the walls. This is called black-body radiation. It may be regarded as an ideal gas of particles having zero rest mass, photons. The photon energy E and momentum p satisfy the relativistic relation that is,

$$E = pc, \qquad (2.3.1)$$

where c is the speed of light. Use this equation, together with an argument of the type given in Section 2.2 or 2.3 to show that in this case

$$PV = \tfrac{1}{3}U. \qquad (2.3.2)$$

2.3.6. For black-body radiation (see question 2.3.5) the Stefan–Boltzmann law applies:

$$U = 4\,\frac{\sigma}{c}\,VT^4 \qquad (2.3.3)$$

where

$$\sigma = \frac{2\pi^5 k^4}{15 h^3 c^2} \qquad (2.3.4)$$

is the Stefan–Boltzmann constant, c is the velocity of light, and T is the wall temperature in kelvin. At what temperature will the pressure of black-body radiation be 1 atmosphere?

2.3.7. For this question refer to (2.1.7) and (2.1.8) for additional information. Using quantum statistics it can be shown that

$$PV = \tfrac{2}{3}U,$$

for a degenerate Fermi gas ($T \ll T_f$). Show that the mean energy of a free conduction electron in a metal is $\tfrac{3}{5}kT_f$.

3
Processes in ideal gas systems

In this chapter we shall explore the notion of a *process* and its connection
with *thermodynamic state* which we introduced in Chapter 2. Remember,
for simplicity in Chapters 2–4 the development is being illustrated using
ideal gas systems only.

3.1 States and processes

First we reconsider the term, *state*. For a monatomic ideal gas we already
know

$$PV = NRT \qquad\qquad (3.1.1a)$$

and

$$U = \tfrac{3}{2}NRT. \qquad\qquad (3.1.1b)$$

It is clear that when three of the variables P, V, T, N, and U are specified
(including P or V) then the other two are also determined. In such a situation we say the *state* of the system is known. In general if we claim to know
the state of a system we must have enough information to determine the
value of all its thermodynamic variables. When the variables do not change
with time, the state is an *equilibrium state*. Normally when we use the word
state we also imply a state of equilibrium.

To give the state of a system is rather like fixing the coordinates of a
point in space. Suppose the state of a system changes as a result of external
intervention. For example, we might change the volume of a cylinder containing a gas. The physical procedure used to impose that change identifies
a *process*. For instance, a certain process might be a constant pressure
(isobaric) process or perhaps a constant volume (isochoric) process. Thus
defining a process is like specifying the path between two points.

In practice we specify a process by identifying as many aspects of the state
change as possible. It is rather like a recipe. For example, the terms **work**
and *heat* define particular types of process. Below we shall see that they
represent two quite different mechanisms for energy transfer. We shall also
find that different processes imply different restrictions on the possible
changes in state parameters. You may find that the thermodynamic
constraints implied by each process condition are difficult to identify
at first. However these concepts represent an important part of basic

thermodynamics. By identifying these constraints it is often possible to write down equations which allow the new state variables to be calculated.

3.2 Quasistatic work

Quasistatic processes

In a *quasistatic process* the change of state is carried out so slowly that the system is effectively in a state of equilibrium throughout. To achieve this the speed of the process must be controlled. It must be slow compared to the rate at which the system would naturally proceed towards equilibrium. It is evident that the equations of state (3.1.1a, b) will continue to apply throughout a quasistatic process. In terms of our geometric analogy above, in a quasistatic process there is a well-defined path between the initial and final state points. This process line is smooth and continuous. A non-quasistatic process, on the other hand, corresponds to a discontinuous path. In this situation there may be no identifiable equilibrium states at all between the initial and final states.

For example, consider an ideal monatomic gas system. Take $N = 10\,\text{mol}$, $T = 300\,\text{K}$, and $V = 10 \times 10^{-3}\,\text{m}^3$. The state of the system is uniquely specified since we can calculate the pressure and energy using (3.1.1a, b). Assume that we are able to vary the volume of the system at will and let us change its state to one having $V = 20 \times 10^{-3}\,\text{m}^3$ also at $T = 300\,\text{K}$. This may be achieved quasistatically by ensuring that the volume increases very slowly while keeping the gas vessel in good contact with a large body of fluid having $T = 300\,\text{K}$.

Such a process would represent a quasistatic isothermal increase in volume. There would be no significant fluctuations in the temperature throughout this change of state and the pressure would vary with the volume according to (3.1.1a).

Alternatively the same final state could be reached by a non-quasistatic processes. A number of such processes are possible. We might, for instance, suddenly increase the volume to $V = 20 \times 10^{-3}\,\text{m}^3$. This change would be accompanied by temperature and pressure fluctuations within the vessel. If the system is then put in contact with a large body of fluid at $T = 300\,\text{K}$, it would eventually reach a state having $V = 20 \times 10^{-3}\,\text{m}^3$ at $T = 300\,\text{K}$. But during this change of state, the system would not be in equilibrium and the equations of state (3.1.1a, b) would not apply. This illustrates how the thermodynamic variables of a system may not all have defined values during a non-quasistatic process. Some do, but not all.

Quasistatic work

Quasistatic work is a particular example of a quasistatic process. In a gas this is a process in which the volume is changed smoothly and slowly. For all practical purposes the gas remains in an equilibrium state, although a changing state, throughout the process. During the process the forces at the walls of the vessel have their equilibrium values. Like many thermodynamic notions the concept of quasistatic work is an idealized one which can be approached only as a limit of real processes. Nevertheless it is a very useful idea.

Besides indicating a particular kind of process, the term *work* is also used to represent the mechanical work done by external forces on the walls of the gas during a quasistatic volume change. In the quasistatic limit, the external work, $đW$, performed on a gas at pressure P in an infinitesimal volume change dV is

$$đW = -P \, dV. \tag{3.2.1}$$

You should try to demonstrate this result yourself. (See the example below if you need help.) It is most important to remember that, if the process is not quasistatic, this equation is inapplicable. In that situation $-P \, dV$ does not represent the work done in an incremental volume change. It may not be possible to calculate the work done at all in such cases.

For a complete quasistatic volume change $a \rightarrow b$, beginning at state a and ending at state b, the total work done on the system is obtained by integrating (3.2.1),

$$(W_{ab})_{\text{quasistatic}} = -\int_{V_a}^{V_b} P \, dV. \tag{3.2.2}$$

Hence W_{ab} is given by the area under the process line in P–V coordinates, as in Fig. 3.1. The sign convention used here is the normal one in physics: the work, W_{ab}, is positive when work is performed *on* the system. Hence if the volume decreases throughout the process W_{ab} is positive.

If the process is cyclic the process line will form a closed loop. In this situation the work done on the gas is positive if the process advances

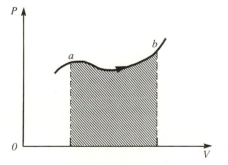

Fig. 3.1   Indicator diagram for a quasistatic work process $a \rightarrow b$. The work done on the system is the negative of the shaded area.

counter-clockwise. Such a representation of a process in P–V coordinates is often referred to as an *indicator diagram*. It was first used during the eighteenth century by James Watt, inventor of the condensing steam engine. He also invented a device for automatically plotting the indicator diagram of a working steam engine in order to obtain the work done per stroke.

It is evident that the work increment, đW, and the total work, W_{ab}, are determined by the particular process of change. Clearly, work is not a state property, such as P, V, U, or T. The increase in the energy of a system in a change of state $a \rightarrow b$ is simply given by the difference

$$\Delta U = U_b - U_a. \tag{3.2.3}$$

But in general the work performed on the system, W_{ab}, may not be calculated using an equation of this form. Instead the work done can be calculated only if the process is known in sufficient detail. When the process is quasistatic, W_{ab} can be calculated using (3.2.2), provided that P can be expressed as a function of V. For these reasons the work increment đW is represented here as a crossed differential. This emphasizes that the work done on the gas is a process, not a state, quantity.

Example

Establish equation (3.2.1). The system is a fixed amount of gas in a variable volume. To be specific we suppose the gas is contained in a cylinder, shown in Fig. 3.2, fitted with a movable piston of cross-sectional area, A. The piston slides without friction or leaking.

Solution

Suppose the system undergoes an infinitesimal quasistatic process in which the piston moves by an increment dx under external control. We shall take forces in the direction of increasing x to be positive. In equilibrium the external force exerted on the piston by the external agent is $F = -PA$. Here P, the pressure of the gas, is taken to be constant during the infinitesimal process. Thus the work done on the gas by the external agent is đ$W = -PA\,dx = -P\,dV$. The quasistatic condition is an important restriction, for otherwise we could not write $F = -PA$. The equation for quasistatic displacement work applies not just to gases, but to fluids in general.

Example

A system composed of 5 mol of an ideal monatomic gas undergoes a quasistatic process at a constant pressure of 1.0×10^6 Pa, in which the temperature changes from 300 K to 500 K.

(a)   Calculate the change in energy, ΔU.

(b)   Obtain the external work done on the gas.

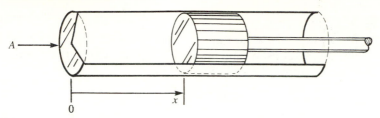

Fig. 3.2   Piston and cylinder of cross-section A.

(c)   The volume is returned to its initial value while keeping the internal energy, U, constant. Calculate the external work done on the gas in this process, assuming it to be quasistatic.

(d)   The gas is finally cooled to 300 K while the volume stays fixed. Determine the work done on the gas in this process.

Solution

(a)   For an ideal gas the energy depends only on N and T (eqn 3.1.1b). Hence

$$\Delta U = \tfrac{3}{2} N R \Delta T = 1.5 \times 5 \times 8.314 \times 200 = 12.471 \text{ kJ}.$$

(b)   Since the process is quasistatic, (3.2.2) is applicable. The pressure is constant, so the equation reduces to

$$W_{ab} = -P(V_b - V_a).$$

Here the volumes are obtained from the ideal gas equation $PV = NRT$. We find $V_a = 12.471 \times 10^{-3} \text{ m}^3$, $V_b = 20.785 \times 10^{-3} \text{ m}^3$ and hence $W_{ab} = -8.314 \text{ kJ}$. Notice, the quantity of work done on the gas is negative, indicating that the gas actually performed work on an external system.

(c)   The process is again quasistatic, and therefore (3.2.2) is applicable. Since both U and N are constant, T must be constant and hence we have

$$P = \frac{NRT}{V} = \frac{20785}{V}.$$

Thus, integrating over V we find

$$W = -20785 \ln (12.471/20.785) = 10.617 \text{ kJ}.$$

Hence work is done on the gas.

(d)   In this process the volume is constant and therefore no displacement work was performed. The system is now restored to its initial state. Notice that more work was done on the gas during compression than was done externally during the expansion. This illustrates how the work done depends on the path of a process.

Questions

3.2.1. Work, W_{ab}, is carried out on N mol of an ideal gas in a quasistatic process at constant pressure, P, during which the temperature changes from T_a to T_b. Show that

$$W_{ab} = NR(T_a - T_b). \tag{3.2.4}$$

3.2.2. A vessel contains N mol of a monatomic ideal gas. The volume, which is initially V_a, changes to V_b in a quasistatic process carried out at a fixed temperature T. Show that the work performed on the system is

$$W_{ab} = NRT \ln \left(\frac{V_a}{V_b} \right). \tag{3.2.5}$$

3.2.3. Use the equations of state for black-body radiation (eqns (2.3.2) and (2.3.3)) to show that the work done in an isothermal quasistatic volume change, from V_a to V_b, in this system is given by

$$W_{ab} = \frac{4\sigma}{3c} T^4 (V_a - V_b). \tag{3.2.6}$$

3.2.4. Calculate the work, W_{ab}, performed on an ideal gas in the quasistatic processes, $a \rightarrow b$, shown in Figs 3.3–3.8. Where necessary, take $N = 1$. Remember to determine the *sign* of W_{ab}.

3.2.5. Refer to Fig. 3.9: $V_a = 0.5 \times 10^{-3} \, \text{m}^3$, $P_b = 100 \, \text{kPa}$ (as in Fig. 2.2).
(a) Calculate the work performed on the gas in the process segments $a \rightarrow b$, $b \rightarrow c$, $c \rightarrow a$, assuming that they are quasistatic.

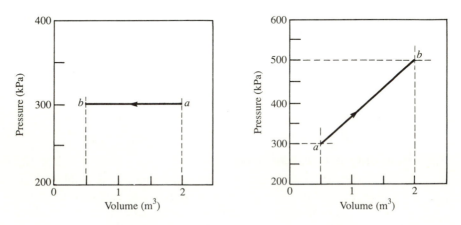

Fig. 3.3   **Fig. 3.4**

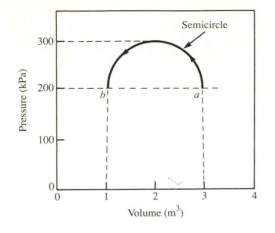

Fig. 3.5

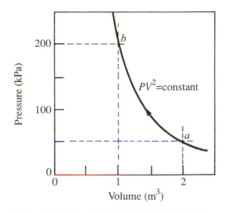

Fig. 3.6

(b)   Compare the work performed on the gas with the increase in
the energy, U, in each case. Is it possible that quasistatic work
is the only process for energy exchange with the system in each
of these process segments?

3.3   Heat-only processes

It is a familiar experience that hot objects tend to cool to the temperature
of their surroundings. On the other hand, objects colder than their surroundings are warmed. These processes represent an exchange of energy
between the objects concerned and their surroundings. In our present

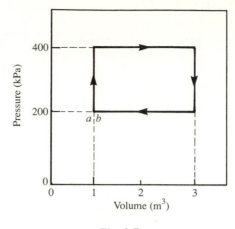

Fig. 3.7

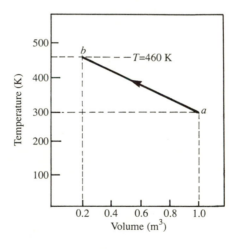

Fig. 3.8

example of an ideal gas we know that the internal energy is directly related
to the temperature (eqn (3.1.1b)), so we can easily accept that temperature
changes are generally accompanied by an energy change.

It is useful to use the term *heat-only process*, or just *heat transfer*, to
denote a process in which the energy of a system changes while no work
is performed on the system. Our familiar experience is that such processes
are always associated with some temperature difference between the system
and its surroundings. (In Chapters 5 and 9 we shall reconsider this link
between heat transfer and temperature.) For an ideal gas it is necessary for

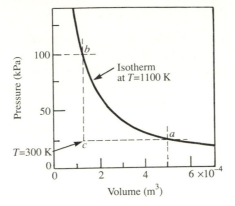

Fig. 3.9 Illustrating quasistatic processes in helium.

the volume to remain constant if a process is to be a heat-only process. For this situation we define the heat $đQ$ transferred to the system by

$$đQ = dU \qquad (3.3.1)$$

where $đQ$, like $đW$, is a process quantity, not an increment in a state variable.

If a heat transfer process is carried out so slowly that the system remains close to equilibrium throughout, then it is a *quasistatic heat transfer* process. However, (3.3.1) applies to any heat-only process, whether or not it is quasistatic.

Like work the term heat is used to convey two ideas: it defines a particular type of process, and it refers to the energy transferred to the system in that process. Since heat is a process quantity a system has no thermodynamic property called 'heat'. It is nonsense to say that a particular system 'has a lot of heat'. At first you may have difficulty accepting this restriction on the use of 'heat' because in everyday language it is quite common to employ the word 'heat' to imply a thermodynamic property. Next time you get rather close to a large bonfire you may well find you say something like, 'Wow—there is a lot of heat in that fire!'. Some authors continue to use the word 'heat' in this sense. We shall avoid this form of expression by using the term *internal energy* where appropriate.

3.4 Adiabatic processes

In an adiabatic process, work is the only means for energy transfer to or from a system, so that heat-only and adiabatic processes are mutually exclusive. For an adiabatic process, therefore, we can express the principle of conservation of energy as

$$đW = dU. \qquad (3.4.1)$$

In the present context this equation may be referred to as the first law of thermodynamics for adiabatic processes.

Adiabatic processes may be quasistatic or non-quasistatic. Equation (3.4.1) is applicable in either case. For a quasistatic adiabatic change of state we have, from (3.2.1) and (3.4.1),

$$(dU)_{\text{quasistatic, adiabatic}} = -P\,dV$$

so that we may write:

$$P = -\left(\frac{\partial U}{\partial V}\right)_{\text{quasistatic, adiabatic}} . \qquad (3.4.2)$$

This formula is mathematically untidy because it mixes the state variables P, U, and V with two process qualifiers, quasistatic and adiabatic. In Sections 4.1 and 6.7 we shall show that these process restrictions imply that the entropy S, which is a function of state, is constant. We will then be able to write (3.4.2) without invoking descriptive process characteristics.

An example of a non-quasistatic adiabatic process is the free expansion of an ideal gas (question 2.3.3). During free expansion no heat is exchanged with the system and no work is done. It is not a quasistatic process because the gas changes from the initial to the final state passing through non-equilibrium states as it does so.

Stirring is another example of a non-quasistatic adiabatic process. Figure 3.10 shows a gas contained in a fixed volume. A mechanical stirrer is mounted on a motor driven shaft which passes through a wall seal. When the stirrer is rotated work is performed on the gas. This is a work process because the mechanism for energy transfer across the system boundary could have been used to do mechanical work, such as lifting a weight or winding a spring.

The process illustrated in Fig. 3.10. does not involve any change in the volume of the gas, unlike quasistatic work (Section 3.2). During this process the system is not in equilibrium. In fact the forces against which work is performed in this process do not exist when the system is in equilibrium. The temperature and pressure of the system are not defined by unique equilibrium values during the process. Instead small elements of the system may have defined values for the pressure and temperature for short periods. Hence we really need to introduce a temperature and pressure field which varies with time and position to represent such processes.

The word *adiabatic* was coined by Rankine (1859) initially to represent 'a curve of no transmission' of heat on the P–V indicator diagram. Evidently he intended adiabatic to mean a quasistatic work-only process, but the quasistatic condition was not stated explicitly. Both Maxwell (1872) and Planck (1897) used adiabatic in this sense. However, it is now normal

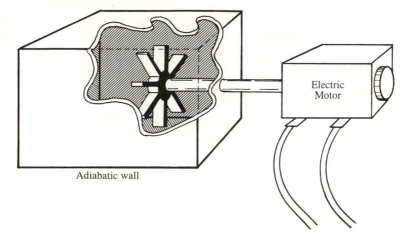

Fig. 3.10 Example of a non-quasistatic work process. Work is done on the gas by the electric motor as a result of vigorous stirring. The walls and drive shaft are adiabatic so that no heat transfer can take place.

in thermodynamics to use adiabatic, as we do here, to denote a work-only process whether or not it is quasistatic. On the other hand, some authors (Landau and Lifshitz (1980), Pippard (1964)) have retained Rankine's intended meaning for this term. It is interesting that in mechanics the term adiabatic has lost its connection with heat transfer; there it is used to identify a quasistatic change.

3.5 The first law

When a system undergoes a process involving both heat and work interactions, the requirements for energy conservation are expressed by the relation:

$$\mathrm{d}U = \mathrm{d}Q + \mathrm{d}W. \qquad (3.5.1)$$

This expression is known as the first law of thermodynamics including both heat and work processes. It applies whether or not the process is quasistatic. As before, the đ symbol is used to indicate that đQ and đW are process-specific quantities. The total work and heat involved may be determined only when the mechanism of the process is specified.

On the other hand, the difference in the internal energy of two states a and b does not depend on the process: U is a state property. We say that dU is an *exact differential* because the integral

$$\int_a^b \mathrm{d}U = U_b - U_a$$

has a unique value, independent of the path of the process a–b. It depends only on the the initial and final states. But the integrals

$$\int_a^b đQ \quad \text{and} \quad \int_a^b đW,$$

cannot be evaluated unless the path of the process by which the system moves from state a to state b is defined. There is no generally applicable expression for these integrals having the form $Q_b - Q_a$ or $W_b - W_a$. For this reason (see Appendix D) we say that $đQ$ and $đW$ are *inexact differentials*.

Example

For an ideal gas obtain an expression for the molar heat capacity at constant volume, defined as

$$c_v = \left(\frac{đQ}{dT}\right)_{V=\text{const},\,N=1}.$$

Solution

Physically the molar heat capacity, c_v, is the quantity of heat required to change the temperature of 1 mol of the gas by 1 K, given that the volume is fixed. Begin by considering N mol. Clearly no displacement work can be done in a constant volume heating process. There is an implied condition that non-quasistatic work processes are excluded too, otherwise the quantity of interest would be undefined. Hence we can write $đQ = dU$ and using $U = \frac{3}{2}NRT$ we have

$$đQ = \tfrac{3}{2} NR \, dT$$

for this process. Putting $N = 1$, it follows that

$$c_v = \tfrac{3}{2} R. \tag{3.5.2}$$

Example

Obtain an expression for the molar heat capacity of an ideal monatomic gas for a heating process at constant pressure, defined as

$$c_p = \left(\frac{đQ}{dT}\right)_{P=\text{const},\,N=1}.$$

Solution

We assume the process is quasistatic. In this case the volume will change in order that the pressure may remain fixed. Thus the work term must be included in the first law,

$$đQ = dU - đW.$$

For N mol we can write

$$dU = \tfrac{3}{2} NR \, dT = Nc_v \, dT.$$

Since P and N are constant, we can take the differential of $PV = NRT$ to obtain

$$đW = -P \, dV = -NR \, dT.$$

Hence

$$đQ = Nc_v \, dT + NR \, dT.$$

But by definition

$$đQ = Nc_p \, dT,$$

and hence it follows that

$$c_p = c_v + R. \qquad (3.5.3)$$

This equation is valid for ideal gases whether monatomic or not. In the particular case of a monatomic ideal gas we obtain, from (3.5.2),

$$c_p = \tfrac{5}{2} R. \qquad (3.5.4)$$

Example

Show that a quasistatic adiabatic process in an ideal gas is characterized by the equation

$$PV^\gamma = \text{const}, \qquad (3.5.5)$$

where $\gamma = c_p/c_v$.

Solution

Consider N mol of an ideal gas. In this process all the variables T, P, V, and U change. First we use the fact that the process is quasistatic: $đW = -P \, dV$, and adiabatic: $dU = đW$. Hence, for an increment of the process,

$$dU = -P \, dV.$$

Second, because the process is quasistatic the equation of state $PV = NRT$ applies throughout; expressed in differential form we have,

$$P \, dV + V \, dP = NR \, dT.$$

Third, we can use the energy equation in the form $dU = Nc_v \, dT$. By eliminating dU and dT we obtain

$$P \, dV + V \, dP = -\frac{R}{c_v} P \, dV.$$

Rearranging, using (3.5.3), and the definition of γ we obtain,

$$\frac{dP}{P} + \gamma \frac{dV}{V} = 0.$$

Now integrate and we have

$$\ln(P) + \gamma \ln(V) = \ln(P V^{\gamma}) = \text{const.}$$

Equation (3.5.5) follows.

Questions

3.5.1.   Suppose an ideal gas is taken through a quasistatic isothermal process from state a to state b. Show that the net work done on the system, W_{ab}, and the net heat input, Q_{ab}, are related by

$$W_{ab} = -Q_{ab}.$$

(a)   Would this equation apply if the process were not isothermal, given that the end states, a and b remain the same?

(b)   Does the result depend on whether or not the process is quasistatic?

(c)   If the system were not an ideal gas, do you think this equation would necessarily apply? Explain your answers.

3.5.2.   A system comprising N mol of a monatomic ideal gas undergoes a quasistatic isothermal process at temperature T, during which the volume changes from an initial value, V_a, to a final value, V_b. Show that the heat transferred to the gas is

$$Q_{ab} = NRT \ln \left(\frac{V_b}{V_a} \right). \qquad (3.5.6)$$

How should this result be modified if the gas were ideal but not monatomic?

3.5.3.   Refer to Figs 3.3–3.8 of question 3.2.4. Calculate the heat, Q_{ab}, transferred to the gas in each of the quasistatic processes $a \rightarrow b$ illustrated. Where the amount of the system is required, but not otherwise defined, take $N = 1$ mol.

3.5.4.   Use eqn (3.5.5), together with the ideal gas equation of state (3.1.1a), to show that the following equations hold for a quasistatic adiabatic process

$$TV^{\gamma-1} = \text{const} \qquad (3.5.7)$$

$$T^{\gamma}P^{1-\gamma} = \text{const.} \qquad (3.5.8)$$

3.5.5.   In a quasistatic adiabatic process an ideal gas changes from an initial state identified by P_a, V_a to a final state, P_b, V_b. Show that the work done on the gas is given by

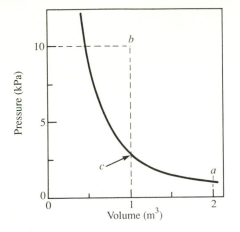

Fig. 3.11 Quasistatic adiabatic (solid line) in a monatomic ideal gas.

$$W_{ab} = \frac{P_a V_a^\gamma}{\gamma - 1}\left(\frac{1}{V_b^{\gamma-1}} - \frac{1}{V_a^{\gamma-1}}\right) \qquad (3.5.9)$$

where $\gamma = c_p/c_v$.

3.5.6. Figure 3.11 shows two equilibrium states, a and b, of a fixed amount of monatomic ideal gas. These states are linked by a two-step process, $a \to c \to b$. You may assume $c_v = \frac{3}{2}R$. For state a, $P = 10^3$ Pa, $V = 2.0\,\text{m}^3$, and $T = 300$ K. For state b, $P = 10^4$ Pa, $V = 1.0\,\text{m}^3$. The process $a \to c$ is a quasistatic adiabatic process and the process $c \to b$ is a constant volume stirring process, which is adiabatic, but non-quasistatic.

(a) Calculate the temperature for the states b and c. Hence calculate the increase in internal energy, U, of the gas for the two processes $a \to c$ and $c \to b$.

(b) Calculate the work done in the process $a \to c$ using (3.5.9). Compare this with the result in part (a).

(c) The process $c \to b$ is an adiabatic process, but (3.5.9) cannot be used to calculate the work performed. Why?

3.5.7. Figure 3.12 shows the same states a and b for the system represented in Fig. 3.11. Here state d is obtained from state a by a quasistatic adiabatic process. It is also reached from b by a quasistatic isothermal process.

(a) Determine P and V for the state d and calculate Q for the process $d \to b$. What is W for this process?

(b) Calculate W for the process $a \to d$, using (3.5.9).

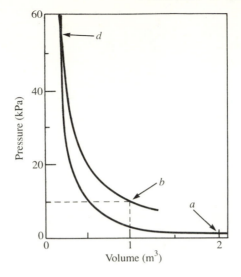

Fig. 3.12 Isotherm (*b–d*) and quasistatic adiabatic (*a–d*) in a monatomic ideal gas.

(c) Use the results from (a) and (b) and question 3.5.6(a) to confirm that

$$\Delta U = Q + W$$

for the complete process $a \rightarrow d \rightarrow b$.

3.5.8. The work performed on the system in the adiabatic quasistatic process $a \rightarrow d$, (Fig. 3.12) can be calculated using either (3.5.9), replacing *b* with *d*, or by using,

$$W_{ad} = \Delta U = Nc_v(T_d - T_a) \qquad (3.5.10)$$

where c_v is the molar heat capacity at constant volume. Show that (3.5.9) and (3.5.10) are equivalent when the process is quasistatic.

3.5.9. Two moles of an ideal gas for which $\gamma = 1.5$ are taken through the cycle shown in Fig. 3.13. Find Q, ΔU, and W for the processes labelled *A*, *B*, and *C*.

3.5.10. (a) A monatomic ideal gas, initially at 17°C, is suddenly, but quasistatically, compressed to one-tenth of its original volume. What is its temperature after compression?
(b) Carry out part (a) for a diatomic gas having $\gamma = 7/5$.
(c) Suppose the process in (a) is not quasistatic; how would you expect this to affect the temperature obtained?

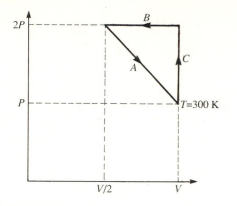

Fig. 3.13   Quasistatic processes in an ideal gas.

3.5.11.   Two moles of helium at 27°C occupies 20 litres. The gas is expanded first at constant pressure until the volume has doubled, and then adiabatically until the temperature returns to its initial value. Treat helium has an ideal monatomic gas and assume that both processes are quasistatic.
(a)   Determine the final volume.
(b)   What is the heat supplied in the overall process?
(c)   What is the change in the internal energy?

3.5.12.   A system comprising a monatomic ideal gas initially occupies $0.5\,m^3$ when the pressure is $200\,kPa$. In a constant pressure quasistatic process the heat input is $50\,kJ$. Determine the final volume and the work done on the gas.

3.5.13.   The speed of sound in a gas (denoted by c to avoid confusion with the symbols for volume) is given by

$$c = \frac{1}{\sqrt{\rho\kappa}},\qquad\qquad(3.5.11)$$

where ρ is the density and κ is the compressibility for a quasistatic adiabatic process, (q-a), defined by

$$\kappa = -\frac{1}{V}\left(\frac{dV}{dP}\right)_{q\text{-}a}.\qquad\qquad(3.5.12)$$

(a)   For an ideal gas show that

$$\kappa = \frac{1}{\gamma P},\qquad\qquad(3.5.13)$$

where $\gamma = c_p/c_v$.

(b)   Hence show that

$$c = \sqrt{\frac{\gamma RT}{M_g}},$$ (3.5.14)

where M_g is the molecular weight expressed in $kg \, mol^{-1}$.
(c)   Obtain γ for air given that the speed of sound at $0°C$ is
        $331 \, m \, s^{-1}$. ($M_g = 0.029 \, kg \, mol^{-1}$; see Section 2.1)

3.5.14.   (Adapted from Endem (1938)). On a wintry day, when the out-
          door temperature is $10°C$, I decide to heat a room of my house
          to $20°C$. The volume of the room is $20 \, m^3$, the initial tempera-
          ture is $10°C$, and atmospheric pressure is $101 \, kPa$. In the follow-
          ing, ignore the heat loss by conduction to or through the walls.
          Take c_v for air to be $3.6R$.
          (a)   Suppose the room is heated from an initial temperature of
                $10°C$ keeping the amount of air in the room constant during
                the process. Determine the heat input required to raise the
                room temperature to $20°C$.
          (b)   Because of incidental leaks to the outside, the room pressure
                actually remains constant during the heating process. Show
                that when the room temperature reaches $20°C$ the energy of
                the remaining air is no greater than that of the air originally
                in the room at $10°C$. (This means, of course, that all the heat
                used to bring the room to $20°C$ is lost to the outside.)
          (c)   Determine the heat input required to bring the room to $20°C$
                in the constant pressure process described in part (b).

3.5.15.   In question 2.1.8 we considered an isothermal model for the
          earth's atmosphere. But the atmosphere is not isothermal,
          primarily because parcels of air undergoing vertical motion are
          subject to compression or expansion processes. To represent these
          we may establish an adiabatic model for the atmosphere in which
          heat transfer and mixing between adjacent elements of the
          atmosphere are ignored. In addition we shall assume that vertical
          transport processes are quasistatic and that air can be treated as
          a mixture of ideal gases. Thus the model ignores the important
          influence of moisture phase change.
          (a)   Use eqn (3.5.8) to show that in a quasistatic adiabatic pro-
                cess, denoted by the subscript, q-a,

$$\left(\frac{\partial T}{\partial P}\right)_{q\text{-}a} = \frac{\gamma - 1}{\gamma}\frac{T}{P} = \frac{\gamma - 1}{\gamma}\frac{M_g}{\rho R}$$ (3.5.15)

where $\gamma = c_p/c_v$, M_g is the mass per mol, ρ is the density
(in $kg \, m^{-3}$), and R is the molar gas constant. ($M_g = 0.029$
$kg \, mol^{-1}$; see Section 2.1.)

(b) Hence, using (2.1.9), show that,

$$\left(\frac{\partial T}{\partial h}\right)_{q\text{-a}} = -\frac{\gamma - 1}{\gamma}\frac{M_g g}{R}. \tag{3.5.16}$$

This is the *adiabatic lapse rate*. By taking $\gamma = 1.4$ for air, show that the adiabatic lapse rate according to this model is $-9.8\,\text{K km}^{-1}$. The actual lapse rate is rather less than this, being typically -5 to $-6\,\text{K km}^{-1}$. The difference is due to mixing, heat transfer, and the influence of atmospheric moisture.

3.6 Analysis of processes

The questions in Section 3.5 provided specific information about the process of interest, and about the initial and final states of the system. However in this section you are asked to identify the equilibrium states and the process attributes yourself. You should then be able to write down the equations of constraint for the process and solve them. Such problems are straightforward, or not, depending on the way you first analyse what is involved physically. If you ignore the analysis stage the thermodynamics of simple processes will become an unfathomable swamp. I suggest that you begin by using the following strategy to help you establish a systematic method of analysis.

1. Define the thermodynamic system of interest. Identify its external interactions, such as heat or work. This may seem to be stating the obvious, but it often leads to surprises.

2. Identify the process characteristics (such as quasistatic, work, isothermal). If possible establish the initial and final states of the system.

3. If you have enough information at this stage, calculate the heat and work inputs by using the process and state information from (2). You may be able to use energy conservation to determine the final state by using the equations of state, such as (3.1.1a, b).

4. If the process is quasistatic, try to use the equations of state to determine the final state, if this is not known already. In this situation you should also be able to calculate the work and heat inputs from the state equations.

5. If the strategy suggested does not provide a solution, reconsider the problem. You may find that there is a basic reason for there being no answer. For instance, there may be insufficient information about a process.

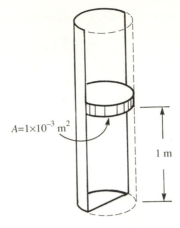

$A = 1 \times 10^{-3} \, \text{m}^2$

1 m

Fig. 3.14 Right circular cylinder fitted with leak-free frictionless piston.

Example

A right circular cylinder containing an ideal monatomic gas is fitted with a frictionless leak-free piston. The axis of the cylinder is vertical (Fig. 3.14) and the system is initially in equilibrium at a temperature of 300 K. The weight of the piston is 150 N and its cross-sectional area is $1 \times 10^{-3} \, \text{m}^2$. Initially the height of the gas column is 1 m. The external pressure on the exterior of the cylinder and piston is zero.

An experimenter places a mass weighing 100 N on top of the piston. This is done in four different ways in separate experiments. The starting condition of the apparatus is the same in each case. For each experiment determine the final height of the piston, the work done on the gas, and the heat input.

(a)   The experimenter gently lowers the added mass so that the piston slowly approaches its new equilibrium state while the gas temperature remains at 300 K throughout the process.

(b)   The added mass is released by the experimenter on to the top of the piston when it is in its initial position. Eventually the piston reaches a new equilibrium state at 300 K.

(c)   The added mass is lowered gently as in experiment (a), but in this experiment the cylinder and piston are insulated to prevent heat transfer into or out of the gas.

(d)   The added mass is released as in (b), but the system is insulated against heat transfer as in (c).

Solution (a)

For this example I will follow the above solution strategy in some detail. The numbers (1)–(5) refer to the steps suggested.

(1) The thermodynamic system is the ideal monatomic gas in the cylinder. Its external interactions are the displacement work, due to movement of the piston, and heat transfer with the surroundings. There are three contributions to the work: the change in potential energy of the piston; the change in potential energy of the added mass; and the work done by the experimenter as the mass is lowered.

(2) The process is quasistatic isothermal compression of the gas at $T = 300\,\text{K}$. The system is in equilibrium when the pressure, P, of the gas satisfies the equation $PA = $ total piston weight, where A is the area of the piston. Hence the initial and final states of the system can be found. We shall label the initial and final states by the letters i and f respectively.

$$\text{Initial pressure,} \quad P_i = \quad 150/(1 \times 10^{-3}) = 150\,\text{kPa.}$$
$$\text{The initial volume,} \quad V_i = \quad 1 \times 10^{-3}\,\text{m}^3,$$

and hence from the ideal gas equation of state (which applies because the initial state is an equilibrium state) we have

$$\text{mol of gas, } N = \frac{P_i V_i}{RT} = \frac{150 \times 10^3 \times 1 \times 10^{-3}}{R \times 300} = 0.0601\,\text{mol.}$$

Similarly the final pressure is

$$P_f = 250\,\text{kPa,}$$

and the final volume is

$$V_f = \frac{150}{250} \times V_i = 0.6 \times 10^{-3}\,\text{m}^3.$$

Hence the final height of the piston is 0.6 m.

(3) Let us see what we can determine from the change of state alone. First, since $\Delta T = 0$, then $\Delta U = 0$, because the system is an ideal gas. Hence, from the first law we have the heat and work related by

$$Q_{if} + W_{if} = 0.$$

Second, from the reduction in the elevation of the piston and the added weight (0.4 m), we can determine the work performed on the gas as a result of their losing potential energy. For the piston this represents 60 J and for the added mass, 40 J. Third, there is a work contribution from the experimenter because the added mass was lowered gently on to the piston. But we cannot determine this quantity of work from the initial and final states alone; for that we shall need to use information about the process itself.

(4) The process is a quasistatic work process at constant temperature. Hence, we may use (3.2.5) we obtain the total work done on the gas

$$W_{if} = NRT \ln\left(\frac{V_i}{V_f}\right) = 0.5 \times 300 \times \ln\left(\frac{1}{0.6}\right) = 76.62 \text{ J},$$

and the heat transferred to the gas, from (3), is $Q_{if} = -76.62$ J.

This is the total work done on the gas. Its contributions are W_e, the work done by the experimenter, and the loss of potential energy by the piston, and the added mass (100 J).

Hence, $W_e + 100 = 76.62$, and we have $W_e = -23.38$. The minus sign simply means that work was done on the arm of the experimenter as the mass was added to the system.

This concludes the first part of this example. I have included detailed comment here in order to clarify the steps involved. It may seem cumbersome to record so much, but you will usually find it is better to do this than to rush headlong into the swamp.

Solution (b)

(1) The system is the same as in part (a), but the process is different because the added mass is placed on top of the piston and then released. We can visualize what will happen here: the mass and the piston will fall until their motion is reversed by the gas pressure in the cylinder. The system will oscillate like a mass on a spring. To analyse what happens in detail here is a complex problem, but we do not need to do that. All we need to know is that the experimenter does no work on the system in this process.

(2) Here the process is not quasistatic because no external control is exerted on the movement of the piston and added mass. Yet the initial and final pressures and temperatures will be the same as in part (a) and hence the initial and final states of the system must be the same. Thus $Q_{if} + W_{if} = 0$ and the final elevation of the piston will be 0.6 m above the base of the cylinder.

(3) The work done on the gas is simply given by the loss of potential energy by the piston and the mass, which is 100 J. Also, because $Q_{if} = -W_{if}$, the heat input to the system is -100 J.

Notice that, while the initial and final states are the same in parts (a) and (b), more work was done on the system in the non-quasistatic process than in the quasistatic process. This is a special case of a general result which we discuss in Chapter 7.

Solution (c)

(1) The system remains the same as in (a) and (b) but there is no heat transfer.

(2) The initial state has been obtained already. Here we also know

$P_f = 250\,\text{kPa}$, but we do not yet know the final temperature or the final elevation of the piston.

(3) Since the process is adiabatic we know $Q_{if} = 0$, and hence $W_{if} = \Delta U$, from the first law. But no further progess is possible here until we introduce the process information.

(4) Since the system is an ideal gas, and the process is both quasistatic and adiabatic, we know $PV^\gamma = \text{const}$ throughout the process (eqn 3.5.5), and since the gas is monatomic, $\gamma = 5/3$. Then we have $P_i V_i^\gamma = P_f V_f^\gamma$, and solving for V_f, we get

$$V_f = \left(\frac{P_i}{P_f}\right)^{1/\gamma} V_i = (0.6)^{0.6} \times 1 \times 10^{-3} = 0.736 \times 10^{-3}\,\text{m}^2.$$

Thus the final elevation of the piston is $0.736\,\text{m}$ relative to the base of the cylinder.

We now use the ideal gas equation of state to find

$$T_f = \frac{P_f V_f}{NR} = 368.0\,\text{K}.$$

We can calculate the work done by knowing

$$\Delta T = 368.0 - 300.0 = 68.0\,\text{K},$$

that is,

$$W_{if} = \Delta U = 1.5 NR\,\Delta T = 1.5 \times 0.5 \times 68.0 = 51.0\,\text{J}.$$

This work is not equal to the reduction in the potential energy of the piston and its added mass because the experimenter intervened.

Solution (d)

(1) As above, the thermodynamic system is the gas. The process is adiabatic and non-quasistatic. The work done is the loss of potential energy by the piston and its added mass. No work is done by the experimenter. In this respect this process is similar to that in part (b).

(2) The initial state has already been obtained in part (a). The pressure in the final state we also know: $P_f = 250\,\text{kPa}$, but at this stage we do not know the temperature.

(3) Since the process is adiabatic, $Q_{if} = 0$. Hence, from the first law, $W_{if} = \Delta U$, where $W_{if} = -mg\,\Delta h$. Here Δh denotes the *increase* in the height of the piston in the process and mg is the weight of the piston and its additional mass; $mg = 250\,\text{N}$.

We shall use the equations of state (3.1.1a, b) to determine the final state of the system. We know $\Delta U = \frac{3}{2} NR\,\Delta T = 1.5 \times 0.5\,\Delta T$ and hence we have

$$1.5 \times 0.5 \, \Delta T = -250 \, \Delta h, \quad \text{or} \quad \Delta T = -\frac{1000}{3} \Delta h.$$

Now the final state is an equilibrium state satisfying $P_f V_f = NR \, T_f$, where

$$V_f = (1.0 \times 10^{-3} + 1.0 \times 10^{-3} \, \Delta h),$$

and

$$T_f = 300 + \Delta T.$$

If we substitute for ΔT in terms of Δh, the equation of state becomes

$$250 \times 10^3 (1.0 \times 10^{-3} + 1.0 \times 10^{-3} \, \Delta h) = 0.5 \left(300 - \frac{1000}{3} \Delta h \right),$$

which yields

$$\Delta h = -0.24 \, \text{m}.$$

Thus we find $\Delta T = 80 \, \text{K}$ and the final temperature is $T_f = 380 \, \text{K}$. The work done on the gas is

$$W_{if} = 0.24 \times 250 = 60 \, \text{J}.$$

Note that the work done on the system in this non-quasistatic process is greater than the work done in the corresponding quasistatic process discussed in part (c). This result is demonstrated in a more formal and general way in Chapter 7.

Questions

3.6.1.   Recalculate the answers in the example based on Fig. 3.14 above, but assume now that the external pressure of the atmosphere is 100 kPa.

3.6.2.   Two gas cylinders, A and B, for which $V_A = 30$ litre, $V_B = 10$ litre, are linked by a narrow pipe, as shown in Fig. 3.15. Initially the valve, V, is closed and cylinder A is charged with an ideal monatomic gas. At this point $P_A = 10 \, \text{MPa}$, $T_A = 300 \, \text{K}$, and $P_B = 0$. The valve, V, is then slowly opened and the gas pressure is allowed to equalize. The final temperatures are $T_A = T_B = 300 \, \text{K}$,
   (a)   Is the process described quasistatic or not? Give reasons.
   (b)   What is the final pressure? What is the change in the internal energy?
   (c)   Calculate the external work performed on the system.
   (d)   Determine the total heat transferred to the system.
   (e)   Would the answers to parts (a)–(d) of this question change if this process occurred as a free expansion rather than as a slow controlled leak? Give reasons.

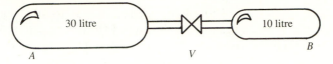

Fig. 3.15 Two cylinders linked by a narrow tube with a valve, V. Initially V is closed and B is evacuated.

3.6.3.   A cylinder, A, having $V_A = 30$ litre, is connected to a second cylinder, B, of circular cross-section, as shown in Fig. 3.16. Cylinder B is fitted with a piston of diameter 10 cm, which slides without friction or leaking. Initially A contains an ideal monatomic gas; in addition, $P_A = 10$ MPa, $T_A = 300$ K, $V_B = 0$ and the compression force on the spring is zero. The spring, which obeys Hooke's law, has a force constant of $1 \times 10^5 \, \text{Nm}^{-1}$.

The valve, V, is opened slowly and the gas leaks through from A to B, compressing the spring. Equilibrium is eventually established again at a temperature of 300 K throughout all parts of the system. Assume the pressure on the right-hand side of the piston can be neglected; neglect the volume of the connecting pipe; and assume the heat capacity of the cylinders can be ignored.
(a)   Calculate the distance by which the spring is compressed in the final equilibrium state.
(b)   Determine the work done on the gas.
(c)   Calculate the total heat transferred to the gas during this process.
(d)   Is the process described quasistatic or not?

3.6.4.   Consider again the system shown in Fig. 3.16. The initial state of the apparatus is described in question 3.6.3. An externally applied

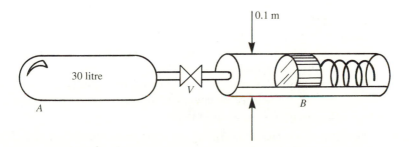

Fig. 3.16 Cylinder of fixed volume (A) coupled to a cylinder fitted with a piston and spring (B). Initially the valve V is closed, the volume of B is zero, and the compression force due to the spring is zero. The volume of the connecting tube is negligible.

force holds the piston hard against the left-hand side of the volume B while the valve V is opened fully. The external force is then gradually reduced allowing the piston to move slowly while the gas expands isothermally, compressing the spring. In the final state the temperature is 300 K throughout the system.

(a) Calculate the amount by which the spring is ultimately compressed.

(b) Establish the work of compression of the spring and the external work done on the gas. Explain the relationship between these quantities.

(c) Determine the heat transferred to the gas during this process.

3.6.5.   Reconsider the system shown in Fig. 3.16. The initial state of the apparatus is as described in question 3.6.3. An externally applied force holds the piston hard against the left-hand side of the volume B while the valve V is opened fully. The piston is then permitted to move slowly, but nevertheless quickly enough that the gas expands adiabatically, compressing the spring. The process is quasistatic. In equilibrium the temperatures in volumes A and B are the same.

(a) Calculate the temperature for the equilibrium state and determine the amount by which the spring is compressed.

(b) Establish the work of compression of the spring and the change in the internal energy of the gas. Explain the relationship between these quantities.

3.6.6.   For the final time, reconsider the system shown in Fig. 3.16. As above, the initial state of the apparatus is correctly described in question 3.6.3. In the process of interest, an externally applied force holds the piston firmly against the left-hand side of the volume B while the valve V is opened fully. The piston is then allowed to move freely so that the gas expands adiabatically, compressing the spring. In the equilibrium state the temperature is the same in both A and B.

(a) Calculate the temperature for the equilibrium state and the amount by which the spring is compressed.

(b) Determine the work of compression of the spring and the change in the internal energy of the gas. Explain the relationship between these quantities.

3.6.7.   Two cylinders, each containing an ideal monatomic gas, are fitted with frictionless pistons as shown in Fig. 3.17. The apparatus is constructed so that the two pistons, and the 10 kg mass they support, always move together and by equal amounts. The weight of the pistons and push rod assembly is negligible, the area of piston

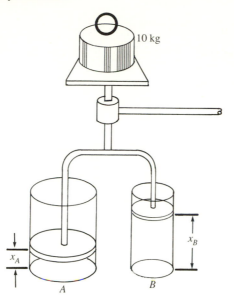

Fig. 3.17 Two cylinders supporting a single mass. The supporting rod allows vertical movement without tilting, even when the forces on the pistons are unequal.

A is $0.02\,\text{m}^2$, and the area of piston B is $0.01\,\text{m}^2$. Take $g = 10\,\text{m s}^{-2}$ and the external pressure to be zero.

Initially the push rod assembly is clamped with $x_A = 0.1\,\text{m}$ and $x_B = 0.3\,\text{m}$. In both cylinders the temperature is $300\,\text{K}$ and the pressure is $1.0 \times 10^4\,\text{Pa}$.

When the clamp is released the system is allowed to move freely until a new state of equilibrium is reached. The temperature of the gas in both cylinders is then $300\,\text{K}$.

(a) Determine the equilibrium pressure in each cylinder and the amount by which the 10 kg mass is elevated.

(b) Calculate the total external work performed on the gas in the cylinders.

(c) Calculate the net quantity of heat transferred to the gas.

(d) Establish, if you can, the external work done on the gas in *each* cylinder. If you think this cannot be done give reasons.

3.6.8. Suppose the rise of the piston yoke in question 3.6.7 is resisted by some external force so that the expansion of the gas takes place quasistatically and isothermally at $300\,\text{K}$. Reconsider parts (a)–(d) of question 3.6.7.

3.6.9. A system is set up as described in question 3.6.7 but the motion

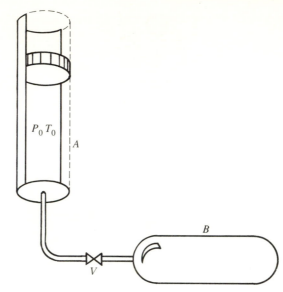

Fig. 3.18 Fixed-volume cylinder (B) linked to a cylinder (A) which is maintained at a constant pressure by the weight of a piston. The volume of the connecting tube is negligible.

of the piston yoke is restrained by some external force so that the expansion process takes place quasistatically and adiabatically.

(a) Determine the equilibrium temperature and pressure in each cylinder and the amount by which the 10 kg mass is elevated.

(b) Calculate the external work done by the gas in each cylinder.

(c) Suppose the yoke is released from its starting point again, but without any external restraint. The process in each cylinder is adiabatic. Establish, if you can, the equilibrium state for each cylinder. If you think this cannot be done on the basis of the information given, explain why not.

3.6.10. A cylinder, A, containing a monatomic ideal gas, is fitted with a frictionless leak-free piston. The axis of the cylinder is oriented vertically (Fig. 3.18) so that the weight of the piston maintains the gas at a constant pressure, P_0. The cylinder is linked to an evacuated vessel, B, by a thin capillary tube. Initially the valve, V, shown in the figure, is closed and the temperature of the gas is T_0. Both A and B are adiabatically isolated from their surroundings and heat conduction along the capillary can be ignored.

The valve, V, is opened and the gas is admitted to B. The system slowly adjusts to a new state of equilibrium, and the final pressure is P_0. Show that the equilibrium temperature of the gas in B is $\frac{5}{3} T_0$.

3.6.11. Consider the system described in question 3.6.10. Initially some monatomic ideal gas is contained in the cylinder B, the pressure and temperature being $32P_0$ and T_0, respectively. The valve, V, is opened and the gas is admitted slowly into the cylinder A. In equilibrium the pressure of the system is P_0. Assume that the material in B remains a monatomic ideal gas and that during the process it expands adiabatically and quasistatically. Show that the equilibrium temperature of the gas in B is $T_0/4$.

4
Entropy of an ideal gas

The concept of entropy evolved first during the 1850s, largely due to Clausius (1850, 1865) who coined the word. Since entropy is arguably the most important thermodynamic state function, the sooner it is introduced the better. Here we shall start with the ideal gas state equations and obtain the entropy function for the gas in order to explore its basic properties. This will be an informal introduction to entropy, but many of the properties we encounter also hold quite generally. A more formal introduction is given in Chapter 6.

4.1 Introduction to the entropy function

Primary properties

We have already seen how the work and heat associated with a change of state of an ideal gas can be calculated, provided the process of change is quasistatic and the process path is known. The work is given by

$$W = - \int P \, dV, \tag{4.1.1}$$

but we have no comparable expression for heat. Here we shall obtain such an expression, by calculating Q using the first law, $Q = \Delta U - W$. First consider an infinitesimal quasistatic process:

$$(đQ)_{\text{quasistatic}} = dU + P \, dV. \tag{4.1.2}$$

For an ideal gas, $dU = Nc_v \, dT$ and $P = NRT/V$, and hence

$$(đQ)_{\text{quasistatic}} = Nc_v \, dT + \frac{NRT dV}{V}. \tag{4.1.3}$$

If we now divide (4.1.3) throughout by T, the two terms on the right-hand side can be integrated directly. The integrals do not depend on the path of integration. Given an initial state, a, and final state, b, we have,

$$\int_a^b \left(\frac{đQ}{T} \right)_{\text{quasistatic}} = Nc_v \int_{T_a}^{T_b} \frac{dT}{T} + NR \int_{V_a}^{V_b} \frac{dV}{V}$$

$$= Nc_v \ln \left(\frac{T_b}{T_a} \right) + NR \ln \left(\frac{V_b}{V_a} \right). \tag{4.1.4}$$

This shows that while dQ is an inexact differential, dQ/T is exact for a quasistatic process, because the integral can be evaluated without specifying the path of the process. We say $1/T$ is an integrating factor for dQ and the integral defines the entropy function, S, for the ideal gas as:

$$\int_a^b \left(\frac{dQ}{T}\right)_{\text{quasistatic}} = S_b - S_a. \tag{4.1.5}$$

Hence, if S_0 is the entropy of N mol of an ideal gas at temperature T_0 and volume V_0, then at temperature T and volume V,

$$S = Nc_v \ln\left(\frac{T}{T_0}\right) + NR \ln\left(\frac{V}{V_0}\right) + S_0. \tag{4.1.6}$$

It is evident from (4.1.6) that S is a function of the state of the system; S is not related to the process of change, unlike Q or W. Since T has a properly defined value only for an equilibrium state of the system, S is also defined only for equilibrium states.

More generally we may use

$$dS = \left(\frac{dQ}{T}\right)_{\text{quasistatic}} \tag{4.1.7a}$$

or

$$(dQ)_{\text{quasistatic}} = T\,dS, \tag{4.1.7b}$$

as the defining relations for the entropy. We have shown that these definitions are useful for the case of an ideal gas and we shall expand the context further in Chapter 6.

Equation (4.1.7b) is analogous to (3.2.1) for work. By integrating over a complete quasistatic process, $a \rightarrow b$, the total heat transferred to a system can be calculated from the entropy function.

$$(Q_{ab})_{\text{quasistatic}} = \int_{S_a}^{S_b} T\,dS. \tag{4.1.8}$$

This equation is analogous to (4.1.1) for the work done on a system in a quasistatic process. Suppose we represent the locus of states in a quasistatic process by a line using (T,S) coordinates. Then (4.1.8) shows that the heat input is simply the area between this process line and the S-axis.

Example

Show that the entropy of an ideal gas is constant in a quasistatic adiabatic process.

Solution

This is simply a matter of revisiting the definitions. Adiabatic means that đ$Q = 0$ throughout the entire process, and since the process is quasi-static, đ$Q = T\,dS$. It follows that S must be constant for the complete process.

Example

A volume contains 4 kg of helium. Initially the temperature is 2000 K and the pressure is 1 MPa. The gas is cooled while the volume changes quasistatically in a process conducted so that $PV^3 = \text{const}$. The final temperature is 500 K. Determine the change in entropy by calculating $\int$ đQ/T over the path of the process.

Solution

Denote the initial and final states by a and b, respectively. From the data given and the ideal gas equation it follows that $N = 1000$ mol and the initial volume, $V_a = 16.63 \text{ m}^3$. Being a quasistatic process the system simultaneously satisfies the two process equations

$$PV = NRT \qquad \text{and} \qquad PV^3 = P_a V_a^3.$$

From the first law we have

$$\text{đ}Q = \tfrac{3}{2}NR\,dT + P\,dV = \tfrac{3}{2}NR\,dT + P_a V_a^3 \frac{dV}{V^3}$$

and eliminating P from the process equations, we have $T T_a^{-1} V_a^{-2} = V^{-2}$. Hence

$$\frac{dV}{V^3} = -\frac{dT}{2T_a V_a^2},$$

and throughout the process

$$\text{đ}Q = \tfrac{3}{2}NR\,dT - \tfrac{1}{2}NR\,dT = NR\,dT.$$

Thus

$$S_b - S_a = \int_{T_a}^{T_b} \frac{\text{đ}Q}{T} = NR\ln(500/2000) = -NR\ln(4) = -11.53 \text{ kJ K}^{-1}.$$

This agrees with $S_b - S_a$ obtained by the direct use of (4.1.4). Note, $V_b = 2V_a$, from the process equations. This result also illustrates that the entropy change does not depend on the particular quasistatic path used to evaluate the integral of đQ/T. Only the initial and final states are important. In contrast the work and heat input in the process do depend on the path chosen. For this process one can show that $W = -6.24$ MJ and $Q = -12.48$ MJ.

Questions

4.1.1. (a) Consider two states, a and b, of a fixed quantity of an ideal gas. Show that the entropy difference, given by (4.1.4), is also given by

$$S_b - S_a = N c_p \ln\left(\frac{T_b}{T_a}\right) - NR \ln\left(\frac{P_b}{P_a}\right) \qquad (4.1.9)$$

$$= N c_v \ln\left(\frac{P_b}{P_a}\right) + N c_p \ln\left(\frac{V_b}{V_a}\right) \qquad (4.1.10)$$

(b) Hence show that the locus of states of an ideal gas which are linked by a quasistatic adiabatic process can be represented by any one of the three equations:

$$TV^{\gamma-1} = \text{const} \qquad (4.1.11a)$$

$$PV^{\gamma} = \text{const} \qquad (4.1.11b)$$

$$T^{\gamma}P^{1-\gamma} = \text{const} \qquad (4.1.11c)$$

where $\gamma = c_p/c_v$.

4.1.2. Figure 4.1 shows five states of a fixed amount of an ideal monatomic gas, for which $c_v = \frac{3}{2}R$. For the states a and b: $V_a = 2\,\text{m}^3$, $P_a = 10^3\,\text{Pa}$, $T_a = 300\,\text{K}$, $V_b = 1\,\text{m}^3$, $P_b = 10^4\,\text{Pa}$. These states are

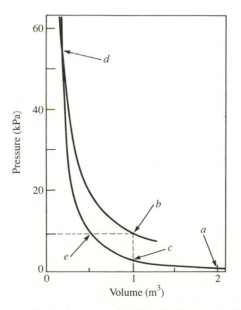

Fig. 4.1 Quasistatic adiabatic (a–c–e–d) and an isotherm (b–d) for a monatomic ideal gas.

linked by three different quasistatic processes. The purpose of this exercise is to calculate S_b-S_a by evaluating $đQ/T$ along these three paths to confirm that the numerical value obtained in each case agrees with (4.1.4). The process segments involved are defined as follows:

$a \to c \to e \to d$	is an adiabatic quasistatic path;
$d \to b$	is an isothermal process;
$e \to b$	is a constant pressure heating process;
$c \to b$	is a constant volume heating process.

Throughout the following take $S_a = 0$.

(a)   Calculate T_b, T_c, T_e, T_d, and T_d.

(b) $S_c = S_d = S_e = 0$. Why? Use (4.1.4) to explicitly confirm one of these values.

(c)   Use (4.1.4) to calculate S_b.

(d)   Calculate S_b by evaluating $\int_c^b \dfrac{đQ}{T}$ along the quasistatic con-

stant volume path $c \to b$. (Note: $đQ = Nc_v \, dT$ for constant volume heating.) This value for S_b should agree with that found in (c).

(e)   Calculate S_b by evaluating $\int \dfrac{đQ}{T}$ along the quasistatic con-

stant pressure path $e \to b$. (Do not just use (4.1.4) again.) Check that your answer agrees with (c).

(f)   Determine the heat Q transferred to the system in the isothermal process $d \to b$. Hence calculate S_b by evaluating

$$\int \frac{đQ}{T} = \frac{Q_{db}}{T}$$ along the quasistatic path $d \to b$.

(g)   The process $c \to b$ could be accomplished adiabatically by means of a stirring process. In this case $\int \dfrac{đQ}{T}$ would be zero.

Yet $S_b \neq S_c$. Why does $\int_c^b \dfrac{đQ}{T}$ not provide the value of $S_b - S_c$ correctly in this case?

4.1.3.   Calculate the heat input to an ideal monatomic gas in the quasistatic processes $a \to b$ shown in Fig. 4.2–4.7. Be sure to use the path shown and to determine the sign for Q. Given that $N = 5 \, \text{mol}$, calculate the work done on the system too.

4.1.4.   At low density (less than $0.1 \, \text{kg m}^{-3}$) water vapour conforms well to the ideal gas equation, $PV = NRT$, provided the temperature is higher than about $320 \, \text{K}$, but the heat capacity is a function of temperature. The following formula gives the specific heat capacity

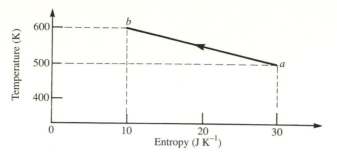

Fig. 4.2

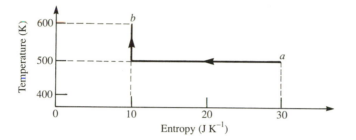

Fig. 4.3

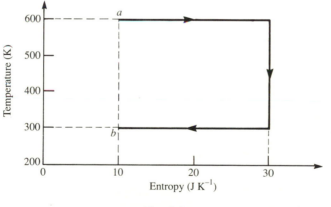

Fig. 4.4

at constant volume, as a function of T (in kelvin):

$$c_v = 1273.0 + 0.3441\, T + 2.833 \times 10^{-4}\, T^2 \,\mathrm{J\,kg^{-1}\,K^{-1}}$$

(a)   Calculate the entropy change for 1 kg of water vapour heated
      from 350 K to 1000 K at constant volume.

(b)   Calculate the entropy change for 1 kg of water vapour heated

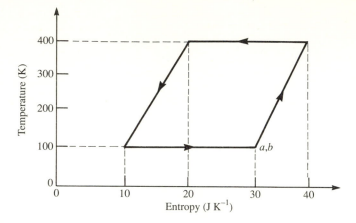

Fig. 4.5

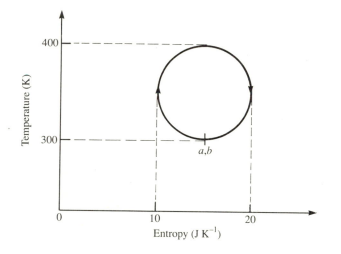

Fig. 4.6

from 350 K to 1000 K at constant pressure. (The molecular
weight of water is 0.018 kg/mol.)

(c) Suppose water vapour, initially at 350 K is compressed
adiabatically and quasistatically to 25% of its original volume.
Obtain an approximate value for $\gamma = c_p/c_v$ and hence
estimate the final temperature, assuming γ is constant. (See
eqn 4.1.11a.)

(d) Obtain an expression for the entropy of water vapour,
using the temperature and the volume as variables. Hence

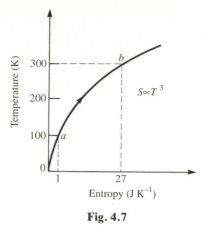

Fig. 4.7

re-estimate the final temperature in part (c) above, using the condition that $S = $ const.

4.1.5. For black-body radiation we know from (2.3.2) and (2.3.3) that

$$U = 4\frac{\sigma}{c} VT^4 \quad \text{and} \quad PV = \tfrac{1}{3}U,$$

where σ is the Stefan–Boltzmann constant given by (2.3.4) and c is the velocity of light. The purpose of this question is to show how the entropy can be calculated for a system other than a simple ideal gas. Use (4.1.7a) to define S.

(a) By applying the first law to a quasistatic process, show that the entropy can be expressed as

$$S = \frac{16\sigma}{3c} VT^3.$$

(b) Show that in a quasistatic adiabatic volume change

$$VT^3 = \text{const.}$$

(c) Given that V_a and V_b are the initial and final volume, respectively, show that the heat input in a quasistatic isothermal change of volume is given by

$$Q = \frac{16\sigma}{3c} T^4 (V_b - V_a).$$

(d) Show that, for a fixed volume, the heat capacity per unit volume defined by

$$c_v = \frac{T}{V}\left(\frac{\partial S}{\partial T}\right)_V$$

is given by

$$c_v = \frac{16\sigma}{c} T^3.$$

Further properties

We show that the energy, U, the temperature, T, and the entropy, S, are formally related by

$$T = \left(\frac{\partial U}{\partial S}\right)_{V,N}. \qquad (4.1.12)$$

This important result derives directly from the defining relation, (4.1.7b). Consider an incremental quasistatic process at constant volume and mole number. Clearly,

$$đW = 0,$$

and hence, from the first law,

$$(đQ)_{\text{quasistatic}} = (dU)_{V=\text{const}, N=\text{const}}.$$

But from (4.1.7b),

$$(đQ)_{\text{quasistatic}} = T\,dS,$$

and hence,

$$T\,dS = (dU)_{V=\text{const}, N=\text{const}},$$

from which (4.1.12) follows. Here we have assumed that U can be expressed as a function of the variables S, V, and N, which is easily justified in the ideal gas case, as we show in question 4.1.8. Notice that (4.1.12) is a relation amongst thermodynamic variables only. Process variables, such as $đQ$, and process descriptive terms, such as *quasistatic*, are not required.

A related result is obtained by combining (4.1.2) and (4.1.7b). For an incremental change of state with N constant,

$$dU = T\,dS - P\,dV. \qquad (4.1.13)$$

This is called the *Gibbs equation* in recognition of its first use by Gibbs (1873). Some authors call this the fundamental relation of thermodynamics, but we shall reserve that name for another equation. The Gibbs equation is also a relation between state variables only. How the change of state occurs is immaterial. But remember, we can interpret the two terms on the right-hand side of (4.1.13) as heat and work only when the change of state occurs quasistatically. So while the equation is expressed using state variables only, it has process implications.

Example

Show that the pressure of an ideal gas is given by

$$P = - \left(\frac{\partial U}{\partial V}\right)_{S,N}. \qquad (4.1.14)$$

Solution

From the first law,

$$dU = đQ + đW.$$

Consider a quasistatic adiabatic process. Here,

$$đW = -P\,dV, \qquad đQ = 0, \qquad S = \text{const}, \qquad N = \text{const}.$$

Thus we can write

$$(dU)_{S=\text{const},\,N=\text{const}} = -P\,dV,$$

from which (4.1.14) follows directly.

Questions

4.1.6.   Equation (4.1.6) was derived on the assumption that N, the number of moles in the system, is constant. We may, however, generalize the equation to include the mole dependence explicitly.

    We know that the energy, U, of a system comprised of two different subsystems is the sum of the energies of the individual subsystems. Energy is said to be an *extensive variable*. Similarly V and N are extensive variables. Suppose we assume that the entropy is extensive also. This is a significant new assumption because it allows us to define the entropy of a composite system, the parts of which need not be in mutual equilibrium.

  (a) Refer to equation (4.1.6). Show that if S_0 and V_0 are constants independent of N, then S defined by (4.1.6) will not be an extensive function. For example, you might show that $S(T, 2V, 2N) \neq 2S(T, V, N)$.

  (b) Show that S will be properly extensive if we define $S_0 = N\,s_0$, $V_0 = N\,v_0$, where s_0 is the entropy of 1 mol at temperature T_0 occupying a volume v_0. Here s_0 and v_0 are called the *molar entropy* and *molar volume*, respectively.

4.1.7.   (a) Use the energy equation (2.2.2) for an ideal gas, together with (4.1.6) to show that

$$U = U_0 \left(\frac{V_0}{V}\right)^{2/3} \exp\left(\frac{2(S - S_0)}{3NR}\right) \qquad (4.1.15)$$

This form of equation is known as a *fundamental relation* relation because all the thermodynamic relations for the system can be obtained from it. We may show this as follows.

(b) Use the derivative equations (4.1.12) and (4.1.14) together with (4.1.15) to re-create the equations for an ideal gas with which we started:

$$PV = NRT, \qquad U = \tfrac{3}{2}NRT.$$

(c) Assume that S is an extensive function of state. (Refer to question 4.1.6.) Show that the dependence of U on N will be correctly presented if we rewrite the fundamental relation (4.1.15) as

$$U = N^{5/3} u_0 \left(\frac{v_0}{V}\right)^{2/3} \exp\left(\frac{2S}{3NR} - \frac{2s_0}{3R}\right), \qquad (4.1.16)$$

where u_0, v_0, and s_0 are the molar energy, volume, and entropy defined in question 4.1.6.

(d) Use equation (4.1.16) to evaluate the chemical potential, μ, defined by

$$\mu = \left(\frac{\partial U}{\partial N}\right)_{S,V}, \qquad (4.1.17)$$

and show that

$$U = TS - PV + \mu N.$$

The chemical potential is an important quantity for identifying the state of equilibrium in chemical reactions. We shall consider the chemical potential again in Chapter 9. Notice that we are able to determine the chemical potential only because (4.1.16) includes the dependence of U on N. This is a consequence of our earlier assumption that S is an extensive thermodynamic property.

4.1.8. (a) Use the results obtained in question 4.1.5 to derive the following fundamental relation for black-body radiation,

$$U = \left(\frac{3}{4}\right)^{4/3} \left(\frac{c}{4\sigma}\right)^{1/3} \frac{S^{4/3}}{V^{1/3}}. \qquad (4.1.18)$$

(b) Use (4.1.12), (4.1.14), and (4.1.18) to reconstruct the original equations of state (2.3.2) (2.3.3). Note that in this case μ, defined in (4.1.17), is zero.

4.2   Entropy creation

Here we shall explore the property of the entropy function, known as entropy creation. We know that the entropy of a gas will increase when we transfer heat to the system, but the entropy of the gas may change even in the absence of a heating process. Indeed we shall see that the entropy also increases in an adiabatic non-quasistatic process. No heat is required. To understand the significance of this effect it will be necessary to introduce the second law, in Chapter 6.

A straightforward illustration of entropy creation is provided by the free expansion of an ideal gas (question 2.3.3). Let us take the gas to be monatomic. The system (Fig. 4.8) initially consists of two volumes separated by a removable partition. One side is evacuated and the other contains N mol of a monatomic ideal gas having volume V_a, pressure P_a, and temperature T_a.

The process of interest is started by removing the partition suddenly. This changes the volume available to the gas from V_a to V_b, where $V_b > V_a$. We may view this as a constraint removal process: a constraint on the state of the gas, the available volume, is replaced by a less restrictive condition.

It is a matter of experience that the gas will expand to fill the larger space, accompanied by disturbances and turbulence which will eventually subside. A new equilibrium state will then be achieved. No work is done during this process and no heat is transferred to or from the system.

We can now calculate the change in entropy due to the change from the initial to the final equilibrium state. Since no work is done, and no heat is exchanged with the surroundings, $Q = 0$, $W = 0$, and therefore $\Delta U = 0$. Hence $T_a = T_b$. Thus, from (4.1.6).

$$S_b - S_a = NR \ln \left(\frac{V_b}{V_a} \right),$$

the temperature term being zero. Now since $V_b > V_a$, the logarithm is positive, and so we have

$$S_b - S_a > 0,$$

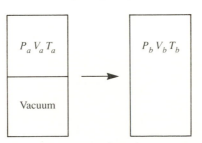

Fig. 4.8   Free-expansion process: initial state a, final state b.

showing that the entropy will have increased. Later we shall see that the entropy always increases whenever an isolated system changes state non-quasistatically. It turns out that such a change of state is also *irreversible*. This means that it is not possible to restore the system to its original state without, at the same time, changing the state of the surroundings. We shall consider this concept in more detail in Chapter 6.

Questions

4.2.1. A composite system consists of an insulated partitioned volume, shown in Fig. 4.9. Side 1 contains 2 mol of a monatomic ideal gas and side 2 has 4 mol of the same gas. Initially the two sides are separated by a fully insulated partition (adiabatic wall) and the temperatures are 500 K and 200 K on sides 1 and 2, respectively. At a particular time the wall is allowed to become conducting while keeping the volumes fixed. The partition remains impervious to the gas. Eventually the system reaches a new state of equilibrium with the same temperature on both sides, T_{12}.
   (a) Calculate T_{12} using conservation of energy.
   (b) Calculate the change of entropy for each side of the composite system, ΔS_1 and ΔS_2. Determine the total change of entropy, ΔS, by assuming (as in question 4.1.6) that S is an additive property and hence that

$$\Delta S = \Delta S_1 + \Delta S_2. \qquad (4.2.1)$$

   You should find that ΔS is positive.

4.2.2. Suppose we change the process described in question 4.2.1 so that instead of replacing the partition by a fixed conducting wall, we now replace it by a conducting frictionless piston. Initially $V_1 = 10$ litre, $V_2 = 5$ litre. Let T_{12} be the common temperature at which equilibrium is obtained, and P_{12} the common pressure.
   (a) Calculate T_{12}.
   (b) Calculate P_{12} and the new values for V_1 and V_2.

Side 1	Side 2
2 moles	4 moles
500 K	200 K

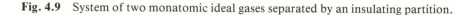

Fig. 4.9 System of two monatomic ideal gases separated by an insulating partition.

(c) Determine ΔS, as in question 4.2.1(b). (Note: ΔS should be positive.)

4.2.3. Consider the free-expansion of an ideal gas shown in Fig. 4.8. Devise an expansion process which would allow the system to change quasistatically from the initial to the final state. You will need to introduce an external heat reservoir with which the system may exchange heat. Take the reservoir to have the same temperature, T, as the gas in its initial state. Show that the external work done on the system in the quasistatic process is $-T(S_b - S_a)$.

4.2.4. Reconsider the process in question 4.2.1. In this question we illustrate how the change in entropy in a process can be used to identify the equilibrium condition.
(a) Suppose the adiabatic wall is restored before the process of heat transfer is complete. The system would then reach a new state which could be called a 'constrained state of equilibrium'. This would, of course, be one consistent with the conservation of energy. Show that conservation of energy implies T_1 and T_2 must satisfy

$$T_1 + 2T_2 = 900. \tag{4.2.2}$$

(b) Plot ΔS, the total change of entropy for the system (using eqn (4.2.1)), as a function of T_1 for the states of equilibrium permitted by (4.2.2). Use a range of values for T_1 which includes T_{12} obtained in question 4.2.1. Compare T_{12} with the value of T_1 for which ΔS is a maximum obtained from the graph. (Note: either plot your graph on graph paper or use a computer plotting routine. Do not just sketch what you imagine the result to be.)

4.2.5. Here we reconsider the process in question 4.2.2. As in question 4.2.4 we shall show how the change in entropy in a process can be used to identify the state equilibrium.
(a) Keeping T_{12} fixed at the equilibrium value found in question 4.2.2, find an expression for ΔS as a function of V_1. That is, assume that the volume re-adjustment does not proceed to completion. We know $V_1 + V_2 = 15$ litre. This allows ΔS to be determined, from (4.2.1), for each assumed value of V_1.
(b) Plot ΔS as a function of V_1 for a range values near the equilibrium value of V_1 obtained in question 4.2.2. Compare the equilibrium value of V_1 with the value for which ΔS is a maximum obtained from the graph.

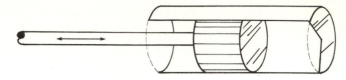

Fig. 4.10 Piston and cylinder for an expansion engine.

4.3 The ideal gas Carnot engine

When heat is transferred between two gases at different temperatures the process is non-quasistatic because the composite system is not in a state of equilibrium. How then can we achieve quasistatic heat transfer between systems at different temperatures? The answer is that we must use a matching device.

The ideal gas Carnot engine is one such device. It consists of a fixed amount of ideal gas in an expansion engine, as shown in Fig. 4.10. The gas undergoes a sequence of quasistatic processes during which it exchanges heat with two heat reservoirs, at temperatures T_1 and T_2. We shall take $T_1 > T_2$.

The term heat reservoir used here indicates a large body at a uniform temperature. The heat conductivity must be so large that the temperature is not locally affected by heat transfer in the processes being considered.

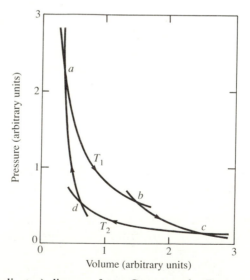

Fig. 4.11 P–V (indicator) diagram for a Carnot cycle. Processes $a \to b$ and $c \to d$ are isothermal; $b \to c$ and $d \to a$ are adiabatic processes. All steps take place quasistatically.

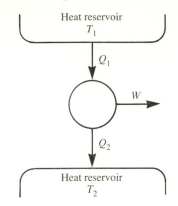

Fig. 4.12 Symbolic diagram for the energy flows in a Carnot engine. The circle symbolizes the cyclic process of the working medium.

A cup of water might be considered to be a heat reservoir for a very small device, but for a thermal power station, which may reject heat at a rate of 2000 MW, a large well-mixed lake might be regarded as a heat reservoir.

The steps of the Carnot cycle are illustrated in Fig. 4.11 in P-V coordinates. The step $a \to b$ is a quasistatic isothermal expansion process during which heat Q_1 is transferred *to the system* at temperature T_1. Similarly, $c \to d$ is a quasistatic isothermal compression process in which Q_2 is transferred *to the reservoir* at temperature T_2. The processes $b \to c$ and $d \to a$ are adiabatic and quasistatic. The quantities Q_1 and Q_2 are illustrated further in Fig 4.12.

The most interesting property of this cycle is the following relationship between the heat transferred and the reservoir temperatures:

$$\frac{Q_1}{Q_2} = \frac{T_1}{T_2}. \tag{4.3.1}$$

This result can be demonstrated simply and directly by considering the Carnot cycle using T-S coordinates. In the T-S plane the cycle is represented by a rectangle, Fig. 4.13. Note that $a \to b$ and $c \to d$ are isothermal, whereas $b \to c$ and $d \to a$ are at constant entropy (isentropic), being quasistatic adiabatic processes. Hence,

$$Q_1 = T_1(S_b - S_a),$$

and

$$Q_2 = - \text{heat input to the system in process } c \to d$$
$$= - T_2(S_d - S_c) = T_2(S_c - S_d).$$

Since $S_b = S_c$ and $S_a = S_d$, eqn (4.3.1) follows. We see that in a quasistatic transfer of energy between two systems at different temperatures,

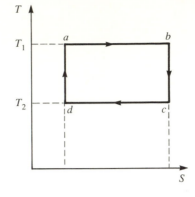

Fig. 4.13   Carnot cycle in T–S coordinates.

$Q_1 \neq Q_2$. Rather, $\Delta S_1 + \Delta S_2 = 0$, as we expect for a quasistatic process in an adiabatically isolated system.

This derivation shows how the use of the most suitable thermodynamic coordinates can make the analysis of a process very simple. Ill-chosen thermodynamic coordinates, conversely, will create untold complications.

Looking ahead we might anticipate that Fig. 4.13 will retain its simple shape even if the working medium of the Carnot engine is not an ideal gas. Indeed this expectation is correct, as we shall show in Chapter 6. On the other hand, we expect that the P–V diagram will normally cease to be applicable if we should replace the ideal gas by some other material.

Questions

4.3.1.   We shall establish (4.3.1) by reviewing the properties of ideal gas processes obtained already.

(a)   Show that the work done on the gas, W_1, in the process $a \rightarrow b$ is given by

$$W_1 = NRT_1 \ln \left(\frac{V_a}{V_b} \right) = -Q_1,$$

and that for the process $c \rightarrow d$,

$$Q_2 = NRT_2 \ln \left(\frac{V_c}{V_d} \right).$$

(Note: Q_2 = heat transferred to the reservoir at temperature T_2.)

(b)   Show that

$$T_1 V_b^{\gamma-1} = T_2 V_c^{\gamma-1},$$

and

$$T_1 V_a^{\gamma-1} = T_2 V_d^{\gamma-1},$$

where $\gamma = c_p/c_v$. Hence show that

$$\frac{V_b}{V_a} = \frac{V_c}{V_d}.$$

(c)   Use (a) and (b) to establish (4.3.1).

4.3.2.   (a)   Carefully plot a P–V diagram for the following Carnot cycle. The initial pressure and temperature are 1 bar and 300 K, respectively, the isothermal expansion is to 0.5 bar, and the temperature after adiabatic expansion is 100 K. Use graph paper or a computer plotting routine. Assume $\gamma = 1.5$ and take the amount of gas to be 1 mol.

(b)   Compute graphically the work done in the various processes and hence find the net work done.

(c)   Find Q_1/Q_2, the ratio of the heat, Q_1, transferred to the gas at 300 K, to the heat, Q_2, transferred to a reservoir at 100 K. Compare Q_1/Q_2 with the ratio of the temperatures.

4.3.3.   Define the thermal efficiency of an ideal gas Carnot engine by

$$\eta = \frac{W}{Q_1}. \tag{4.3.2}$$

Use (4.3.1) to show that

$$\eta = \frac{T_1 - T_2}{T_1}. \tag{4.3.3}$$

Here W, Q_1, T_1, and T_2 are defined in Fig. 4.12.

4.3.4.   The cycles of two different engines are represented in T–S coordinates in Fig. 4.14. Define the thermal efficiency for an engine using eqn (4.3.2). Here W is the work done by the engine for each complete cycle and Q_1 is the heat transferred to the system by a reservoir at temperature T_1. For both engines show that

$$\eta = \frac{T_1 - T_2}{2T_1}.$$

Sketch the corresponding cycles in P–V coordinates.

4.3.5.   The Stirling cycle is similar to the Carnot cycle, except that the adiabatic processes are replaced by constant volume quasistatic heat transfer processes. Figure 4.15 illustrates a Stirling cycle in P–V coordinates.

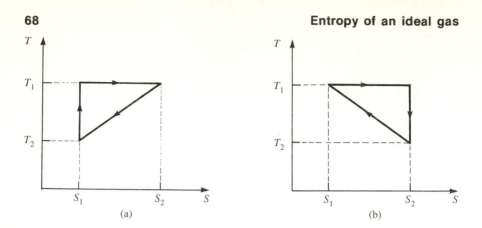

Fig. 4.14 Two hypothetical engine cycles in T–S coordinates.

(a) Show that the work done on the gas, W_1, in the process $a \to b$, is given by

$$W_1 = NRT_1 \ln\left(\frac{V_a}{V_b}\right) = -Q_1,$$

and that for the process $c \to d$, the heat, Q_2, transferred to the reservoir at temperature T_2 is

$$Q_2 = NRT_2 \ln\left(\frac{V_c}{V_d}\right).$$

(b) Hence show that

$$\frac{Q_1}{Q_2} = \frac{T_1}{T_2},$$

and establish that the thermal efficiency (defined by eqn (4.3.2)) of the quasistatic Stirling cycle is the same as that of a Carnot cycle having the same source and sink temperatures (eqn (4.3.3)).

(c) Devise an adapted form of Fig. 4.13 to represent the Stirling cycle in T–S coordinates. Use this diagram to confirm your result in part (b).

(d) Let Q_x be the heat transferred *to an external system* in the constant volume cooling process $b \to c$, and Q_y the heat transferred *to the gas* in the corresponding heating process, $d \to a$. Show that

$$Q_x = Q_y = Nc_v(T_1 - T_2),$$

given the processes are quasistatic.

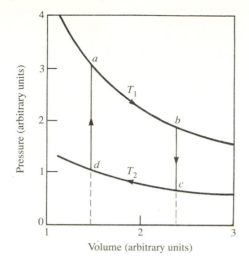

Fig. 4.15 *P–V* indicator diagram for a Stirling engine cycle.

It is evident that Q_x could be stored and then recovered subsequently, as Q_y, to reheat the gas from T_2 to T_1. Such a distributed heat store, which is called a regenerator, is a particular feature of Stirling cycle engines and refrigerators generally. Because the processes $b \rightarrow c$ and $d \rightarrow a$ effectively cancel each other, they make no net contribution to the heat input or work output. The corresponding processes in an ideal gas Carnot cycle involve equal and opposite work quantities instead of heat quantities.

This example of a Stirling cycle illustrates a general principle: the thermal efficiency of an engine which operates in a quasistatic cycle is the same as a Carnot cycle machine having the same source and sink reservoir temperatures. We shall demonstrate this result using the second law in Chapter 6. The ideal gas Ericsson cycle (see next question) provides another illustration of this principle.

4.3.6. The Ericsson cycle, shown in Fig. 4.16, is a quasistatic ideal gas representation for a regenerative gas turbine cycle. In common with the ideal gas Carnot and Stirling cycles, the system exchanges heat with two reservoirs in the isothermal expansion and compression processes, $a \rightarrow b$ and $c \rightarrow d$, respectively.

The processes linking the terminal states of the isothermal steps are constant pressure processes in the Ericsson cycle. Hence, both work and heat are exchanged in the steps $b \rightarrow c$ and $d \rightarrow a$. In the Carnot and Stirling cycles the corresponding processes are constant S and constant V processes, respectively.

Entropy of an ideal gas

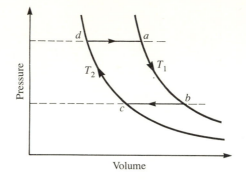

Fig. 4.16 *P–V* indicator diagram for an Ericsson engine cycle.

(a) Show that the heat and work quantities for the constant pressure processes $b \to c$ and $d \to a$ are equal and opposite, and hence that these steps do not contribute to the net input of heat and work for the cycle.

(b) Using the same notation as in questions 4.3.1 and 4.3.5, evaluate Q_1 and Q_2 in terms of T_1, T_2, V_a, V_b, V_c, and V_d. Show that

$$\frac{V_b}{V_a} = \frac{V_c}{V_d}$$

and confirm that

$$\frac{Q_1}{Q_2} = \frac{T_1}{T_2}.$$

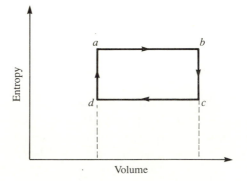

Fig. 4.17 Otto engine cycle in *V–S* coordinates.

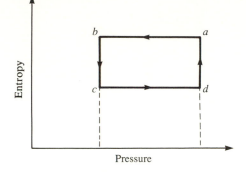

Fig. 4.18 Brayton engine cycle in P–S coordinates.

(c) Establish that the thermal efficiency (defined by eqn (4.3.2)) of the ideal gas Ericsson cycle is the same as that of a Carnot cycle having the same source and sink temperatures. Note that both cycles are quasistatic.

4.3.7. Figures 4.17 and 4.18 show ideal gas representations of the Otto and Brayton cycles, respectively. Assume the gas is monatomic.

(a) Demonstrate that the thermal efficiency of the Otto cycle is given by

$$\frac{W}{Q_{da}} = 1 - \left(\frac{V_a}{V_b}\right)^{R/c_v}.$$

(b) Show that the thermal efficiency of the Brayton cycle is given by

$$\frac{W}{Q_{da}} = 1 - \left(\frac{P_a}{P_b}\right)^{R/c_p}.$$

(c) These cycles involve quasistatic processes only. Suggest reasons why the thermal efficiency of these cycles is different from that of the Carnot cycle. (See the comments at the end of question 4.3.5.)

Part II
The Classical Laws

5
Zeroth and first laws

5.1 Generalizing

In Chapters 2–4 we introduced a number of thermodynamic functions for ideal gases. Because we knew the primary equations of state, such as (2.1.1) and (2.2.2), it was not necessary to introduce the underlying laws of thermodynamics in a formal way. Instead we explored some basic thermodynamic concepts without the need to establish the general theory first.

In this chapter we shall develop the notions of states and processes further. The aim is to enlarge upon ideas that are already familiar. Although we might proceed in a number of ways, we shall follow a traditional route which is based on the classical laws of thermodynamics. This approach acknowledges how firmly the classical laws are embedded in the literature and language of physics.

5.2 Systems, states, and variables

We will briefly reconsider some of the basic ideas already introduced about systems and thermodynamic states, and their associated variables.

System

First, a system is something we choose to consider separately from its surroundings. It will usually have a well-defined boundary so that there will be no ambiguity about what is part of the system and what is not. But a system need not be identified by a boundary, provided, for practical purposes, it can be isolated from adjacent systems.

Second, we assume the system must not be so small that atomic fluctuations become important. Thus we shall concentrate on macroscopic bodies, being systems which have very large quantities of matter compared with the atomic scale.

Our third expectation for a thermodynamic system is that, however it is disturbed or prepared, it will eventually reach a state of equilibrium when isolated from its surroundings. We shall call this the *equilibrium assumption*.

Equilibrium states

A system in an equilibrium state normally has a number of useful macroscopic physical variables (such as volume, pressure, potential energy, momentum, electric charge, and magnetization). The thermodynamic state of the system is identified by specifying parameters such as these. As we noted before, if we claim to know the state of the system, we imply that the values of all the relevant thermodynamic variables are known.

The principal feature of an equilibrium state is that it will remain constant with time, unless the system is deliberately disturbed by some external agent. Equilibrium thermodynamics is concerned with how to represent these states, and with the choice of the most useful parameters to represent the processes by which a system changes from one state to another.

It will be evident that equilibrium states must have an associated timescale. This arises in the following way. If the time after the preparation of the system is too short, the system may not have settled down to a state of equilibrium. So there is a lower limit to the time-scale for which the term equilibrium state is relevant. But if the time-scale is too long, the system constraints, which are essential for maintaining equilibrium, may have changed due to uncontrollable influences. Normally, however, it is not necessary to consider the question of time explicitly in equilibrium thermodynamics. Accordingly we shall generally assume that the system constraints, which define the equilibrium state of a system, can be fixed indefinitely.

State variables

For an ideal gas we have already encountered the variables U, V, P, N, S, and T. We noted that three of these, U, V, and N, could be regarded as constraints for the thermodynamic state of an isolated ideal gas system. These variables have externally imposed limits and so we might think of them as the independent variables in this case. On the other hand, P, S, and T, which are defined only for equilibrium states, are considered to be the dependent variables.

There is, in addition, another distinction between U, V, and N (the three constraint variables) and P and T. The variables U, V, and N are proportional to the size of the system and are called *extensive variables* for this reason. If we take a system in equilibrium and subdivide it, the value of the extensive variables for the subsystems will be proportional to the quantity of matter they each contain. Or, if we have a composite system made up of two subsystems having energy U_1 and U_2, the total energy of the composite system will be the sum, $U_1 + U_2$, and similarly for the volume. On the other hand, when an equilibrium system is subdivided the value of P and T for each subsystem is independent of its size. Accordingly, P and T are said to be *intensive variables*.

5.3 Zeroth law and thermal equilibrium

The zeroth law allows us to show formally that the temperature is a property of the thermodynamic state of a system. This principle was recognized by Maxwell (1872) who referred to it as the 'law of equal temperatures'. The modern name was adopted much later in response to a suggestion by Fowler and Guggenheim (1939) who offered the view that 'This postulate of the "existence of temperature" could with advantage be known as the zeroth law of thermodynamics'.

Since we have already been using the notion of temperature, it is reasonable to ask, what is the purpose of the zeroth law? The answer is that the zeroth law identifies the operational basis for the concept of temperature, a concept which we adopted previously in an implicit way only.

Consider three bodies Σ_1, Σ_2, and Σ_3 and assume that, given time, they can reach a state of internal equilibrium, as we would expect from the equilibrium assumption (Section 5.2). Similarly we shall accept that if any two of these bodies are put together, while isolated from the surroundings, then the composite system will also eventually reach a state of equilibrium. The process of reaching equilibrium is heat transfer. Generally speaking two bodies are said to be in *thermal equilibrium* if no heat transfer occurs when they are brought into thermal contact.

It would be useful to extend the notion of thermal equilibrium to include more than two systems. An example of this situation occurs when Σ_1, say, is in equilibrium with Σ_2 while at the same time Σ_2 is in equilibrium with Σ_3. But what happens if Σ_1 and Σ_3 are then brought into contact is not clear. This problem is addressed by the zeroth law which can be stated:

> If the bodies Σ_1 and Σ_2 are in thermal equilibrium, and bodies Σ_2 and Σ_3 are in thermal equilibrium too, then Σ_1 and Σ_3 will also be in thermal equilibrium.

This relationship, illustrated in Fig. 5.1, implies that the three bodies have some property in common. This property is the temperature. We shall now demonstrate, using the zeroth law in a formal way, that the temperature is a function of the state of the system.

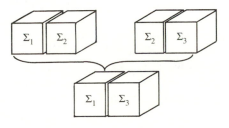

Fig. 5.1 Illustrating the zeroth law.

Consider the three systems shown in Fig. 5.1. When two systems are put in contact the resulting composite system evolves to a particular thermodynamic state of equilibrium. This implies that the equilibrium state of the composite system can be represented by fewer thermodynamic variables than are needed when the systems are separated. We shall assume that one variable can be eliminated. Denote the thermodynamic variables of Σ_1 by $p_1, q_1, r_1, \ldots$ and similarly for Σ_2 by $p_2, q_2, r_2 \ldots$. There must exist a functional relation between these variables when Σ_1 and Σ_2 are in thermal equilibrium. To be definite, we assume that this relation can be expressed in the form

$$p_1 = f(q_1, r_1, \ldots, p_2, q_2, r_2, \ldots). \qquad (5.3.1)$$

Hence, when this equation is satisfied, Σ_1 and Σ_2 will be in thermal equilibrium. Similarly, when Σ_1 and Σ_3 are in thermal equilibrium, there must exist a relation

$$p_1 = g(q_1, r_1, \ldots, p_3, q_3, r_3, \ldots). \qquad (5.3.2)$$

Thus, when Σ_1 is in equilibrium separately with both Σ_2 and Σ_3 we must have

$$f(q_1, r_1, \ldots, p_2, q_2, r_2, \ldots) = g(q_1, r_1, \ldots, p_3, q_3, r_3, \ldots). \qquad (5.3.3)$$

But the zeroth law requires that Σ_2 and Σ_3 must also be in thermal equilibrium in this situation. Consequently, there is a functional relation between $p_2, q_2, r_2, \ldots$ and $p_3, q_3, r_3, \ldots$ similar to (5.3.1). Thus the variables of Σ_1, namely $q_1, r_1, \ldots$ which appear in (5.3.3), must do so in a way which permits them to be eliminated. Hence (5.3.3) can be written in the form,

$$\theta_2(p_2, q_2, r_2, \ldots) = \theta_3(p_3, q_3, r_3, \ldots) \qquad (5.3.4)$$

where θ_2 and θ_3 are functions of the state variables for Σ_2 and Σ_3, respectively. The functions θ_2 and θ_3 are called the *empirical temperature functions* for the systems Σ_2 and Σ_3. This equation provides both necessary and sufficient conditions for Σ_2 and Σ_3 to be in thermal equilibrium.

The zeroth law has allowed us to show that there must exist a temperature function of state. Yet the zeroth law seems to be an unexceptional statement and one might ask the question: could it possibly have been different? After all, the zeroth law seems merely to imply that, if two things are each equal to a third then they are equal to each other. Is there more to it than this? Is there an analogous situation where the corresponding form of the zeroth law does not hold? In fact there are such examples, electrical equilibrium being an obvious one. If we construct a conducting ring from Cu, aqueous $CuSO_4$, aqueous $ZnSO_4$, and Zn, an electrical current will flow. While the parts of the system are in equilibrium when connected in pairs, the composite system is not in equilibrium. Hence the zeroth law has no analogy in electrical equilibrium.

Example

Figure 5.2 shows isobaric (constant pressure) equilibrium states for water, using the specific volume and the Celsius temperature as thermodynamic coordinates. These states are near the well-known anomalous point for water, where the density is a maximum with respect to changes in the temperature. It is evident that if we try to use P and V as the state coordinates in this region, two values of the temperature will be obtained for a given pair (P, V). Does this imply that the temperature is not a function of state, contrary to the zeroth law?

Solution

Figure 5.2 appears to cast doubt over the theorem we have just established. However, the problem is not in the zeroth law, but in our interpretation of it.

While it may seem to be improper to enlarge upon the interpretation of laws when inconsistencies arise, physical laws should be discarded only when there is no avenue for resolving them. In effect the focus of a problem moves from the law itself to an understanding of the 'objects' to which it applies. Here the difficulty occurs because (P, V) does not constitute a proper pair of state variables. Usually the coordinates (P, V) are quite satisfactory in practice: Watt used them for his indicator diagam, discussed in Section 3.2. But (P, V) cannot be used in all situations, although it is commonly assumed otherwise.

The selection of a consistent set of thermodynamic variables is rather like the choice of coordinates for a point in Euclidean space. For instance, we

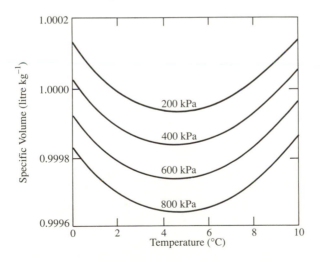

Fig. 5.2 Isobars for water near the anomalous point at 4°C. In this region the system has two possible temperatures for given coordinates $(P-V)$.

may use Cartesian coordinates, (x, y, z), or spherical polar coordinates, (ρ, θ, ϕ), to represent a given point. The choice depends on the special needs of the application, such as the geometric symmetry of the system of interest. But points in Euclidean space cannot be represented in general by an arbitrary selection of three coordinates chosen from these six. For example, use of the coordinates (ρ, x, y) would normally identify two points. Similarly, a consistent set of thermodynamic coordinates for a system cannot be established by selecting variables from a menu in an arbitrary way. In Chapters 7–10 we shall establish procedures to determine the correct variables to represent the thermodynamic state of a number of systems. We show why the state of a fluid cannot be represented by (P, V) in general and that other coordinates, such as (V, T), (P, T), or (U, V) should be used instead.

Questions

5.3.1. Adapted from Redlich (1968). Three bodies consist of Σ_1, a γ-ray emitting body, Σ_2, a γ-ray absorbing body, and Σ_3, which is made of a material transparent to γ-rays. In an experiment Σ_1 is in thermal equilibrium with Σ_3 and Σ_2 is also in thermal equilibrium with Σ_3, but it is found that when Σ_1 and Σ_2 are brought into contact the two bodies are not in thermal equilibrium, because the energy of Σ_2 increases due to γ-ray absorption. In a superficial way this example seems to be inconsistent with the zeroth law. Suggest how the zeroth law can be clarified in order to avoid this difficulty.

5.3.2. Consider two similar balloons, both inflated to the same degree. Each may individually be in equilibrium when connected with a small fixed volume of gas at the same pressure (Fig. 5.3). In addition, this equilibrium condition will be a state of stable equilibrium, that is, if the pressure in one volume or the other is disturbed slightly, then the composite system will readjust afterwards so that equilibrium will be restored again in a short time.

But when the two balloons are linked together by a small tube, the state of equilibrium would be unstable, as illustrated in Fig. 5.4.

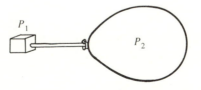

Fig. 5.3   Inflated balloon in equilibrium with a small volume to which it is linked by a narrow bore tube.

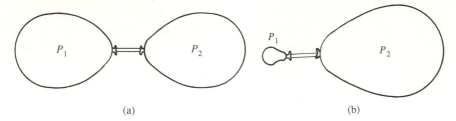

Fig. 5.4 (a) Initial state of two identical inflated balloons linked by a narrow bore tube. (b) Final state of two linked balloons after a perturbation to one of them.

The reason is that the pressure of a balloon generally decreases as it inflated, so if the pressure in one of the balloons is temporarily increased, some air would flow into the other, thus increasing the pressure of the first balloon. Eventually the first balloon would discharge all its contents into the second.

Given that the equilibrium assumption applies, the analogous effect is excluded in heat transfer by the zeroth law. What does this imply about the bodies referred to in formulating the zeroth law?

5.4 First law: energy

An adiabatic process is one in which only work interactions operate on a system. This was the approach we took in Section 3.4. Alternatively, we can define adiabatic in terms of the notion of thermal equilibrium: an adiabatic process is one performed in a time which is short compared to the time a system takes to reach a state of thermal equilibrium with its surroundings. This definition ensures that adiabatic and heat transfer processes are disjoint.

The first law, which summarizes observations showing that the internal energy is a thermodynamic function of state, can be expressed in terms of adiabatic processes:

The work W_{ab} is the same for all adiabatic processes linking two equilibrium states, a and b, of a thermodynamic system.

Thus W_{ab} is path independent for all adiabatic processes $a \rightarrow b$. It is a function of the states a and b alone. Now if we have an adiabatic process $a \rightarrow b$ made up of two subprocesses $a \rightarrow c$ and $c \rightarrow b$, then

$$W_{ab} = W_{ac} + W_{cb}.$$

Since W_{ab} does not depend on the state c, it follows that W_{ab} can be expressed as

$$W_{ab} = U_b - U_a, \tag{5.4.1}$$

for an adiabatic process. Here U_a and U_b represent the internal energy for the states a and b, respectively. Hence the adiabatic expression of the first law implies that there is an internal energy function, U, for the system.

In Section 2.2 we accepted this result for a monatomic ideal gas, U being the sum of the kinetic energies of the individual atoms. For other systems the microscopic origin of the internal energy may be less obvious. Nevertheless the first law shows that a function, U, must exist and that it depends only on the thermodynamic state variables. Normally the internal energy does not include the potential or kinetic energy of a system, viewed as a macroscopic body, because these, if relevant, can usually be expressed as separate terms.

When a process is not adiabatic, U will change due to energy transfer processes other than work. For the moment we shall exclude processes involving matter transfer. Then the additional change in U, due to a non-work energy interaction, is called heat, Q. If we now combine this definition with the adiabatic expression of the first law we have

$$\Delta U = Q + W. \tag{5.4.2}$$

This equation represents the first law in its non-adiabatic form. As we know already, Q and W are process-specific quantities; they are positive when they contribute to the energy of the system. Note that (5.4.1) and (5.4.2) are applicable to both quasistatic and non-quasistatic processes, but we can evaluate W and Q in terms of state variables only in specific cases, such as when the process is quasistatic or adiabatic.

Recognition of the first law evolved during the period 1840–50 in an extended debate in which the contributors included Joule, Helmholtz, and Meyer. Subsequently clear statements of the first law were put forward by Clausius (1850) and by Thomson (1851) who also suggested that U should be called the· *energy*.

5.5 First law processes in fluids

Here we shall consider processes in fluids, including liquids, gases, and mixtures of the two. The distinguishing feature of a fluid is its ability to be deformed quasistatically without work being performed. We shall assume that the viscosity and surface tension of the fluid can be ignored in the processes of interest here. In this situation, the only way to execute work quasistatically on the fluid is by volume displacement.

Liquids and gases are distinguished primarily by their density and compressibility. The density of a liquid is high and its compressibility (see eqn (3.5.12)) is low compared with a gas at the same temperature and pressure. In the case of an ideal gas we have simple equations of state which are directly useful for analysing processes. However, for fluids the available thermodynamic data is normally expressed in the form of empirical and somewhat complicated equations. It is therefore convenient to summarize this data in the form of a diagram.

In this section we shall use graphical representations of the thermo-dynamic properties of water and steam given in Appendix B. Figures B.1–B.3 are particularly relevant and an explanatory introduction to these diagrams is given in Appendix B. From time to time, I have used '*water*' to denote both the liquid and vapour phases as well as liquid–vapour mixtures. Inevitably some of the examples and questions have a technological flavour, which is appropriate given the influence of the practical applications of steam on the early development of thermodynamics.

Example

Calculate the latent heat of vaporization of water at 300°C.

Solution

Assume the process is a quasistatic constant pressure process. From Fig. B.3 we can establish the following state properties for 1 kg of the fluid. Denote the initial and final states by the subscripts i and f, respectively. We use lower case symbols to denote energy, volume, and so on, per unit mass.

Initial state: $u_i = 1.33\,\mathrm{MJ\,kg^{-1}}$ $\quad P_i = 8.5 \times 10^6\,\mathrm{Pa}$ $\quad v_i = 1.4 \times 10^{-3}\,\mathrm{m^3\,kg^{-1}}$;

Final state: $\quad u_f = 2.56\,\mathrm{MJ\,kg^{-1}}$ $\quad P_f = 8.5 \times 10^6\,\mathrm{Pa}$ $\quad v_f = 22 \times 10^{-3}\,\mathrm{m^3\,kg^{-1}}$.

Being a quasistatic process, we have for 1 kg,

$$w = -P\Delta v = -8.5 \times 10^6 (0.022 - 0.0014) = -0.175\,\mathrm{MJ\,kg^{-1}},$$

and hence,

$$q = \Delta u - w = 1.23 + 0.175 = 1.405\,\mathrm{MJ\,kg^{-1}}.$$

Example

Calculate the mean molar heat capacity of water at constant volume, taking $v = 50\,\mathrm{m^3\,kg^{-1}}$, between 50 and 250°C.

Solution

From Fig. 5.5 (or Fig. B.3) we obtain

$$u_i = 2.445\,\mathrm{MJ\,kg^{-1}}, \qquad u_f = 2.73\,\mathrm{MJ\,kg^{-1}}.$$

Thus

$$q = \Delta u = 0.285\,\mathrm{MJ\,kg^{-1}}.$$

Now $\Delta T = 200\,\mathrm{K}$ and hence

$$\frac{q}{\Delta T} = \frac{2.85}{200}\,\mathrm{MJ\,kg^{-1}\,K^{-1}} = 1.425\,\mathrm{kJ\,kg^{-1}\,K^{-1}}.$$

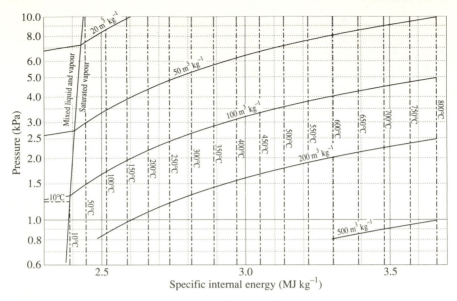

Fig. 5.5 Isotherms (dashed) and isochores (solid) for water in the low-pressure region. This figure represents detail of the bottom right-hand corner of Fig. B3.

The molar mass of water is $0.0180 \text{ kg mol}^{-1}$. Therefore the molar heat capacity is

$$c_v = 0.0180 \times 1.425 \text{ kJ mol}^{-1} \text{K}^{-1} = 25.7 \text{ J mol}^{-1} \text{K}^{-1} \approx 3.1R,$$

where R is the molar gas constant.

Example

Calculate the work and heat input when 2 kg of water undergoes an isothermal process in which the volume changes quasistatically from 200 m^3 to 0.2 m^3 at $100°C$.

Solution

From Fig. B.3 we see that the initial state is in the vapour region and the final state is in the mixed two-phase region. We consider the process in two steps, denoted by a and b below. The specific volume of saturated vapour is $1.7 \text{ m}^3 \text{kg}^{-1}$ at $100°C$.

(a)   For isothermal compression in the vapour phase we have:

$$V_i = 200 \text{ m}^3, \qquad V_f = 2 \times 1.7 \text{ m}^3 = 3.4 \text{ m}^3.$$

For the vapour phase compression, we see from Fig. B.3 that u, the specific internal energy, is approximately constant, and for the purposes of this calculation we shall treat the vapour as an ideal gas.

Now, from Fig. B.3, we have $P = 8.8 \times 10^3$ Pa when $v = 20$ m³ kg⁻¹ at 100°C. Hence for 2 kg of water vapour at 100°C,

$$PV = 8.8 \times 10^3 \times 20 \times 2 = 3.52 \times 10^5 \text{ J},$$

and hence

$$W_a = -\int P\,dV = -3.5 \times 10^5 \ln\left(\frac{V_f}{V_i}\right) = -3.52 \times 10^5 \ln\left(\frac{3.4}{200}\right) = 1.43 \text{ MJ},$$

and since $\Delta U_a = 0$, $Q_a = -1.43$ MJ.

(b) For isothermal, isobaric compression in the two-phase region we have:

$$V_i = 3.4 \text{ m}^3, \qquad V_f = 0.2 \text{ m}^3, \qquad P = 1 \times 10^5 \text{ Pa}.$$

The work is

$$W_b = (3.4 - 0.2) \times 1 \times 10^5 = 0.32 \text{ MJ}.$$

Now,

$$U_i = 2 \times 2.510 \text{ MJ}, \qquad U_f = 2 \times 0.550 \text{ MJ},$$

and so

$$\Delta U_b = -3.92 \text{ MJ}.$$

Hence,

$$Q_b = -4.24 \text{ MJ},$$

and for the complete process,

$$W = 1.75 \text{ MJ} \quad \text{and} \quad Q = -5.67 \text{ MJ}.$$

Example

A vessel of volume 15 m³ contains liquid water and steam in equilibrium at 50°C. The mass of water in the liquid phase is 1.0 kg. Determine the mass in the vapour phase.

Solution

Let m be the mass of vapour phase steam. We know the volume is an extensive variable, and we shall assume that the total volume is the sum of the volumes of the individual phases, whether they are mixed or not. From Fig. B.3 we find that the specific volume of saturated water at 50°C is 0.001 m³ kg⁻¹, and for saturated steam, 12 m³ kg⁻¹. Hence the total volume is given by

$$V = 0.0010 \times 1.0 + 12m = 15.$$

Thus $m = 1.25$ kg.

Notice that the mass of steam is not sensitive to the mass of liquid water in the vessel because the volume occupied by the liquid phase is only a small fraction of the vessel. Until the mass of water is of the order of 1000 kg or more, the volume and hence the mass of steam will not be affected significantly.

Example

A vessel containing 8 kg of saturated steam at 250°C is connected by a closed valve to a long cylinder containing a well-fitting frictionless piston (Fig. 5.6). The weight of the piston, plus the weight of the load it supports, is 1×10^4 N and the area of the piston is 0.1 m^2. Atmospheric pressure is 100 kPa. The valve is opened and the steam is slowly admitted to the cylinder. The process is adiabatic, and at its conclusion the temperature is the same in both chambers. Determine the height to which the piston is raised.

Solution

Initial state: $V_i = 8 \times 0.05 = 0.4$ m^3 (from Fig. B.3)
    $U_i = 8 \times 2.6 = 20.8$ MJ;

Final state: P_f = atmospheric pressure + $1 \times 10^4/0.1 = 2 \times 10^5$ Pa.

To determine the final state, we must use the first law. The pressure in the cylinder is constant during the process, hence

$$W = 2 \times 10^5 \times 0.1 \times h \ (\text{J}),$$

and so the specific internal energy of the final state is

$$u_f = \frac{U_i - 2 \times 10^4 h}{8} \ (\text{J}) = 2.6 - 0.0025h \ (\text{MJ}).$$

In addition the specific volume of the fluid in the final state also involves h:

$$v_f = (0.4 + h \times 0.1)/8 = 0.05 + 0.0125 h \ (\text{m}^3 \text{kg}^{-1}).$$

To determine h we use an iterative process. First select a value for h, calculate u_f and v_f. Then check whether these are consistent with $P = 0.2$ MPa.

At $h = 50$, $v_f = 0.67$, $u_f = 2.48$ (h is too small);
At $h = 70$, $v_f = 0.925$, $u_f = 2.425$ (h is too high);
At $h = 60$, $v_f = 0.8$, $u_f = 2.45$ (which is about right using Fig. B.3).

Hence the equilibrium elevation of the piston and its load will be approximately 60 m.

Questions

Where you require data on the thermodynamic properties of water in the following questions, refer to Appendix B.

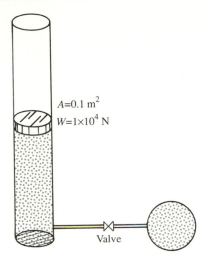

Fig. 5.6 Steam vessel linked to a cylinder fitted with a piston. Initially the valve is closed and the cylinder is empty.

5.5.1. Figure 5.5 shows in detail the low-pressure vapour region of the $\log(P)-u$ diagram for water (see Figs B.1 and B.3). The lines of constant temperature are seen to be parallel to the $\log(P)$-axis, showing that u depends on T only and indicating that steam behaves as an ideal gas in this region. (Note however, the molar heat capacity at constant volume is not constant.)

   (a) To check this suggestion, take $v = 100 \, \text{m}^3\text{kg}^{-1}$ and construct a graph of P versus T, where T is in kelvin. Is the data consistent with the expectation that $P \to 0$ as $T \to 0$? Use the gradient of your graph to calculate the molar gas constant. Compare your value with the accepted value, given in Appendix A. The molar mass of water is $0.018 \, \text{kg mol}^{-1}$.

   (b) Use the data in Fig. B.3 to check that steam obeys the ideal gas equation along a constant temperature line at 700°C from $500 \, \text{m}^3 \, \text{kg}^{-1}$ to $0.05 \, \text{m}^3 \, \text{kg}^{-1}$.

5.5.2. A vessel contains a mixture of liquid water and steam in equilibrium at 300°C. The mass of steam in the vapour phase is 1 kg and the vapour fraction, by mass, is 20%. Determine the volume of the vessel.

5.5.3. A quantity of water, initially saturated liquid, is heated in a quasistatic constant pressure process until it is in the saturated vapour state. The mass is 20 kg and the pressure is 0.5 MPa.

   (a) What is the initial temperature? Does the temperature vary during this process?

(b) Obtain the initial and final volume. Determine the volume when 50% of the fluid is in the vapour phase.

(c) Calculate the external work performed on the fluid during this process.

(d) Determine the change in the internal energy of the system in this heating process.

(e) What is the heat input?

(f) Instead of the heating process described, the volume is suddenly increased from its initial to final value (see part (b)) in a free-expansion process (that is, the process is adiabatic and no work is done). What is the final temperature? Determine the mass of water remaining in the liquid phase.

(g) At the conclusion of the free-expansion process described in (f), the system is heated at constant volume until the saturated vapour phase is reached. Determine the heat input to the system in this process and the work done.

5.5.4. The total mass of water in a volume of $0.1\,m^3$ is $2.0\,kg$. The initial temperature is $50°C$. The vessel is heated at constant volume until all the liquid is vaporized, but no further.

(a) Show that initially the fluid is a mixture of liquid and vapour phases. What fraction of the volume of the vessel is occupied by the liquid phase? What fraction of the fluid mass is in the liquid phase?

(b) What is the final temperature and pressure? Determine the heat and work input to the system.

5.5.5. An equilibrium mixture of liquid water and steam is contained in a volume of 10 litres. The initial pressure is $3.0\,MPa$ and 70% of the mass is in the vapour phase.

(a) Starting from the initial state, the vessel is cooled to $50°C$ keeping the volume fixed. Determine the heat and work input.

(b) Starting from the initial state, the vessel is cooled to $50°C$ keeping the pressure fixed. The process is quasistatic. Determine the heat and work input.

5.5.6. A pressure vessel of volume $2\,m^3$, has a maximum pressure for normal operation of $20\,MPa$. It is to be used as an energy store for heating a building. The minimum useful storage temperature in this application is $100°C$.

(a) Show that the amount of useful energy stored is maximized when the vessel contains $1220\,kg$ of water and the useful (that is, above $100°C$) stored energy is then $1.44\,GJ$.

(b) As the vessel ages the maximum operating pressure is reduced to $2\,MPa$. Determine the optimum quantity of water to maximize the stored energy for this application. Show that the capacity for energy storage is approximately 58% of the original value.

(c) Finally, the maximum permitted operating pressure of the vessel is de-rated to 0.2 MPa by the pressure vessel regulatory authority. Show that the maximum storage capacity is now approximately 0.16 GJ.

5.5.7. An evacuated vessel, of volume 40 m³, is connected through a closed valve to a second vessel, of volume 0.004 m³, which is filled with saturated liquid water at a pressure of 20 MPa. The valve is opened and the system reaches equilibrium, the temperature being then the same in both vessels. The process is adiabatic.

(a) What is the equilibrium temperature?

(b) The composite system is cooled until the temperature is 10°C. Determine the heat input to the system.

5.5.8. A vessel, *A*, of volume 2 m³, contains 2 kg of water at a temperature of 50°C. A second vessel, *B*, of volume 0.5 m³, contains 5 kg of water at 500°C. The two vessels are brough into thermal contact. Determine the equilibrium temperature of the composite system.

5.5.9. The first working steam engine with a piston was constructed by Thomas Newcomen between 1705 and 1711 (Jones 1970).

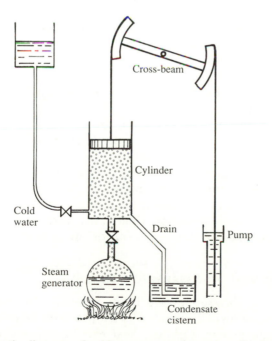

Fig. 5.7 Schematic diagram of a Newcomen engine set-up for pumping water.

Newcomen's engines were amazingly successful, some remaining in use for over 100 years. Most were used for pumping water, particularly for draining mines.

Figure 5.7 shows a sketch of a Newcomen engine. The cylinder was first filled with steam from the boiler. During this process the weight of the piston was supported by the cross-beam, balanced by the weight of the pump side. Hence the steam pressure was essentially atmospheric and no useful work was done. At the top of the stroke cold water was admitted to the cylinder, thus creating a partial vacuum. In this process, the power stroke, the piston was forced back into the cylinder by the external pressure of the atmosphere. Excess water drained from the cylinder to the condensate cistern and subsequently fresh steam was admitted to the cylinder to begin the cycle again.

The following example is based on Newcomen's later engines which were quite large. The cylinder diameter is 2 m and the working stroke of the piston is 3 m. Take the total length of the cylinder when the piston is at its maximum height to be 3.1 m. Cold water is supplied to the steam generator at 10°C at atmospheric pressure, 100 kPa.

(a)   Calculate the mass of steam required to fill the cylinder at 100°C at a pressure of 100 kPa. Determine the corresponding heat input to the steam generator.

(b)   Water, initially at 10°C, is admitted to the cylinder until the pressure drops to 50 kPa. Assume the piston does not move significantly in this process and ignore heat transfer to the walls. Determine the amount of water required to achieve this result. (In practice the heat capacity of the walls was significant.)

(c)   In the subsequent power stroke, the piston drops by 3 m. Assume this takes place quasistatically at constant pressure and determine the external work done.

(d)   The engine executes ten strokes per minute. What is the average rate of work output? Obtain an approximate value for the rate of working as a fraction of the rate of heat input to the steam generator. (Ignore the work required to pump the cooling water into and out of the cylinder.)

5.6   Steady flow processes: the enthalpy

Many practical processes involve fluids entering and leaving a process region as steady flows. Examples of such processes include fluid streams passing through pumps, turbines, heat exchangers, and mixing vessels. Since the notion of a fixed fluid system is not very useful in this situation,

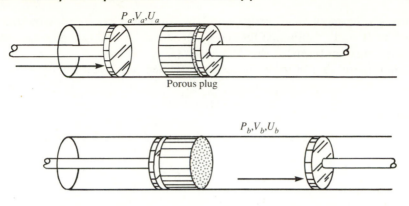

Fig. 5.8 Illustrating a Joule–Thomson (throttling) process using two pistons to maintain fixed pressures at the entry and exit of a porous plug.

the existing expression of the first law is inconvenient. Here we shall show how a new function, called the *enthalpy*, avoids this difficulty. It enables us to recast the first law in a form that can be applied easily to steady flow problems.

Consider a fluid flow in a pipe of uniform cross-section. We suppose the fluid passes through some form of restriction, such as a narrow orifice (throttle) or a porous plug. This is called a throttling, or Joule–Thomson, process. Denote the entry side by a, and the exit side by b.

We can analyse this flow as a two-state process by imagining that the flow is controlled by two pistons, as shown in Fig. 5.8. These move so as to maintain constant pressures, P_a and P_b, on each side of the restriction. Suppose the volumes of sides a and b are initially V_a and zero, respectively. Finally, the corresponding volumes are zero and V_b. Assume the process is adiabatic. Then,

$$\Delta U = U_b - U_a$$
$$W = P_a V_a - P_b V_b$$

and since

$$Q = 0 = \Delta U - W$$

then

$$U_a + P_a V_a = U_b + P_b V_b,$$

assuming changes in the kinetic and potential energy of the fluid can be ignored.

We now define the *enthalpy function, H*:

$$H = U + PV. \tag{5.6.1}$$

This is a thermodynamic property of the fluid. It is not a process variable, such as W and Q, although for certain processes W and Q can be expressed conveniently in terms of the change in H.

Hence, for the throttling process illustrated in Fig. 5.8,

$$H_a = H_b, \tag{5.6.2}$$

and, since

$$\text{(initial mass at side } a) = \text{(final mass at side } b) = M,$$

then

$$h_a = h_b \tag{5.6.3}$$

where

$$h = \frac{H}{M}, \tag{5.6.4}$$

is the *specific enthalpy* for the fluid (units: J kg^{-1}).

Thus, in an adiabatic throttling process the specific enthalpy is the same for the initial and final states of the fluid. But notice that the process is not quasistatic and so the thermodynamic state is not defined throughout the process. We cannot say that the enthalpy is constant *during* the throttling process. Rather, the specific enthalpy has the same value for the equilibrium states of the fluid at the beginning and end of the process.

Questions

5.6.1. (a) Adapt the above argument to show that if heat Q is exchanged with the fluid in the steady flow throttling process illustrated in Fig. 5.8, then

$$Q = H_b - H_a$$

or

$$q = h_b - h_a,$$

where q is the heat input per unit mass of fluid.

(b) Does this result depend on whether the heat is exchanged with the fluid on side a or b (or both)?

(c) Show that

$$q = \Delta h$$

for a constant pressure heating process. (When a substance changes phase at a fixed pressure, the corresponding latent heat is also known as the *specific enthalpy of phase change*.)

5.6.2. Suppose that the flow restriction in Fig. 5.8 is replaced by a work device. It might be a work input device, such as a compressor, or an output device, such as a turbine, or it might simply be a stirring device. This work input or output is in addition to the piston displacement work shown.

(a) Demonstrate that, provided the process is performed with constant fluid pressures P_a and P_b on either side of the device,

$$W_{rs} + Q = H_b - H_a \qquad (5.6.5)$$

where W_{rs} denotes *rotating shaft work* input. This indicates a work process in which work transfer across the system boundary utilizes a rotating shaft or its equivalent. In such a process work is performed without doing displacement work.

(b) In terms of the shaft work input per unit mass flow, w_{rs}, and the heat input per unit mass flow, q, show that

$$w_{rs} + q = h_b - h_a \qquad (5.6.6)$$

5.6.3. Show that the molar heat capacity of a fluid at constant pressure can be expressed as

$$c_p = \frac{1}{N} \left(\frac{\partial H}{\partial T} \right)_{P,N}, \qquad (5.6.7)$$

where H is the enthalpy of N mol of the fluid $\left(\text{cf. } c_v = \frac{1}{N} \left(\frac{\partial U}{\partial T} \right)_{V,N} \right)$.

5.6.4. (a) Show that the enthalpy of a fixed amount of a monatomic ideal gas can be written

$$H = \tfrac{5}{2} NRT. \qquad (5.6.8)$$

(b) For an infinitesimal change of state in a fixed amount of an ideal gas (for which we have already defined the entropy in Chapter 4) show that

$$dH = T\,dS + V\,dP. \qquad (5.6.9)$$

(Hint: use $H = U + PV$ together with (4.1.13).)

5.7 Use of the enthalpy function for water

In a fluid process where the initial and final pressures are fixed the enthalpy function takes on a role analogous to the internal energy for systems having defined initial and final volumes. Because practical processes often involve heat and work exchange with fluids in steady flow, the enthalpy functions for fluids such as water and air are extensively tabulated, much more so than the internal energy. The following questions refer to Figs B.6–B.10

which show the thermodynamic properties of water using $\log(P)$ and h as the thermodynamic coordinates. These figures are similar to the $\log(P)-u$ figures, B.1–B.5, used already.

Questions

5.7.1.   A steady flow of water, initially saturated liquid, at a pressure of 2.0 MPa, passes through a throttle. The final pressure is 0.1 MPa.
    (a)   What state variable is the same for the initial and the final states? Identify the initial and final states on the $\log(P)-h$ chart.
    (b)   What is the final temperature of the fluid?
    (c)   Suppose the fluid entering the throttle is saturated vapour at 2.0 MPa. Locate the initial and final states on the $\log(P)-h$ chart. Determine the final temperature of the fluid at 0.1 MPa in this case.

5.7.2.   The flow rate of steam entering the high-pressure stage turbine of a thermal power station is 209 kg s^{-1}. The entering condition is 16.2 MPa at 541°C and the exhaust is 4.1 MPa at 347°C.
    (a)   Determine the shaft power of the turbine, assuming that the process is adiabatic.
    (b)   Suppose that the steam passes through a throttle before entering the turbine. The turbine entry pressure is now 10 MPa, but the exhaust pressure is the same as in (a). Assume that the process is adiabatic and that the gradient of the turbine process line on the $\log(P)-h$ diagram is unchanged. Determine the reduction in the output power of the turbine.

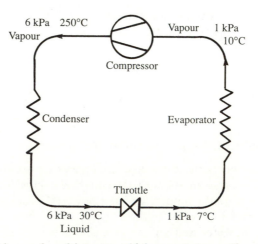

Fig. 5.9   A Rankine cycle refrigerator which uses water as the working fluid.

5.7.3. In order to avoid the use of halogenated hydrocarbons an industrial chiller employs water as the working fluid. The process cycle, called a Rankine vapour compression cycle, is represented schematically in Fig. 5.9. The flow rate of water in the circuit is $1 \, \text{kg s}^{-1}$.
   (a) Determine the rate of heat input to the low-pressure side in order to evaporate the water stream.
   (b) Calculate the required compressor shaft power.
   (c) Calculate the volume of water vapour the compressor must take in per second. Determine the minimum diameter of the compressor intake duct (assumed to have circular cross-section) if the vapour velocity is not to exceed $100 \, \text{m s}^{-1}$.

5.7.4. A water purification plant produces $10^4 \, \text{kg}$ of distilled water per hour. The process is similar to the refrigeration cycle shown in Fig. 5.9, except that the heat input to the evaporator is derived directly from the condenser and the cycle is open, rather than closed. This system is illustrated in Fig. 5.10. Assume that the supply water is at 7°C, 0.1 MPa, and that the condensate is rejected at 17°C, 0.1 MPa.

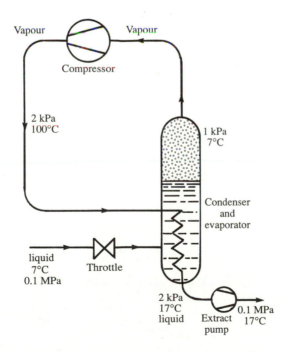

Fig. 5.10 Vapour recompression process for distilling water. By compressing the steam, the latent heat of condensation can be recovered for the evaporation process.

(a) Determine the rate of heat input to the evaporator from the condenser.

(b) What is the required shaft power of the compressor and extract pump? (Hint: see question 5.6.4.)

(c) Determine the rate of heat rejection to the surroundings.

(d) Suppose the distillation is performed instead as a simple single evaporation process at 0.1 MPa in which the latent heat of condensation is used only to preheat the incoming cold water. Compare the required rate of heat input in this process with the power requirement determined in (b).

5.7.5. In a simple distillation plant it is not possible to use the latent heat of the condensing stream to evaporate the incoming stream because the evaporating and condensing temperatures are the same. This difficulty can be avoided in a multi-effect evaporator, in which the process takes place at several different pressures, the higher-pressure stages feeding heat to the lower stages. Figure 5.11 illustrates a two-effect evaporator. In practice five or more stages might be used.

(a) Calculate the required rate of evaporation in each stage, given that all the steam generated in the high-pressure evaporator must condense in the low-pressure evaporator.

(b) Determine the rate at which heat must be supplied to the system. Compare this with the heat demand for a single-effect evaporator operating at 0.1 MPa.

(c) Calculate the required pumping power.

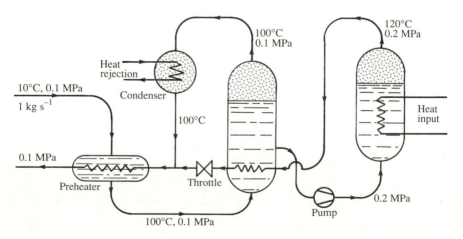

Fig. 5.11 A two-effect water evaporation system. The latent heat of condensation of the high-pressure vapour (0.2 MPa) maintains evaporation in the low-pressure stage (0.1 MPa).

5.8   Potential and kinetic energy in steady fluid flow

Equation (5.6.6), which expresses the first law for a fluid in steady flow, does not include the kinetic and potential energy of the medium. Here we show how these terms are introduced.

Consider a fluid in steady flow through a region of constant volume, R, shown in Fig. 5.12. Let δM be the mass of fluid entering and leaving R in a time interval $(t, t + \delta t)$ and let δW and δQ be the corresponding work and heat inputs to R. Note, δW cannot include volume displacement work because the volume of R is constant: it must represent rotating shaft work, or some equivalent process, such as electrical work transfer. We shall use lower case symbols, u, v, and h, to denote the specific energy, volume, and enthalpy. We suppose the fluid enters and leaves R by way of smooth ducts, at the points labelled a and b, respectively. It is assumed that the velocity of the fluid, denoted c, is axial along these ducts. The height of the duct above a particular datum level will be denoted by z, and the fluid properties at the points a and b will be labelled by the subscripts a and b.

Initially we consider the energy balance equation for the fluid system which consists of the fluid inside R at time t, together with an amount, δM, outside R at the point a. At time $t + \delta t$ the fluid element which was outside R at point a will be inside R, but there will then be an equal amount, δM, outside R at the point b. The work performed on this system in the time δt is $\delta W + P_a \delta M v_a - P_b \delta M v_b$, and the heat input is δQ. The corresponding increase in the energy of the system is $\delta M(u_b - u_a) + \frac{1}{2} \delta M(c_b^2 - c_a^2) + \delta M g(z_b - z_a)$. Hence,

$$\delta W + P_a \delta M v_a - P_b \delta M v_b + \delta Q = \delta M(u_b - u_a) + \tfrac{1}{2} \delta M(c_b^2 - c_a^2)$$
$$+ \delta M g(z_b - z_a). \qquad (5.8.1)$$

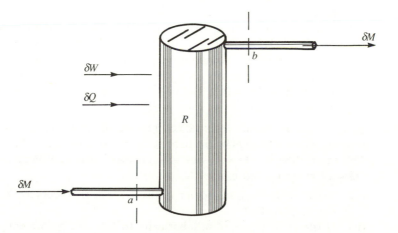

Fig. 5.12   Energy and fluid inputs and outputs for a control volume, R.

We have then, after rearranging and dividing by δM,

$$w + q = h_b - h_a + g(z_b - z_a) + \tfrac{1}{2}(c_b^2 - c_a^2) \tag{5.8.2}$$

where

$$w = \frac{\delta W}{\delta M}, \tag{5.8.3}$$

is the shaft work input to the control volume, R, per unit mass of fluid, and

$$q = \frac{\delta Q}{\delta M}, \tag{5.8.4}$$

is the heat input to the control volume, R, per unit mass of fluid. In addition, h is the specific enthalpy function, defined in (5.6.4). Equation (5.8.2), which includes the potential and kinetic energy terms, is the required generalization of (5.6.6).

Questions

5.8.1. (a) For a process involving an ideal gas show that

$$dh = c_p\, dT, \tag{5.8.5}$$

where c_p is the specific heat capacity (per kg) at constant pressure. If the process is also adiabatic and quasistatic, show that

$$dh = v\, dP = \frac{RT}{M_g}\frac{dP}{P}, \tag{5.8.6}$$

where R is the ideal gas constant (per mol), and M_g is the molar mass. (Hint: see question 5.6.4.)

(b) The pressure of air entering a nozzle is 700 kPa, the temperature is 700 K, and the velocity is 50 m s^{-1}. The pressure at the exit is 500 kPa. Assuming the process is quasistatic and adiabatic what is the exit velocity? Take $c_p = 1.00$ kJ kg^{-1} K^{-1} and $R/M_g = 0.287$ kJ kg^{-1} K^{-1}; assume the air behaves as an ideal gas.

5.8.2. Suppose the air stream referred to in question 5.8.1(b) enters a turbine in which the expansion process is quasistatic and adiabatic. The mass flow is 1 kg s^{-1}. For this problem you may neglect the velocity at entry and exit.

(a) What is the turbine power output, given the pressure at the exit is 350 kPa?

(b) If the air is throttled adiabatically to 500 kPa before entering the turbine what is the change in power output?

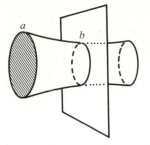

Fig. 5.13   Nozzle with a convergent entry at the cross-section a.

5.8.3.   A fluid in steady flow passes through a simple adiabatic nozzle, as in Fig. 5.13. Given that the velocity of the fluid at cross-section a is negligible, show that c_b, the axial velocity at b, is given by

$$c_b = \sqrt{2(h_a - h_b)}. \qquad (5.8.7)$$

If, in addition, the fluid is incompressible show that

$$c_b = \sqrt{2v(P_a - P_b)} \qquad (5.8.8)$$

where v is the specific volume and P is the pressure. (Note, for a compressible fluid, the state of the fluid at point b cannot be determined from P_b alone unless we have some further information about the process. We return to this difficulty in question 5.8.6.)

5.8.4.   Suppose a mass δM of a fluid enters a fixed volume at point b, as shown in Fig. 5.14, instead of continuing in steady flow. It enters through a pressure regulating valve, so that the pressure at point a is constant. In this process the internal energy of the gas contained within the fixed volume increases by δU_b. Let W be the shaft work performed on the fluid and Q the heat input between the points a and b. Show that the energy conservation relation can be expressed as:

$$W + Q = \delta U_b - \delta M\, h_a + \delta M\, g\,(z_b - z_a) - \tfrac{1}{2}\,\delta M\, c_a^2. \qquad (5.8.9)$$

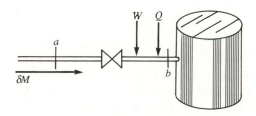

Fig. 5.14   Heat, work, and fluid inputs for a fixed volume fed from a constant pressure fluid source.

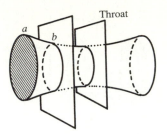

Fig. 5.15   Nozzle with a convergent entry at the cross-section a.

5.8.5.   Compressed helium, supplied through a throttle at a pressure of 10 bar and a temperature of 350 K, is used to charge a 50 litre gas cylinder. Initially the cylinder contains helium at a pressure of 2 bar and the temperature is 290 K.
   (a)   Given that the heat loss from the cylinder is negligible during this process, calculate how many mol of helium are added to the contents of the cylinder. (Refer to eqn (5.8.9).)
   (b)   What is the final temperature?

5.8.6.   (a)   A steady flow of an ideal gas enters a convergent adiabatic nozzle, as in Fig. 5.15, the entry velocity being negligible. Use equation (5.8.7) to show that the velocity, c_b, at some cross-section, b, of the nozzle is given by

$$c_b = \sqrt{2\left(\frac{\gamma}{\gamma-1}\right)(P_a v_a - P_b v_b)},$$

where $\gamma = c_p/c_v$, and v denotes the specific volume.
   (b)   The flow is adiabatic and quasistatic, so that the state of the gas obeys $Pv^\gamma = \text{const}$. Hence, show that:

$$c_b = \sqrt{2\left(\frac{\gamma}{\gamma-1}\right)P_a v_a (1 - (P_b/P_a)^{1-1/\gamma})},$$

and also that

$$\frac{c_b}{v_b} = \sqrt{\frac{2P_a}{v_a}\left(\frac{\gamma}{\gamma-1}\right)\left(\frac{P_b}{P_a}\right)^{2/\gamma}\left(1 - \left(\frac{P_b}{P_a}\right)^{1-\frac{1}{\gamma}}\right)}. \quad (5.8.10)$$

   (c)   Consider c_b/v_b as a function of P_b for a given value of P_a. Use (5.8.10) to show that c_b/v_b will be a maximum at a point b when P_b has the critical value

$$P_b = P_a \left(\frac{2}{\gamma+1}\right)^{\gamma/\gamma-1}. \quad (5.8.11)$$

(d)   Let $\dot{m}$ be the mass flow rate of the gas in a cross-section of area A. Show that

$$\frac{\dot{m}}{A} = \frac{c}{v}, \tag{5.8.12}$$

where c is the corresponding velocity and v the specific volume. Note that this ratio will be a maximum at the throat of the nozzle, shown in Fig. 5.15. Thus the mass flow rate through the nozzle will be a maximum when P_t, the pressure at the throat, has the critical value given by (5.8.11).

(e)   When the mass flow rate through the nozzle is maximized, the flow is said to be *choked*. Show that the temperature of the gas at the throat, T_t, is then given by

$$T_t = \left(\frac{2}{\gamma + 1}\right) T_a, \tag{5.8.13}$$

and that the corresponding throat velocity is given by

$$c_t = \sqrt{\frac{\gamma R T_t}{M_g}}, \tag{5.8.14}$$

which is the velocity of sound (eqn (3.5.14)). Here M_g is the molar mass, in kg per mol, and R is the molar gas constant.

What happens downstream of the throat in choked flow depends on the shape of the nozzle and the pressure at the exit of the nozzle. If the nozzle subsequently diverges and the exit pressure is less than the critical pressure given in (5.8.11), then the velocity may continue to increase in the diverging region of the nozzle. The flow will then become supersonic. Under certain conditions there may occur an abrupt change from supersonic to subsonic flow within a very short transition region. In this non-quasistatic process, which is called a *shock*, flow kinetic energy is converted into internal energy.

6

The second law

6.1 The role of the second law

In Section 4.1 we considered some properties of the entropy function for ideal gases using the definition

$$S_b - S_a = \int_a^b \left(\frac{\text{d}Q}{T} \right)_{\text{quasistatic}} \tag{6.1.1}$$

$$= Nc_v \ln \left(\frac{T_b}{T_a} \right) + NR \ln \left(\frac{V_b}{V_a} \right). \tag{6.1.2}$$

We showed that the integral is independent of the path of integration provided the process is quasistatic and we found that the entropy has some interesting properties:

1. It is useful for representing adiabatic quasistatic processes, S being constant.

2. S always increases in an adiabatic non-quasistatic process (Section 4.2).

3. When an isolated system is allowed to change state, as in heat transfer between two bodies, the equilibrium state is associated with the state of maximum entropy (Section 4.2).

4. The ideal gas Carnot cycle takes on a simple rectangular shape in (T, S) coordinates (Section 4.3).

To understand these properties we need a general theory for entropy. Equation (6.1.1) appears to be a promising starting point, but we cannot presume that the integral will always be independent of the path $a \rightarrow b$. Thus we need to know whether the quantity

$$\left(\frac{\text{d}Q}{T} \right)_{\text{quasistatic}}, \tag{6.1.3}$$

is an exact differential in general. It turns out that we are unable to decide this question on the basis of the zeroth and first laws alone and a new principle is required. Recollect that we found it necessary to employ the equations of state in order to calculate the entropy of an ideal gas. Hence for systems in which the equations of state are unknown we shall need additional information. Here we shall see that this is provided by the second law of thermodynamics. Formally, the role of the second law can be linked to the existence of an integrating factor for the differential

$$đQ = dU - X\,dx - Y\,dy - Z\,dz - \cdots, \tag{6.1.4}$$

which is called a Pfaffian form (Appendix D). Equation (6.1.4) is a generalized form for the first law for quasistatic processes in which the terms $X\,dx$, $Y\,dy,\ldots$, represent generalized work terms. When there is only one such work term it can be shown that an integrating factor can always be found, as we showed for an ideal gas in Section 4.1. The form given in (6.1.3) is then an exact differential. But when there are more work terms an integrating factor will exist only under certain conditions. According to the formulation of the second law by Carathéodory (1909) these conditions are satisfied by thermodynamic systems. (For simplified treatments see Pippard (1964), Adkins (1983), or Zemansky and Dittman (1981).) We shall take a less formal approach to the second law in this chapter, but in Chapter 7 we shall establish the entropy function as a state property in a more general way.

The second law has a pivotal role in thermodynamics. How then have we managed to proceed up to this point without making explicit use of the second law? The answer is that we have used various heuristic arguments and principles, such as the equilibrium assumption (Section 5.2) and other appeals to familiar experience. In this chapter we shall clarify these assumptions.

6.2 Irreversibility and the second law

The calculation of the entropy for an ideal gas, equation (6.1.1), requires that the process concerned should be quasistatic. Operationally 'quasistatic' is understood to be a restriction on the rate at which a process is permitted to proceed. Now the spirit of thermodynamics is to express process conditions in terms of states, or state properties. But the quasistatic condition concerns process rates and so it is not of this type. Hence we require a new concept which encompasses the quasistatic notion whilst expressing a state constraint rather than a process condition. We shall now introduce the notion of *irreversibility* to meet this need.

The word 'irreversible' is part of everyday language: in relation to past occurrences it means 'unalterable'. We are all familiar with events which fit this meaning, such as a child's balloon bursting or the burning of an art treasure. In thermodynamics the notion of an irreversible process appears to have been used first by Carnot (1824) and in this context it has a specific meaning which we may express in the following way.

Consider a particular process, x. Define an *effacing process*, $\bar{x}$, to be one which can cancel out all the thermodynamic effects

of x. Thus, if x is followed by $\bar{x}$, it would then be as if no process
had occurred at all. Such a process is said to be *reversible*.
Conversely, x is *irreversible* if no effacing process, $\bar{x}$, can
exist.

We shall illustrate the meaning of this concept by using the free expansion
of an ideal gas, a process which we introduced in question 2.3.3. The system
consists of some gas contained in a cylinder. Initially the cylinder is divided
by a partition having the gas on one side while the other side is evacuated.
When the partition is removed the gas enters the evacuated space without
any work or heat transfer to the surroundings. In this example the process
x is the free-expansion process leading ultimately to the gas occupying the
entire cylinder in a new state of equilibrium.

Any method by which we successfully restore the gas back into one side
of the cylinder, whilst leaving no evidence of that achievement, would
represent an effacing process, $\bar{x}$. There may be many such processes since
the path of $\bar{x}$ need not retrace the steps involved in x. After all, $\bar{x}$ need only
annul all the thermodynamic changes due to x; no other relationship between x and $\bar{x}$ is required. Thus the notion of irreversibility is not a process
attribute. Instead it expresses a relationship between the initial and final
states of the system.

In principle it is easy to establish that a particular process is reversible,
since we need only to identify a single effacing process. But if we want to
say that a process is irreversible we must show that all efforts to find an
effacer will fail, so how can we make such an assertion? After all, our
failure to find an effacer might just be due to ignorance. Thus an assertion
that a particular process is irreversible must always remain vulnerable to
challenge. In isolation the statement cannot be verified in an operational
way. Nevertheless an assertion that a certain process is irreversible must be
taken seriously if it has not been discredited despite our best efforts.
Accepted statements that some processes are irreversible are therefore given
special status: they represent the second law of thermodynamics.

6.3 Non-quasistatic processes are irreversible

As it happens we believe most natural processes are irreversible and so there
are countless ways to express the second law. They can be summarized by
the statement that *non-quasistatic processes are irreversible*, and we shall
take this to be an expression of the second law. It provides a link between
the process characteristic, non-quasistatic, and the state relationship,
irreversible.

There are particular forms of the second law which have historical significance. Carnot first recognized the second law, but he did so before the law
of conservation of energy was understood. Later Clausius (1850) clarified

the significance of the first law and gave an expression of the second law, known as the Clausius statement:

> No engine working in a cycle can transfer energy from a cold
> to a hot body and have no other effect.

By hot and cold we shall mean an ordering of body temperature measured using ideal gas thermometers. Now spontaneous heat transfer from a hot to a cold body is an observed process. Hence the Clausius statement means that heat transfer between bodies at different temperatures is irreversible. We can understand this, since heat transfer between bodies at different temperatures is a non-quasistatic process.

Thomson (1851) offered a different expression for the second law, now called the Kelvin–Planck statement:

> It is impossible for an engine, working in a cycle, to exchange
> heat with a single reservoir, produce an equal amount of work,
> and have no other effect.

It is a matter of experience that a real engine can be used to perform stirring or frictional work processes on a heat reservoir and thereby increase its energy, so we can transfer energy from a work store (such as an elevated weight) to a heat reservoir without producing any additional external effect. The Kelvin–Planck statement means that this process cannot be effaced. Thus the Kelvin–Planck statement asserts that stirring and frictional work processes are irreversible, which is reasonable since they are non-quasistatic processes. Hence there can be no 'inverse stirring process' in which a stirrer immersed in a stationary fluid spontaneously rotates so as to perform work on an external system. But the second law is a stronger assertion than this: it holds that the effect which this process would achieve cannot be achieved by any other means.

The fact that non-quasistatic processes are irreversible does not imply that a quasistatic process will be reversible. The reason is this: in order to establish that a process x is reversible it is necessary to find an effacing process $\bar{x}$. The operational test is clear. But the test for a quasistatic process is a qualitative one. It is necessary for the process always to be slow and the system always to be close to equilibrium. But is this enough to ensure the process is reversible?

Suppose we stretch a piece of plastic material, such as a length of plasticine. No matter how slowly we do this process, it is irreversible. The reason is that this process is much the same as a frictional process. When a surface is moved while in friction contact with another surface work is done on the system irrespective of the direction of motion. Similarly, in a plastic deformation process, work is done on the material in both the stretch and compression process. Thus the quasistatic condition is insufficient to ensure that it is reversible. Of course, one might add to the requirements for a quasistatic process. For instance, we could say that in a quasistatic process the forces acting on a system should have their equilibrium values. But again the question arises: is this sufficient to ensure that the process will be reversible in all situations?

Ultimately the difference between quasistatic and reversible is that one is a process description and the other is a state relation. The quasistatic condition is about how a process must be performed. The reversible condition, on the other hand, is concerned with the relationship between the initial and final states for a process.

To summarize, if a process is non-quasistatic, then it is irreversible. Equivalently, if a process is reversible then it must be quasistatic. But the quasistatic condition does not imply that a process is reversible; nor does irreversible imply non-quasistatic. However, for the processes in fluids we have considered previously it is reasonable to equate the terms quasistatic and reversible.

Finally, be aware that the term 'quasistatic process' is interpreted in other ways by others; it is quite common to simply take reversible and irreversible to be equivalent, respectively, to quasistatic and non-quasistatic.

Questions

6.3.1. Demonstrate that the Clausius and Kelvin–Planck statements of the second law are equivalent. You may assume that two-reservoir heat engines exist because they can be constructed in fact. (This is a machine, which operates in a cycle, and has heat interactions with a hot heat source reservoir and a cold heat sink reservoir. As a result the machine produces a work output.)

A useful procedure is to show that the contrary assumption for one statement refutes the other, as follows:

(a) Show that if the Clausius statement (CS) is true then the Kelvin–Planck statement (KPS) must be true. Do this by showing that, if the KPS is false then one could violate the CS, by using a frictional process driven by the hypothetical one-reservoir engine referred to in the KPS.

(b) Show that the KPS implies the CS. This may be done by showing that if the CS is false then one would be able to violate the KPS using a real two-reservoir heat engine, together with the hypothetical device which violates the CS.

6.3.2. Consider the following processes:

(a) the free expansion of a compressible fluid (for example, question 2.3.3 and Section 4.2);

(b) the heating of an electrical resistor as a result of the passage of an electrical current;

(c) a waterfall splashing into a pool below.

Demonstrate that each process is irreversible by showing that if it

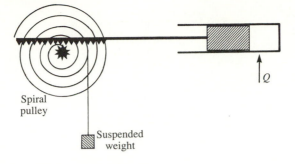

Spiral
pulley

Suspended
weight

Fig. 6.1 Mechanism for compressing a gas which maintains $P \propto 1/V$.

were reversible, another process could be devised that would violate either the Clausius or Kelvin–Planck statements of the second law.

6.3.3. Show that where a process, x, has identified subprocesses, $x_1, x_2, x_3, \ldots, x_n$, then x is reversible if and only if each subprocess x_i is reversible, $i = 1, \ldots, n$.

6.3.4. Figure 6.1 shows a fixed mass of gas being heated in a cylinder fitted with a movable piston. The spiral mechanism from which the weight is suspended ensures that the restoring force on the piston decreases as the volume expands. Hence the system is in equilibrium at any position of the piston when the gas is at constant temperature.

This device permits a quasistatic isothermal process in which the gas absorbs heat isothermally and does work at the same time, by lifting the weight. The work done would then be equal to the heat absorbed. Would such a process represent a violation of the second law? Explain your answer.

6.4 The Carnot theorem

The thermal efficiency of a cyclic ideal gas expansion engine, operating quasistatically, is given by equation (4.3.3)

$$\frac{W}{Q_1} = \frac{T_1 - T_2}{T_1}. \qquad (6.4.1)$$

The quantities W, Q_1, T_1 and T_2 are identified in Fig. 6.2, the temperatures T_1 and T_2 being measured using an ideal gas thermometer.

Below we shall use the second law to demonstrate that this equation holds for all cyclic machines which operate between the same reservoirs and employ reversible processes only. Such a machine is called a *Carnot machine*. A Carnot machine is an engine or refrigerator depending on the

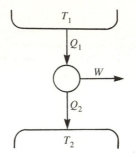

Fig. 6.2 Direction of heat and work inputs in a Carnot cycle engine.

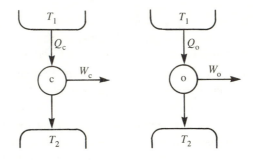

Fig. 6.3 Carnot cycle, c, and another cycle, o.

direction of heat transfer. The example presented in Section 4.3 is a particular case of a Carnot engine.

Consider two cyclic heat engines operating between two reservoirs at temperatures T_1 and T_2 as shown in Fig. 6.3. One of the machines is a Carnot engine, denoted by c, and the other we denote by o. Define the thermal efficiency, η_c and η_o, by

$$\eta_c = \frac{W_c}{Q_c}, \tag{6.4.2}$$

$$\eta_o = \frac{W_o}{Q_o}. \tag{6.4.3}$$

Now it is always possible to choose the machine o to ensure that η_o is very small, or indeed zero. (A thermally conducting rod connecting the two reservoirs would represent a zero-efficiency engine.) So arrangements can always be made to ensure that $\eta_o \leq \eta_c$. But can η_o exceed η_c? The answer is no. We can demonstrate this by showing that if $\eta_o > \eta_c$ we could construct a device to violate the second law.

Suppose we arrange that $W_o = W_c$. Since c is reversible, we can envisage

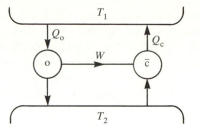

Fig. 6.4 Cycle o engine coupled to a Carnot engine in reverse, $\bar{c}$.

a coupled device in which the output work, W_o, of o is used to drive c in reverse, as in Fig. 6.4. Then,

$$\frac{Q_c}{Q_o} = \frac{Q_c}{W_c} \cdot \frac{W_o}{Q_o} = \frac{1}{\eta_c} \cdot \eta_o.$$

Hence if $\eta_o > \eta_c$, then $Q_c > Q_o$. At the end of any completed cycle there would be a net transfer of heat to the upper reservoir from the lower, without any change occurring elsewhere external to the system. This process would violate the second law (Clausius statement).

We conclude that

$$\eta_o \leqslant \eta_c. \tag{6.4.4}$$

Now if o is also reversible, then we can swap the labels o and c and apply the same argument, leading to the conclusion

$$\eta_c \leqslant \eta_o.$$

Hence, if both o and c are reversible, $\eta_o = \eta_c$.

Conversely, if $\eta_o = \eta_c$, then o must be reversible because there would exist a device, namely c, which could annul all the effects of the processes due to o.

Further, if $\eta_o < \eta_c$, then o cannot be reversible. For otherwise we could again create a result contrary to the second law by operating o in reverse, as we did with c in Fig. 6.4.

These results, which represent the *Carnot theorem*, were established by Sadi Carnot (1824) in a paper which effectively founded thermodynamics. The Carnot theorem can be summarized as follows:

$$\eta_o = \eta_c \quad \text{if o is reversible, otherwise} \quad \eta_o < \eta_c.$$

Clearly no engine can be more efficient than a Carnot engine and in this sense it represents a limiting thermodynamic ideal. All practical engines are less efficient than a Carnot engine because they are irreversible to some extent. Similar theorems apply to refrigeration and heat pump cycles (see questions 6.5.2 and 6.5.3).

The Carnot theorem shows that the efficiency of a reversible engine depends on the temperatures of the reservoirs alone. Hence (6.4.1) holds for all reversible cyclic engines. If the engine is reversible the efficiency does not depend on the internal construction, the materials employed, or the details of the cycle process, in any way at all. Only the temperatures of the reservoirs are relevant. Thus the equations we obtained for an ideal gas Carnot engine (Section 4.3):

$$\frac{Q_1}{Q_2} = \frac{T_1}{T_2} \tag{6.4.5}$$

and

$$\eta_c = \frac{T_1 - T_2}{T_1} \tag{6.4.6}$$

apply to all reversible engines. We can now see why the quasistatic Stirling and Ericsson cycles also satisfy these equations (see questions 4.3.5 and 4.3.6). The simplicity and generality of these results derive directly from the concept of irreversibility and the second law.

6.5 Thermodynamic temperature

From the Carnot theorem we know that the ratio Q_1/Q_2 (see Fig. 6.2) is the same for all reversible cyclic machines operating between the same heat reservoirs. Since the only identifying characteristic of a reservoir is its temperature, this result suggests that the Carnot cycle may be used to define a scale of temperature in a very general way.

Let θ_1 and θ_2 be the empirical temperatures for the upper and lower reservoirs, respectively (to be distinguished from the ideal gas temperatures, see Section 5.3). Then the Carnot theorem implies that there exists some function, f, of θ_1 and θ_2 alone, such that

$$\frac{Q_1}{Q_2} = f(\theta_1, \theta_2). \tag{6.5.1}$$

Because the left-hand side is in the form of a quotient, we must be able to express f as

$$f(\theta_1, \theta_2) = \frac{g(\theta_1)}{g(\theta_2)}, \tag{6.5.2}$$

where $g(\theta)$ is a function of the empirical temperature, θ, of the reservoir (see also question 6.5.4). There are two choices available to us for the function g.

1. Define a practical temperature scale, θ, and then establish, by measurement, an empirical function $g(\theta)$ which satisfies (6.5.1) and (6.5.2).

2.   Choose a particular expression for $g(\theta)$ and then use (6.5.1) and (6.5.2)
     to determine the temperature, θ, of a real system.

Both procedures have been used in the past. The first option leads to
the definition of *practical temperature* scales while the second defines the
thermodynamic temperature. Here we shall consider only the thermodynamic temperature. For this case we adopt the simplest possible non-
trivial expression for g by taking $g(\theta)$ and θ to have the same value, a definition first proposed by Thomson (1848). We shall consistently denote this
quantity, called the thermodynamic temperature, by T. Thus,

$$g(T) = T, \qquad (6.5.3)$$

and hence,

$$\frac{Q_1}{Q_2} = \frac{T_1}{T_2}, \qquad (6.5.4)$$

is the defining relation for the thermodynamic temperature. Of course
(6.5.4) does not identify the thermodynamic temperature function completely, being defined only by a ratio. It is therefore necessary to define
the thermodynamic temperature of a particular reference reservoir.
We shall use the triple point of water, as in Section 2.1, taken to be
273.16 K. The reasons for this choice are discussed further in Section
12.1.
   It is clear that the thermodynamic temperature function coincides with
the ideal gas temperature, because the defining relation, (6.5.4), is also
satisfied by the ideal gas temperature, (4.3.1). Thus the definition of the
thermodynamic temperature represents, for practical purposes, a change in
viewpoint only. It is a non-trivial change, however, because the definition
of the thermodynamic temperature is completely independent of any
physical detail of the thermometer selected.

Questions

6.5.1. (a)   Given a heat source at 1500°C which can provide heat at a
              rate of 500 MW, and a heat sink at 20°C, what is the max-
              imum rate of performing mechanical work on an external
              system? What are the maximum and minimum rates at which
              heat may be input to the sink at 20°C?

        (b)   Repeat (a) for a source temperature of 50°C. What do you
              infer about the effect of the temperature of the source of heat
              on the opportunity to perform work?

        (c)   Given source and sink temperatures of 327°C and 27°C,
              classify the following cycles (referring to Fig. 6.2) as revers-
              ible, irreversible, or unphysical, giving reasons. All quantities
              are in joule.

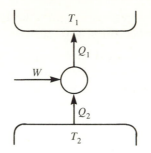

Fig. 6.5 Direction of heat and work inputs in a Carnot cycle refrigerator.

$$
\begin{array}{llll}
\text{(i)} & Q_1 = 1000, & W = 300, & Q_2 = 700; \\
\text{(ii)} & Q_1 = 500, & W = 250, & Q_2 = 350; \\
\text{(iii)} & Q_1 = 700, & W = 400, & Q_2 = 300; \\
\text{(iv)} & Q_1 = 800, & W = 400, & Q_2 = 400; \\
\text{(v)} & Q_1 = 400, & W = 800, & Q_2 = -400.
\end{array}
$$

6.5.2. Figure 6.5 shows schematically the energy inputs and outputs for a refrigerator. Define the *coefficient of performance* for the refrigerator as

$$
\varepsilon = \frac{Q_2}{W}. \tag{6.5.5}
$$

(a) Let ε_c be the coefficient of performance for a reversible, or Carnot, refrigerator. Noting that (6.5.4) holds, show that

$$
\varepsilon_c = \frac{T_2}{T_1 - T_2}, \tag{6.5.6}
$$

where T_1 and T_2 are the temperatures of the hot and cold reservoirs, respectively.

(b) Denote the coefficient of performance for another refrigerator, o, by ε_o. Use the second law to show that

$$
\varepsilon_o = \varepsilon_c \quad \text{if o is reversible, otherwise} \quad \varepsilon_o < \varepsilon_c.
$$

(c) The source and sink temperatures of a refrigerator are $-73°\text{C}$ and $27°\text{C}$, respectively. Classify the following cycles (referring to Fig. 6.5) as reversible, irreversible, or unphysical, giving reasons. All quantities are in joule.

$$
\begin{array}{llll}
\text{(i)} & Q_1 = 1000, & W = 300, & Q_2 = 700; \\
\text{(ii)} & Q_1 = 500, & W = 250, & Q_2 = 350; \\
\text{(iii)} & Q_1 = 700, & W = 400, & Q_2 = 300; \\
\text{(iv)} & Q_1 = 800, & W = 400, & Q_2 = 400; \\
\text{(v)} & Q_1 = 400, & W = 800, & Q_2 = -400.
\end{array}
$$

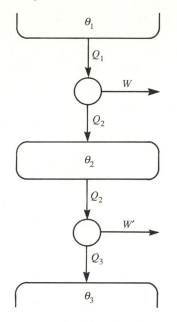

Fig. 6.6 Two Carnot cycle engines linked through an intermediate reservoir. The empirical temperatures of the reservoirs are labelled θ_1, θ_2, and θ_3.

6.5.3. The functioning of a heat pump is similar to a refrigerator, but the normal purpose is to transfer heat to the high-temperature reservoir, rather than to cool the low-temperature reservoir. With reference to Fig. 6.5 the coefficient of performance of a heat pump is defined as

$$\kappa = \frac{Q_1}{W}. \tag{6.5.7}$$

(a) Let κ_c be the coefficient of performance for a reversible (Carnot) heat pump. Use (6.5.4) to show that

$$\kappa_c = \frac{T_1}{T_1 - T_2}. \tag{6.5.8}$$

(b) Denote the coefficient of performance for another heat pump, o, by κ_o, and use the second law to show that

$$\kappa_o = \kappa_c \quad \text{if o is reversible, otherwise} \quad \kappa_o < \kappa_c.$$

(c) The source and sink temperatures for a heat pump are 6°C and 37°C, respectively. Classify the following cycles (referring to Fig. 6.5) as reversible, irreversible, or unphysical, giving reasons. All quantities are in joule.

(i) $Q_1 = 1000$, $W = 100$, $Q_2 = 900$;
(ii) $Q_1 = 1500$, $W = 10$, $Q_2 = 1490$;
(iii) $Q_1 = 90$, $W = 5$, $Q_2 = 85$;
(iv) $Q_1 = -1000$, $W = -100$, $Q_2 = -900$;
(v) $Q_1 = -90$, $W = -5$, $Q_2 = -85$.

6.5.4. Consider two reversible cyclic machines linking two reservoirs at empirical temperatures θ_1 and θ_3, as in Fig. 6.6. Show that f, defined in (6.5.1), must satisfy

$$f(\theta_1, \theta_3) = f(\theta_1, \theta_2)f(\theta_2, \theta_3) = \frac{f(\theta_1, \theta_2)}{f(\theta_3, \theta_2)} ,$$

where θ_2 is the empirical temperature of some intermediate reservoir. Hence show that f must take the form of a quotient, as in eqn (6.5.2).

6.5.5. Assume the coefficient of performance of a particular refrigerator (question 6.5.2) is 25 per cent of that for a Carnot refrigerator. Given the ambient temperature is 20°C calculate the work input to the refrigerator in the following processes:
(a)   freeze 1 kg of water (0°C, latent heat 334 kJ kg^{-1});
(b)   liquefy 1 kg of N_2 (-196°C, latent heat 199 kJ kg^{-1});
(c)   liquefy 1 kg of H_2 (-252.8°C, latent heat 456 kJ kg^{-1});
(d)   liquefy 1 kg of He (-268.9°C, latent heat 20.6 kJ kg^{-1}).

6.5.6. A body is cooled from an initial temperature, T_a, to a final temperature, T_b. The mass of the body is m and the specific heat of the body for this process is constant, c. The temperature of the heat sink for the refrigerator is also T_a. Show that the minimum work input to the refrigerator for this process is given by

$$W_{\min} = mc \left\{ T_b - T_a - T_a \ln \left(\frac{T_b}{T_a} \right) \right\} . \tag{6.5.9}$$

6.5.7. A heat pump is used to heat a body from an initial temperature T_a, to a final temperature, T_b. The mass of the body is m, the specific heat of the body for this process is a constant, c, and the temperature of the heat source for the heat pump is also T_a. No external work is performed on the body.
(a)   Show that the minimum input of work to the heat pump in this process is given by (6.5.9).
(b)   Show that (6.5.9) is also the maximum work output of an engine in a process for which the initial and final temperatures of the body are T_b and T_a, respectively. Here $T_b > T_a$. The engine rejects heat to a reservoir at temperature T_a.

6.6 The inequality of Clausius

For two reservoirs

When we analysed refrigerators and engines in the preceeding section we defined the heat quantities so that they were always positive. In this section we shall focus on the heat and work inputs to the system itself. To do this we shall take heat and work interactions with the system to be positive when they are inputs to the system.

First we shall re-express the Carnot theorem for an engine in terms of Q_1, Q_2, and W, defined in Fig. 6.7. In this case $W = -|W|$, $Q_1 = |Q_1|$, and $Q_2 = -|Q_2|$. Hence the Carnot theorem can be expressed as

$$\frac{|W|}{Q_1} = \frac{-W}{Q_1} \leq \frac{T_1 - T_2}{T_1}.$$

Since the process is cyclic, the first law requires

$$W + Q_1 + Q_2 = 0,$$

and hence,

$$\frac{Q_1 + Q_2}{Q_1} \leq \frac{T_1 - T_2}{T_1},$$

which, on rearranging yields,

$$\frac{Q_1}{T_1} + \frac{Q_2}{T_2} \leq 0. \tag{6.6.1}$$

Here the equality holds if the cycle is reversible, otherwise the inequality applies. This equation is the inequality of Clausius (1865, 1867) for a process in which a system interacts with two reservoirs only. The same expression is obtained for a refrigerator cycle, but in that case the correct form of the Carnot theorem is that given in question 6.5.2. In addition, for a refrigerator, $|W| = W$, $Q_1 = -|Q_1|$, $Q_2 = |Q_2|$.

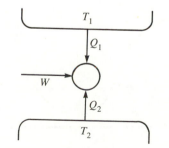

Fig. 6.7 Direction of heat and work inputs for a Carnot cycle to demonstrate the inequality of Clausius.

Systems with two state coordinates

Consider a system which interacts (i) with a set of heat reservoirs at different temperatures, sequentially, and (ii) with a work reservoir. The thermodynamic state of the system moves in a cyclic path, c, as in Fig. 6.8. We label the coordinates used to represent the state of the system X and Y. For an ideal gas we could use P and V.

We assume that we can regard the cycle c as if it were made up of a large number of elementary Carnot cycles, as indicated in Fig. 6.9. To be definite, we suppose the system is a fluid for which we can use P and V as thermodynamic coordinates. Each elementary cycle is constructed so that the isothermal heat transfer processes occur at the boundary of the process c.

Consider one of these elementary cycles. Let T_1 and T_2 to be the temperatures of the upper and lower boundaries, as shown in Fig. 6.9. The corresponding heat inputs to the system are Q_1 and Q_2. Then from (6.6.1),

$$\frac{Q_1}{T_1} + \frac{Q_2}{T_2} \leq 0. \tag{6.6.2}$$

Since this relation applies to each elementary Carnot cycle embedded in the boundary of the cycle c, we have

$$\sum_i \frac{Q_i}{T_i} \leq 0, \tag{6.6.3}$$

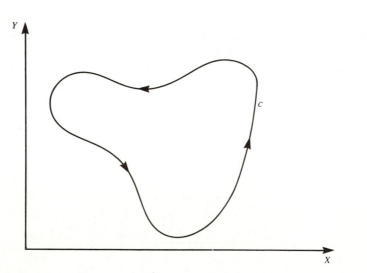

Fig. 6.8 An arbitrary cyclic process, c, which can be represented using only two state coordinates.

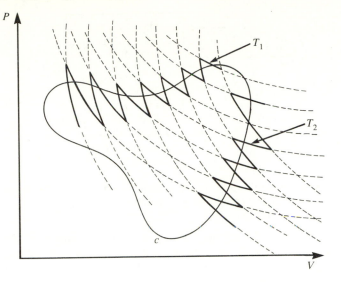

Fig. 6.9 Decomposition of an arbitrary cycle, represented in two coordinates, (P–V) into Carnot subcycles.

where the sum is over the discrete isothermal increments, i, in the cycle c. In the infinitesimal limit, the sum becomes an integral and we have for the cyclic path c,

$$\oint_c \frac{\text{d}Q}{T} \le 0. \tag{6.6.4}$$

The equality applies if the process c is reversible and the inequality if c is irreversible. Equation (6.6.4) is the inequality of Clausius for this class of system.

General systems

A difficulty with the above demonstration for systems with two state coordinates is that we assumed the cycle, c, may be represented as a superposition of simple Carnot cycles. Figure 6.9 makes this assumption look reasonable, but only because we assumed that the cycle could be represented using two state variables. We cannot therefore rely on this demonstration of the inequality of Clausius for systems requiring more than two thermodynamic coordinates. The following demonstration avoids this difficulty.

The process, c, may be represented as an interaction between a single heat reservoir, Σ_{res}, at temperature T_{res} and the system, Σ, as shown in Fig. 6.10. Here the transfer of heat between Σ_{res} and Σ is accomplished reversibly using a Carnot cycle refrigerator. The system is quite separate

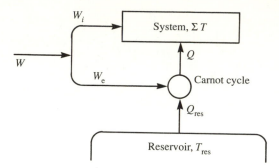

Fig. 6.10 Heat and work inputs to a system Σ which is taken through a cyclic process whilst interacting reversibly with a single heat reservoir.

from the Carnot cycle refrigerator. We could equally well have used a Carnot cycle engine and the following discussion would still apply.

Consider an increment in the cycle of Σ, taking the temperature of the boundary of Σ as T and the heat input to Σ as δQ. δQ_{res} is the heat transferred from Σ_{res} to the Carnot machine. From the Carnot theorem we have

$$\frac{\delta Q}{\delta Q_{\text{res}}} = \frac{T}{T_{\text{res}}}.$$

Hence when Σ has a change of state $a \rightarrow b$,

$$\sum_a^b \frac{\delta Q}{T} = \sum \frac{\delta Q_{\text{res}}}{T_{\text{res}}}.$$

Taking the limit in which the steps in this processes are infinitesimal, we have for a cyclic process, c, which begins and ends with Σ in the same state,

$$\oint_c \frac{\mathrm{d}Q}{T} = \frac{Q_{\text{res}}}{T_{\text{res}}}, \tag{6.6.5}$$

where

$$Q_{\text{res}} = \oint_c \mathrm{d}Q_{\text{res}} \tag{6.6.6}$$

is the total heat transferred to the Carnot machine from Σ_{res} at temperature T_{res} in the complete cycle c.

Let W be the total work done on Σ and the Carnot refrigerator in a whole cycle of the system, Σ, as in Fig. 6.10. Now the work, W, performed on a system which exchanges heat with a single reservoir, must be positive or zero. For if W were negative, there would be a net output of work in a complete cycle, contrary to the Kelvin–Planck statement of the second law.

Since the internal energy of Σ returns to the same value at the end of the cycle, c, the net heat, Q_{res}, transferred to the refrigerator over a complete cycle must be either negative or zero.

$$Q_{res} \leq 0.$$

Moreover, if c is reversible, then the sign of Q_{res} will change when the cycle c is traversed in the reverse direction. But Q_{res} must remain negative or zero. Thus,

$$Q_{res} = 0, \quad \text{if the cycle } c \text{ is reversible, otherwise} \quad Q_{res} < 0,$$

and hence from (6.6.5),

$$\oint_c \frac{\mathrm{d}Q}{T} \leq 0, \tag{6.6.7}$$

where the equality is necessary and sufficient for the cyclic process to be reversible. This is the inequality of Clausius in a general context.

Question

6.6.1. Equation (6.6.1) was derived above on the assumption that the cycle being analysed was an engine cycle. Show that the same result is obtained using a refrigerator cycle.

6.7 The entropy function

Equilibrium systems

We shall use the inequality of Clausius to show how the equation for calculating entropy differences, (6.1.1), can be adapted for any thermodynamic system. Consider the cyclic process shown in Fig. 6.11. The cycle passes through two equilibrium states, a and b, which are connected by two labelled paths, r and o. The path $b \rightarrow a$, labelled r, is reversible; the path $a \rightarrow b$, labelled o, may be either reversible or irreversible. The inequality of Clausius requires that

$$\oint \frac{\mathrm{d}Q}{T} = \int_a^b \left(\frac{\mathrm{d}Q}{T}\right)_{\text{path o}} + \int_b^a \left(\frac{\mathrm{d}Q}{T}\right)_{\text{path r}} \leq 0. \tag{6.7.1}$$

In particular, if o is reversible, then the complete cycle is reversible and the equality holds. It then follows that

$$\int_a^b \left(\frac{\mathrm{d}Q}{T}\right)_{\text{path o}} = \int_a^b \left(\frac{\mathrm{d}Q}{T}\right)_{\text{path r}},$$

and therefore $\int_a^b (\mathrm{d}Q/T)_{\text{rev}}$ is the same for all reversible paths $a \rightarrow b$. Hence $\mathrm{d}Q/T$ is an exact differential for a reversible process. We may therefore

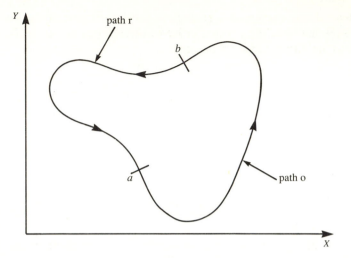

Fig. 6.11 Cyclic process which passes through equilibrium states a and b. Path r involves only reversible processes; path o may be reversible or irreversible.

generalize the defining equation for the entropy function which we first introduced in Section 4.1:

$$dS = \left(\frac{\mathrm{d}Q}{T}\right)_{rev},$$ (6.7.2)

and

$$S_b - S_a = \int_a^b \left(\frac{\mathrm{d}Q}{T}\right)_{rev},$$ (6.7.3)

where the subscript rev denotes that the process is reversible. Clearly in a reversible process the heat transferred to a system can be calculated in terms of the entropy as

$$(Q_{ab})_{rev} = \int_a^b T\,dS.$$ (6.7.4)

This equation is analogous to (3.2.2). Note that the qualification *quasi-static*, which we used in Chapter 4, has been replaced here by the requirement that the process should be *reversible*. As in the work examples, the heat input in a reversible process can be obtained as the area between the process line and the $T = 0$ axis on a (T, S) graph.

Composite systems

Equation (6.7.3) allows us to determine the difference in the entropy of two states of a system in equilibrium. In applying this equation it is natural to think of T as the temperature of the system and $\text{d}Q$ as the heat transferred to the system. This would suggest that we may define the entropy only for systems which are in equilibrium at a well-defined temperature. However, this restriction is unnecessary. We shall now show that the definition of entropy includes composite systems, the separate parts of which are in *internal* equilibrium, whilst not necessarily being in *mutual* equilibrium with each other. We shall also show that the change in the entropy of a composite system is the sum of the changes in the entropy of its parts. We have already used this result from time to time; here we shall demonstrate it.

Suppose a composite system, Σ_c, shown in Fig. 6.12, has two adiabatically separated subsystems, Σ_1 and Σ_2, each of which is in equilibrium internally. They need not be in equilibrium with each other and so their temperatures will generally be different. We may change the entropy of each subsystem by reversible exchange of heat with a reservoir, Σ_{res}, at a temperature T_{res}. In this process a Carnot cycle machine would be required, in the way described in Section 6.6. The entropy changes for Σ_1 and Σ_2, ΔS_1 and ΔS_2, may be expressed in terms of the heat transferred from the reservoir to the Carnot machine, Q_1 and Q_2, as in Fig. 6.12. Hence, using (6.6.5) and (6.7.3) we have

$$\Delta S_1 = \frac{Q_1}{T_{\text{res}}} \qquad \text{and} \qquad \Delta S_2 = \frac{Q_2}{T_{\text{res}}} \ .$$

Since the system temperature does not appear here, we may apply the same form of expression to the composite system, Σ_c. Let ΔS_c denote the entropy change for Σ_c. The total heat transferred from the reservoir to the

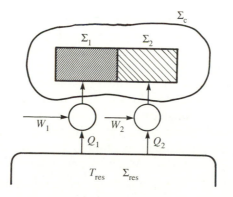

Fig. 6.12 Reversible heat transfer to two systems, Σ_1 and Σ_2, from a heat reservoir, Σ_{res}. Σ_1 and Σ_2, which are not in thermal equilibrium, are separated by an adiabatic wall.

Carnot machine is Q_c. Then we may define the change in the entropy of Σ_c:

$$\Delta S_c = \frac{Q_c}{T_{\text{res}}}. \tag{6.7.5}$$

Since $Q_c = Q_1 + Q_2$ we have

$$\Delta S_c = \frac{Q_1 + Q_2}{T_{\text{res}}} = \Delta S_1 + \Delta S_2. \tag{6.7.6}$$

Hence the entropy change of a composite system, defined by (6.7.5), is the sum of the entropy changes of its subsystems. If we integrate (6.7.6) we would obtain a relation between S_c, S_1, and S_2 which includes undefined constants of integration. We shall further assume that we may impose the condition

$$S_c = S_1 + S_2. \tag{6.7.7}$$

This means that the constants of integration cannot be chosen arbitrarily. Ultimately this assumption is justified by the third law of thermodynamics, discussed in Section 9.2.

Questions

Where necessary in the following questions you should refer to the thermodynamic data on water and steam in Appendix B.

6.7.1. A solid of mass m has specific heat c. Show that the change in the entropy of the mass when it is heated from T_a to T_b is

$$S_b - S_a = mc \ln(T_b/T_a) \tag{6.7.8}$$

where the temperature is in kelvin. Ignore volume changes.

6.7.2. (a) 10 kg of steam at 400°C at a pressure of 50 MPa is expanded reversibly at constant temperature until the volume is 200 m³, and then adiabatically and reversibly until the volume is 500 m³. Determine the final temperature, and the heat and work input to the system.

(b) The process described in part (a) is changed. First the steam expands adiabatically and reversibly until the volume is 500 m³. The system is then heated at constant volume until the temperature is the same as the final state obtained in part (a). Obtain the heat and work input to the system.

6.7.3. In Section 4.2 we obtained the temperature and pressure of an ideal gas as partial derivatives of U with respect to S and V, respectively. Here we illustrate the use of these relations numerically. To obtain approximate values for these partial derivatives use the

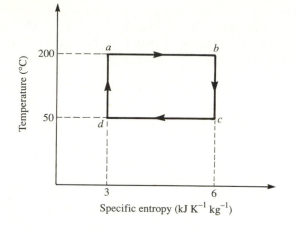

Fig. 6.13   States of Carnot cycle in which steam is the working fluid.

graphical data given in Fig. B.5. In the following the lower case
symbols refer to the specific quantities.
(a)   Evaluate $(\partial u/\partial s)_v$ along the line $v = 0.2\,\mathrm{m^3\,kg^{-1}}$ between
$s = 4$ and $s = 5\,\mathrm{kJ\,kg^{-1}\,K^{-1}}$. Compare with the corresponding temperature.
(b)   Evaluate $-(\partial u/\partial v)_s$ along the line $s = 7\,\mathrm{kJ\,kg^{-1}\,K^{-1}}$ between $v = 0.05$ and $0.1\,\mathrm{m^3\,kg^{-1}}$. Compare with the corresponding pressure.

6.7.4.   A heat pump (see question 6.5.3) is used to heat a body in a reversible process. Let T_0 be the temperature of the cold reservoir for
the heat pump. Show that the heat provided to the heat pump
from the reservoir is $T_0 \Delta S$, where ΔS is the increase in the
entropy of the body due to the process.

6.7.5.   In an ideal gas Carnot cycle the net work done on the system in
the two reversible adiabatic processes is zero. This does not necessarily apply to other systems. Consider the following reversible
cycle in which the working substance is 1 kg of water. The T–S
coordinates of the cycle $a \rightarrow b \rightarrow c \rightarrow d \rightarrow a$ are shown in Fig. 6.13.
(a)   Determine the heat and work input to the system in each
subprocess.
(b)   Confirm that

$$\frac{|W|}{Q_1} = \frac{T_1 - T_2}{T_1}$$

where W is the net work input in the cycle and Q_1 is the
heat input in the subprocess $a \rightarrow b$.

6.7.6.   A body of mass m undergoes a heating process. The initial and
final temperatures are T_a and T_b respectively. Consider the limit
of a very large mass m so that $T_b \approx T_a$. Show that (6.7.8) reduces
to the equation for the entropy increase of a reservoir:

$$\Delta S = \frac{Q}{T_a}$$

where Q is the heat input to the mass. You may use $\ln(1 + x) \approx x$
when $x \ll 1$.

6.7.7.   One kg of water (liquid and vapour phases) at a pressure of 2 kPa
occupies a volume of 50 m^3. The system is compressed reversibly
and adiabatically until the pressure is 10 MPa. Obtain the final
volume and temperature and the work done on the system.

6.7.8.   A cylinder contains 0.1 kg of steam at 450°C, as shown in
Fig. 6.14. The cross-sectional area of the cylinder is 0.01 m^2 and
the steam is contained by a well-fitting, frictionless piston of mass
100 kg. The axis of the cylinder is vertically oriented and the height
of the volume enclosed is 1 m initially. Atmospheric pressure is
0.1 MPa. Take $g = 10 \, \text{m s}^{-2}$.

(a)   The motion of the piston is controlled by an external
restraint, so that the steam expands quasistatically and adiabatically. What is the equilibrium height of the plug? Determine the work done on the external restraint mechanism.

(b)   The piston is released from its initial position. Assume the
subsequent expansion is quasistatic and adiabatic. Calculate
the maximum upward velocity of the plug.

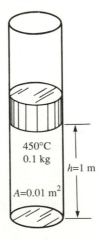

Fig. 6.14   Initial state of a cylinder containing 0.1 kg of steam at 450°C. The
piston weights 100 kg and the external atmospheric pressure is 0.1 MPa.

6.7.9.   At sufficiently low temperatures a gas composed of the free conduction electrons in a metal (see question 2.1.7) satisfies the
equations

$$PV = \tfrac{2}{3}U \tag{6.7.9}$$

and

$$U = \tfrac{3}{5}a\,\tilde{N}^{5/3}V^{-2/3} \tag{6.7.10}$$

where

$$a = \left(\frac{3}{\pi}\right)^{2/3}\frac{h^2}{8m}, \tag{6.7.11}$$

$\tilde{N}$ is the number of electrons in volume V, h is Planck's constant,
and m is the electron mass. Show that the entropy of this system
is constant over the states represented by these equations.

6.7.10.   In question 6.7.9 the temperature was effectively set to zero.
A better approximation, which is applicable provided $T \ll T_{\mathrm{f}}$,
employs an equation for U containing the temperature explicitly:

$$U = \tfrac{3}{5}a\tilde{N}^{5/3}V^{-2/3}\left(1 + \frac{5}{12}\left(\frac{\pi T}{T_{\mathrm{f}}}\right)^2\right). \tag{6.7.12}$$

Here

$$T_{\mathrm{f}} = \frac{a}{k}\left(\frac{\tilde{N}}{V}\right)^{2/3},$$

is the Fermi temperature, as in (2.1.7), and a is given in (6.7.11).
Starting from (6.7.12) show that, under this approximation, S can
be expressed as

$$S = \frac{\pi^2\tilde{N}k}{2}\frac{T}{T_{\mathrm{f}}}. \tag{6.7.13}$$

6.8   Entropy and irreversibility

It might appear as if we have generalized the results we established in
Section 4.1 by merely replacing the quasistatic process requirement with the
reversible condition. But the inequality of Clausius, eqn (6.6.7), is a new
result. In this section we shall use this relation to show how the change in
the entropy of a system can provide a measure of the irreversibility of a
process.

Consider path o for the process $a \rightarrow b$, shown in Fig. 6.11. Suppose this
process takes place in an adiabatically isolated system. Since $đQ = 0$
throughout,

$$\int_a^b \left(\frac{\text{d}Q}{T}\right)_{\text{path o}} = 0,$$

and since path r, $b \to a$, is reversible,

$$\int_b^a \left(\frac{\text{d}Q}{T}\right)_{\text{path r}} = S_a - S_b,$$

and hence,

$$\oint \frac{\text{d}Q}{T} = \int_a^b \left(\frac{\text{d}Q}{T}\right)_{\text{path o}} + \int_b^a \left(\frac{\text{d}Q}{T}\right)_{\text{path r}} = S_a - S_b.$$

It then follows, using the inequality of Clausius,

$$\oint \frac{\text{d}Q}{T} \leq 0,$$

that

$$S_b \geq S_a, \tag{6.8.1}$$

where, as before, the inequality applies when the process is irreversible. This is an important result which it is worthwhile to emphasize: in an adiabatic process,

$$\Delta S = 0 \quad \text{if the process is reversible, otherwise} \quad \Delta S > 0.$$

This result, called the *principle of increasing entropy*, shows that states of lower entropy cannot be reached in an adiabatic process. The Carathéodory form of the second law is a related expression of this principle:

> In the neighbourhood of any state of a system in equilibrium there exist states which are inaccessible by adiabatic processes alone.

We simply note here that the Carathéodory statement can be used as the foundation expression of the second law and from it the entropy and the thermodynamic temperature may be developed. An attraction of this approach is that it focuses on states rather than processes, thus avoiding the need for cyclic machines to introduce the entropy. However desirable this may be, many accounts of the Carathéodory theory are flawed because it is presumed that pressure and volume may always be used as state coordinates for a fluid. This assumption does not always hold (see Section 5.3).

We are now in a position to appreciate the link between the entropy function and irreversibility in a general way. Consider an isolated system, Σ, which undergoes an irreversible process, x, beginning at state a and ending at state b. The surroundings bear no record of the event. But since the process is irreversible, Σ cannot be restored to state a except by a process which will change the state of the environment. This information must be contained in the thermodynamic variables for states a and b since there is no thermodynamic evidence elsewhere that the process $a \rightarrow b$ occurred.

Thus the fact that a certain adiabatic change, $a \rightarrow b$, is irreversible must be ascertainable from the states a and b only. Since details of the process are not required, the term *irreversible change* is a change-of-state attribute. The very existence of irreversible processes implies there must be a thermodynamic state function which provides a record of irreversibility in an isolated system. There must be some specific property difference. The entropy is that property.

Equivalently, the entropy allows us to say which of two states precedes the other in time. In effect the entropy establishes a direction for the arrow of time in physical processes. This is an interesting result because our understanding of time asymmetry from the quantum, or microscopic, point of view is incomplete. In statistical mechanics the entropy of a system is related to the disorder, which may be represented by the number of quantum states, Ω, accessible to a system given its macroscopic constraints, such as its volume and energy. The entropy is then given by $S = k \ln \Omega$, a well-known equation due to Boltzmann. An irreversible process in an isolated system is therefore associated with an increase in the disorder. We do not address this aspect of thermodynamics further in this book, but see, for instance, Callen (1985), Riedi (1988), Waldram (1985), and Zemansky and Dittman (1981).

Questions

6.8.1. Vessel A contains 10 kg of water at 50°C at a pressure of 1 MPa. Vessel B contains 5 kg of steam at 500°C at a pressure of 1 MPa. The vessels are connected and the fluids mix whilst the total volume is kept constant. Determine the increase in the entropy of the system. The vessels have the same final temperature.

6.8.2. Two kg of water, a mixture of liquid and vapour, occupies a volume of 20 litres at a pressure of 10 MPa. The vessel is connected by a closed valve to a second vessel of volume 100 m³, which is initially evacuated. The valve is opened and the system comes to equilibrium at a uniform temperature. Determine the increase in the entropy of the system.

6.8.3. Two solid masses, A and B, each of 1 kg, are placed in thermal contact while remaining adiabatically isolated from their surroundings. The initial temperatures are $T_A = 400$ K, $T_B = 300$ K, and the specific heats, in kJ kg^{-1} K^{-1}, are

$$C_A = 1 \qquad C_B = 1 + 2 \times 10^{-3} T.$$

(a)   Calculate the final common temperature, assuming that the process is one of heat transfer only.

(b)   Calculate the change in the entropy of each body and the change in the entropy of the composite system.

6.8.4.   Two identical bodies, x and y, of mass m and specific heat c, are at temperatures T_x and T_y, respectively. The bodies are located in an adiabatic enclosure. At a particular time the bodies are placed in themal contact.

(a)   Determine the final common temperature, T_w.

(b)   Show that the change in the entropy of the composite system is given by

$$\Delta S = mc \ln \left[\frac{(T_x + T_y)^2}{4 T_x T_y} \right]. \tag{6.8.2}$$

(c)   Show that ΔS given by (6.8.2) must be positive. Does this mean that the process is irreversible?

6.8.5.   Consider the system described in question 6.8.4. Suppose the two bodies, x and y, are brought to the same temperature, T_w, obtained in that question in a process carried out reversibly using Carnot machines which exchange heat with an external reservoir at temperature T_{res}, as shown in Fig. 6.15.

(a)   Show that the external work performed in this process is given by

$$W_e = T_{res} \Delta S, \tag{6.8.3}$$

where ΔS is given in (6.8.2).

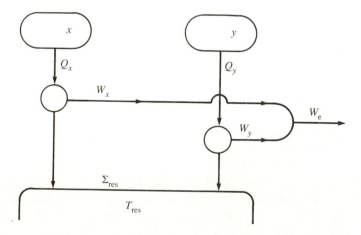

Fig. 6.15   A reversible process for bringing two bodies to a common temperature, T_w, whilst interacting with a single heat reservoir, Σ_{res}.

   (b)   Since the total energy of the bodies x and y does not change
in this process where did the energy represented by this work
come from?

   (c)   The entropy of the composite system, x plus y, increases in
this process. Does this mean that the process is irreversible?
Explain.

6.8.6.   Two streams of water merge and mix while remaining adiabatically
separated from the surroundings. One, which is initially at a temperature of 5°C, has a flow rate of $2\,\text{m}^3\,\text{s}^{-1}$. The other is initially
at 25°C and has a flow rate of $1\,\text{m}^3\,\text{s}^{-1}$. What is the rate of
increase of entropy due to this process? (Density of water is
$1000\,\text{kg}\,\text{m}^{-3}$; specific heat of water is $4.18\,\text{kJ}\,\text{K}^{-1}\,\text{kg}^{-1}$.)

6.8.7.   In a Simon liquefier, helium is allowed to leak slowly from a pressure vessel which is otherwise adiabatically isolated from its surroundings. If the initial pressure is sufficiently high, and the initial
temperature low enough, the helium remaining is in the liquid phase
when the system reaches atmospheric pressure.

   Suppose the initial temperature is 8 K, the initial pressure is
5 MPa, and the final pressure is 101 kPa. At the final pressure the
enthalpy of phase change of liquid helium is $20.6\,\text{kJ}\,\text{kg}^{-1}$ at
4.22 K. Assume that the remaining helium behaves as an ideal
monatomic gas until it liquefies, and that the heat capacity of the
vessel can be ignored. Determine the percentage of the remaining
mass of helium that will be in the liquid phase in the final state.
(Refer to question 3.6.11 for hints.)

7

Reversible and irreversible work processes

In this chapter we introduce some basic results deriving from the second law. These concern the work performed on systems undergoing reversible and irreversible work processes. We shall also introduce work processes on systems other than fluids.

7.1 Reversible process theorems

The reversible process theorems (Haywood 1980) are analogous to the Carnot theorem, but they apply to non-cyclic processes, whereas the Carnot theorem applies to cyclic processes. We shall use the Kelvin–Planck expression of the second law in order to establish these theorems.

Adiabatic processes

Consider the work performed on a system, Σ, when it has a change of state as a result of a purely adiabatic process. We suppose that the system is subject to a particular set of external constraints at the beginning and end of the process. For example, the process might be the expansion of a fixed amount of fluid, with given initial and final volumes. The fixed mass and specified volumes represent the constraints referred to. We shall take the view that the energy, together with the constraints, uniquely identify the state of the system. This is an important assumption which effectively defines what we mean by constraint. (If you find that the state of a system is not uniquely identified by the constraints selected, you have not identified the constraints correctly.)

Let a be the initial state of the system Σ, and let b be the final state following a particular reversible process, r. Let c be the final state when another process, o, is used. The final states of Σ, b, and c, will generally be different, as indicated schematically in Fig. 7.1. Let W_r be the work performed on Σ in the reversible process, r, and let W_o be the work for the process o.

We shall show that

$$W_o = W_r \quad \text{if o is reversible, otherwise} \quad W_o > W_r. \quad (7.1.1)$$

Suppose $W_r > W_o$. Then $U_b > U_c$. The states b and c may differ only in

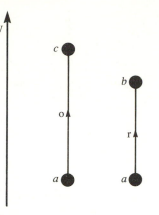

Fig. 7.1 Two possible adiabatic processes of a system, r and o. The initial state, a, is the same for the processes, but the final states, b and c, will be different. These states will normally have different energy, U.

the energy of the system, because the constraints are otherwise the same. Hence we could change the state of the system from c to b by heat transfer to Σ from a single external reservoir. In this situation we could perform a cycle: $a \to c$, by process o; $c \to b$, by heat input; and $b \to a$, by process $\bar{r}$, the inverse of r. In this cycle the total work input to the system is $W_o - W_r$, which is negative. Since the system interacts with only one reservoir, the cycle would violate the Kelvin–Planck statement. We conclude that $W_r \le W_o$.

It is now easy to show that if o is reversible, and only then, $W_r = W_o$. This concludes the demonstration of the result we sought. In summary, when a system undergoes an adiabatic process, subject to particular initial and final state constraints, the work done on the system is a minimum when the process is reversible; it is the same for all reversible changes.

Non-adiabatic processes

The reversible process theorem for non-adiabatic processes is quite different to the adiabatic form, but the theme is the same. Here we consider the work performed on a system Σ when it undergoes a particular change of state $a \to b$, during which it may exchange heat with another system Σ', as shown in Fig. 7.2. Note that Σ' could be a reservoir, meaning that Σ' could be a very large system compared to Σ, but it need not be. We do not exclude the possibility that the temperature of Σ' might change during the process.

Suppose the change of state, $a \to b$, is effected by a reversible process, r, in which work W_r is performed on Σ, by some external agent which we need not specify, and heat Q_r is transferred to Σ from Σ'. Let o be another process for which Σ has the same end states. The corresponding

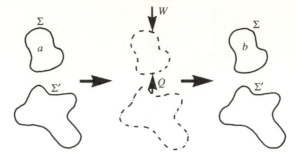

Fig. 7.2 A process $a \rightarrow b$ in which work W is performed on Σ and heat Q is transferred from Σ' to Σ. If the process is reversible it may be necessary to use a Carnot cycle machine to ensure the heat transfer from Σ' to Σ is reversible, as illustrated in Figs 6.10 and 7.4. In that case W then includes the work input to the Carnot machine.

work and heat inputs to Σ are W_o and Q_o, respectively. Then from the first law,

$$\Delta U_{ab} = W_o + Q_o = W_r + Q_r$$

and hence

$$W_o - W_r = Q_r - Q_o. \tag{7.1.2}$$

We shall show that

$$W_o = W_r \quad \text{if o is reversible, otherwise} \quad W_o > W_r. \tag{7.1.3}$$

Consider a cyclic process in which Σ undergoes the state change $a \rightarrow b$, by process o, and the state change $b \rightarrow a$, by process $\bar{r}$, the effacing process for r, as shown in Fig. 7.3. Such an effacing process must exist since r is reversible. The corresponding heat and work inputs to Σ in process $\bar{r}$ will be $-Q_r$ and $-W_r$, respectively, and in the complete cycle the work input to Σ will be $W_o - W_r$. Now Σ exchanges heat with one other system only, and in this circumstance the net work input must be positive or zero, by the Kelvin–Planck statement. Hence $W_o \geq W_r$.

If $W_o = W_r$ then (7.1.2) shows $Q_o = Q_r$. Hence $\bar{r}$ is an effacing process for o, and so o is reversible. Conversely, if o is reversible, then the argument used above shows that $W_r \geq W_o$ as well as $W_o \geq W_r$. Hence $W_o = W_r$.

In summary, the relation (7.1.3) has been established. Equivalently, from (7.1.2) and (7.1.3),

$$Q_o = Q_r \quad \text{if o is reversible, otherwise} \quad Q_o < Q_r. \tag{7.1.4}$$

These results comprise the reversible process theorem in its non-adiabatic form. The theorem is analogous to the Carnot theorem in that it relates the work and heat for a process to the reversible character of the process.

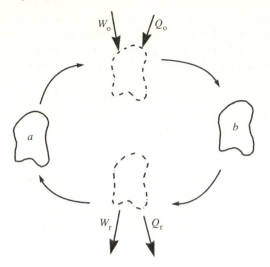

Fig. 7.3   A cyclic process $a \to b \to a$ in which $a \to b$ is performed by the process labelled o, and $b \to a$ is performed by the process $\bar{r}$.

Note that the reversible process theorem is expressed in terms of states, whereas the Carnot theorem concerns cyclic processes, so that the initial and final states need not be specified. In the following examples we show how the reversible process theorem may be used to demonstrate the existence of the entropy function and to establish its properties, without recourse to arguments which employ Carnot cycles in an explicit way. But remember that a Carnot cycle machine must be used as a matching device if we wish to have a reversible heat interaction between two systems at different temperatures. In this respect the Carnot machine is fundamental.

Questions

7.1.1.   For a reversible process involving heat exchange with one heat reservoir, we know, from the reversible process theorem (non-adiabatic form), that Q depends only on the initial and final states, a and b, of the system and the state of the reservoir. Now the only defined thermodynamic parameter of a heat reservoir is its empirical temperature, θ (see Section 5.3). Denote the state variables which identify a and b by $\{a\}$ and $\{b\}$, respectively. The reversible process theorem shows that there must exist a function ψ of the state variables and θ such that

$$Q_{ab} = \psi(\theta, \{b\}) - \psi(\theta, \{a\}). \tag{7.1.5}$$

Consider a process $a \to b$ which is both adiabatic and reversible. Then,

$$\psi(\theta, \{b\}) = \psi(\theta, \{a\}).$$

But since $Q_{ab} = 0$, the reservoir temperature has no physical role in this process and hence the relationship between $\{a\}$ and $\{b\}$ does not involve θ. Hence we must be able to express $\psi(\theta, \{a\})$, in a form which allows θ to be cancelled. Thus we can write

$$\psi(\theta, \{a\}) = \phi(\theta)\sigma(\{a\}) \qquad (7.1.6)$$

where $\phi(\theta)$ is a function of the reservoir temperature only, and σ is a function of the state of the system such that

$$\sigma(\{a\}) = \sigma(\{b\}) \qquad (7.1.7)$$

for states linked by adiabatic reversible processes. This is the *empirical entropy* function.

(a) Show that no pair of states having the same empirical entropy can differ by the energy alone. (Hint: if they did, then a reversible heat interaction could be used to link them; this possibility is excluded by the reversible process theorem.)

(b) Use the result from part (a) to show that in a heat interaction the value of σ must change. It follows that σ must be a monotonic function of the internal energy, U. The surfaces in state space defined by the set of states x satisfying the equation

$$\sigma(\{x\}) = \sigma_0,$$

must therefore represent a set of non-intersecting surfaces, each identified by a value of σ_0. We may choose here whether σ should be an increasing or decreasing function of U. Since it is the usual choice, we shall take σ to increase with heat input.

(c) Show, from the reversible process theorem (non-adiabatic form), that if an adiabatically isolated system has an irreversible change of state $a \rightarrow b$, then,

$$\sigma(\{b\}) > \sigma(\{a\}). \qquad (7.1.8)$$

This result represents the entropy increase theorem. (Take σ to be an increasing function of U, as noted in (b).)

7.1.2. Two heat reservoirs Σ_1 and Σ_2, having empirical temperatures θ_1 and θ_2, have reversible heat interactions with a system, Σ. Suppose that heat Q_1 is first transferred from Σ_1 whilst Σ undergoes a state change $a \rightarrow b$. Subsequently Σ undergoes a state change $b \rightarrow a$ whilst heat Q_2 is transferred to Σ_2. Use (7.1.5) and (7.1.6) to show that

$$\frac{Q_1}{Q_2} = \frac{\phi(\theta_1)}{\phi(\theta_2)}. \qquad (7.1.9)$$

By defining the function $\phi(\theta) = T$ to be the thermodynamic temperature we have reproduced the results obtained in Chapter 6 without explicitly invoking a Carnot cycle. Nevertheless where two systems are not in thermal equilibrium the reversible exchange of heat must involve an auxiliary device, effectively a Carnot engine or refrigerator, as indicated in Fig. 6.10.

7.2 The availability function

Consider a system Σ which undergoes a reversible process while exchanging heat with a heat reservoir Σ_{res} at temperature T_{res}. According to the reversible process theorem the work and heat input to Σ, W_r and Q_r, depend only on the initial and final states of the system, a and b. Here we shall derive expressions for W_r and Q_r.

Isothermal processes

Let us begin with the simplest situation: we suppose the process is isothermal such that Σ and Σ_{res} are in mutual thermal equilibrium throughout. From the first law we have

$$U_b - U_a = Q_r + W_r,$$

where U is the internal energy of Σ. Since the process is isothermal and reversible,

$$S_b - S_a = \frac{Q_r}{T_{res}},$$

where S is the entropy of Σ. Hence,

$$W_r = U_b - U_a - T_{res}(S_b - S_a). \tag{7.2.1}$$

This result is consistent with our expectations from Section 7.1. We are led to define a new function, called the *Helmholtz free energy*, F,

$$F = U - TS. \tag{7.2.2}$$

It is evident that when a system undergoes a reversible isothermal process, $a \to b$, while exchanging heat with a reservoir at temperature T_{res}, the work input is

$$W_r = F_b - F_a = \Delta F, \tag{7.2.3}$$

where ΔF is evaluated at the temperature T_{res}. The name *free energy* is applied to F because it allows us to express the work done in an isothermal process in terms of the change in a state function. Such functions are also called thermodynamic potentials and we shall consider them in more detail in Chapter 10. (The terms *Helmholtz free energy, Helmholtz function*, and *Helmholtz potential* may be used interchangeably.)

Non-isothermal processes

Equation (7.2.1) is more generally applicable than this derivation suggests since the system Σ need not be in thermal equilibrium with the reservoir, Σ_{res}. It is necessary only that the exchange of heat between Σ and Σ_{res} should be reversible. For our purposes we shall use a Carnot refrigerator (or heat pump), as in Fig. 7.4, as an auxiliary device.

In this situation there are two contributions to the external work when Σ changes state, $a \rightarrow b$: the work performed on Σ directly and the work done on the Carnot refrigerator. Since the refrigerator does not have a net change of state in the process we obtain from the first law

$$W_r + Q_{res} = U_b - U_a. \tag{7.2.4}$$

The Carnot relation applies for each increment of the process:

$$\frac{dQ_{res}}{dQ} = \frac{T_{res}}{T},$$

or, rearranging and integrating,

$$Q_{res} = T_{res} \int_a^b \frac{dQ}{T},$$

and since the process is reversible,

$$\int_a^b \frac{dQ}{T} = S_b - S_a.$$

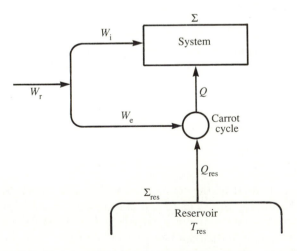

Fig. 7.4 Reversible heat transfer process when Σ and Σ_{res} are not in thermal equilibrium.

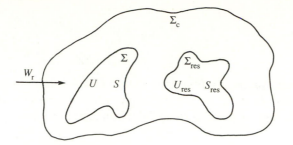

Fig. 7.5   Relationships between Σ, Σ_{res}, and Σ_c in a general reversible process.

Hence,

$$Q_{\text{res}} = T_{\text{res}}(S_b - S_a). \tag{7.2.5}$$

This result may also be obtained directly from (6.6.5). Equation (7.2.1) now follows from (7.2.5) and (7.2.4). Alternatively we can write

$$W_r = A_b - A_a, \tag{7.2.6}$$

where

$$A = U - T_{\text{res}}S, \tag{7.2.7}$$

is the *non-flow availability function* for the system, or simply the *availability function*. Notice that A, unlike F, is not a thermodynamic function of the state of Σ alone. The availability function involves state variables of both Σ and Σ_{res}.

General derivation

The existence of the availability function was suggested in a general way by the reversible process theorem introduced in Section 7.1. The form of eqn (7.2.6) for the work W_r, is fully consistent with the non-adiabatic reversible process theorem. In addition the reversible process theorem shows that the work input, W_o, for any irreversible process, o, must be greater than W_r.

Equations (7.2.6) and (7.2.7) may also be obtained in a direct way as follows. Consider the composite system Σ_c comprised of Σ and Σ_{res}, as shown in Fig. 7.5. In the reversible process $a \rightarrow b$, the only energy input to Σ_c is the work input, W_r. From the first law

$$W_r = \Delta U + \Delta U_{\text{res}} \tag{7.2.8}$$

where ΔU_{res}, ·the increase in the energy of the reservoir, Σ_{res}, is

$$\Delta U_{\text{res}} = Q_{\text{res}} = T_{\text{res}} \Delta S_{\text{res}}. \tag{7.2.9}$$

Here Q_{res} is the heat input to Σ_{res} in the reversible process. Now for Σ_c the process is adiabatic and reversible, and hence

$$\Delta S_{\mathrm{res}} + \Delta S = 0. \tag{7.2.10}$$

It follows that

$$W_{\mathrm{r}} = \Delta U - T_{\mathrm{res}}\,\Delta S = \Delta A,$$

where A is the availability function. This demonstration illustrates how we can move away from a specific representation of a reversible process, as in Fig. 7.4, to use of just the thermodynamic relations for such processes, as in (7.2.9) and (7.2.10).

The Gouy–Stodola theorem

Suppose the system Σ in Fig. 7.5 undergoes an irreversible change of state, $a \to b$. Then the entropy of the composite system Σ_{c} will increase, by the entropy principle, Section 6.8. We shall now show that the additional work input to Σ is directly related to the entropy created due to irreversibility.

Let r be a reversible process in which Σ changes state $a \to b$, and let o be another process having the same end states. During the process the heat input from Σ_{res} to Σ, measured at the boundary of Σ_{res}, is Q_{r} and Q_{o} for processes r and o, respectively. The additional work input to Σ (and to the auxiliary devices) compared with a reversible process, is given by (7.1.2):

$$W_{\mathrm{o}} - W_{\mathrm{r}} = Q_{\mathrm{r}} - Q_{\mathrm{o}}$$

Here

$$Q_{\mathrm{r}} = T_{\mathrm{res}}(S_{\mathrm{b}} - S_{\mathrm{a}}) = T_{\mathrm{res}}\,\Delta S, \tag{7.2.11}$$

and hence

$$W_{\mathrm{o}} - W_{\mathrm{r}} = T_{\mathrm{res}}\,\Delta S - Q_{\mathrm{o}}. \tag{7.2.12}$$

Now the increase in the entropy of the composite system Σ_{c}, due to the process o being irreversible, is given by,

$$\Delta S_{\mathrm{i}} = \Delta S - \frac{Q_{\mathrm{o}}}{T_{\mathrm{res}}}. \tag{7.2.13}$$

The irreversibility may be due to the process being irreversible within Σ, or to the heat transfer between Σ and Σ_{res} being irreversible. In both cases the additional work required due to irreversibility is

$$W_{\mathrm{o}} - W_{\mathrm{r}} = T_{\mathrm{res}}\,\Delta S_{\mathrm{i}}. \tag{7.2.14}$$

This result, called the *Gouy–Stodola theorem*, is useful for analysing loss mechanisms in complex processes. We can interpret the theorem in two ways: consider a system undergoing a process $a \to b$ in which it may exchange heat with a reservoir, Σ_{res}.

1. If W_r is positive, $T_{res} \Delta S_i$ is the *additional* input of work needed to accomplish the change of state $a \to b$ compared with a reversible process having the same end states. The reversible work *input* is given by the *increase* in availability A in the process.

2. For a process in which W_r is negative (a work output process), $T_{res} \Delta S_i$ is the *loss* of work output compared with a reversible process having the same end states a and b. The reversible work *output* is given by the *decrease* in A.

To illustrate the requirement that the process must be reversible, consider the potential energy, mgh, of an elevated mass, m, at a height h. If h changes the decrease in potential energy equals the work which can be performed during the change, provided it takes place reversibly. But if an elevated weight is lowered whilst the descent is resisted by a frictional brake, then the potential energy will be dissipated into internal energy at the brake. The opportunity to do external work will be lost because the process is irreversible.

Example

Two bodies, x and y, of identical composition, have masses 1 kg and 5 kg, respectively. The specific heat capacity of the material is 1 kJ kg^{-1} and both bodies are initially in thermal equilibrium at 300 K.

(a) What is the minimum input of work in order to change the temperature of x to 330 K, given that a heat reservoir is available at 300 K?

(b) What is the minimum input of work in order to change the temperature of x to 330 K, given that a heat reservoir is available at 280 K?

(c) What is the minimum input of work in order to change the temperature of x to 330 K, given that the composite system comprised of x and y is adiabatically enclosed?

Solution (a)

In this process only the state of body x need change. The minimum work input is the increase in the availability of x, ΔA, where

$$\Delta A = \Delta U - T_{res} \Delta S.$$

Here

$$\Delta U = mc \, \Delta T = 1 \times 1 \times 30.0 = 30.0 \text{ kJ}$$

$$\Delta S = mc \ln \left(\frac{T_2}{T_1} \right) = 1 \times 1 \ln \left(\frac{330}{300} \right) = 0.09531$$

$$T_{res} = 300.$$

Hence

$$\Delta A = 30.0 - 28.593 = 1.407 \text{ kJ}.$$

Solution (b)

If we allow the state of x alone to change, the minimum work input is calculated in the same way as in part (a), except that $T_{res} = 280$ K. We get

$$\Delta A_x = 3.313 \text{ kJ}.$$

Note that there is an increase in the minimum work because the temperature lift from the reservoir to the final temperature of x has increased.

But the temperature of y is not restricted here. Consider then the minimum work input if in the same process the temperature of y is allowed to reach the reservoir temperature. Then

$$\Delta A_y = -1 \times 5 \times 20 - 280 \times 5 \times \ln\left(\frac{280}{300}\right) = -3.410 \text{ kJ},$$

and hence the minimum net work input is

$$W_{min} = -0.097 \text{ kJ}.$$

Hence, by allowing the body y to reach equilibrium with the reservoir at 280 K reversibly, the work output from that process exceeds the work input required to reversibly heat the body x to 330 K. This illustrates how temperature differences may be exploited to perform external work. In practice, however, it is difficult to make effective use of small temperature differences, such as those in this example.

Solution (c)

The work input is minimized when the process is reversible, which means that $\Delta S = 0$ for the composite system, since the process is adiabatic. The work input is the increase in the energy of the system. We shall therefore determine the state of the system for which $T_x = 330$ K, and $\Delta S = 0$. Let T_y be the temperature of body y. Then,

$$\Delta S_x + \Delta S_y = 1 \times 1 \times \ln\left(\frac{330}{300}\right) + 5 \times 1 \times \ln\left(\frac{T_y}{300}\right) = 0.$$

Hence,

$$(1.1)^{0.2}\frac{T_y}{300} = 1.0,$$

yielding

$$T_y = 294.335 \text{ K}.$$

The work input is now obtained by determining the increase in the energy of the composite system in this isentropic process:

$$\Delta U = \Delta U_x + \Delta U_y = 30.0 - 1 \times 5 \times (300 - 294.335) = 1.678 \text{ kJ}.$$

Questions

7.2.1. Two mol of a monatomic ideal gas initially occupy a volume of 1 litre at a temperature of 300 K, as shown in Fig. 7.6. The partition is suddenly removed and the gas undergoes a free-expansion process. The final volume, when the system reaches equilibrium again, is 2 litres. Label the initial state a and the final state b. A heat reservoir is available at a temperature of 300 K, as shown. The change of state in the expansion process could be carried out reversibly and isothermally with the heat input being provided directly by the reservoir.

    (a)   Calculate the external work performed on the system in the process $a \rightarrow b$, under the condition that the process is performed reversibly. (Devise a convenient reversible process to do this.) Does your choice of process path affect the result? Show that your result agrees with the value obtained by using (7.2.6).

    (b)   Repeat part (a) but now assume that the only reservoir available is at 250 K. Devise a suitable reversible expansion path linking the initial and final states, a and b. You will need to use a Carnot machine to provide heat to the system, before, during, or after the expansion, depending on how you choose to do it. Some of the work of expansion will be required to run the Carnot machine.

7.2.2. A monatomic ideal gas is initially in the state shown in Fig. 7.6. The system, comprising 2 mol, undergoes a process to a state of final volume 2 litres and temperature 250 K. The temperature of the only heat reservoir available is 250 K.

    (a)   Determine the work done on the system when this process is performed reversibly by: (i) adiabatic reversible expansion

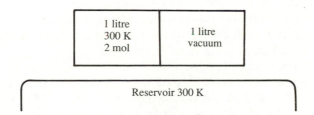

Fig. 7.6 Initial state of a system and reservoir.

until the temperature is 250 K, followed by, (ii) isothermal expansion until the volume is 2 litres. Compare with the change in the availability for the system, obtained from (7.2.7).

(b) Another reversible path for the state change described in part (a) is: (i) cool the gas reversibly, using a Carnot engine which rejects heat to the reservoir, followed by, (ii) expand the volume isothermally to 2 litres. Calculate the work done on the system in this process and compare with your answers from part (a).

7.2.3. A vessel contains water at $T = 273.15$ K. It is desired to freeze 1 kg of water, the latent heat of fusion being 334 kJ kg^{-1}. The heat rejected by the refrigerator goes to warm up a second vessel containing 1 kg of water, which is initially at the same temperature as the first. Determine the minimum amount of heat that the refrigerator must reject to the second vessel in this process. (The specific heat of water is 4.18 kJ K^{-1} kg^{-1}.)

7.2.4. Consider two bodies, x and y, of mass m and specific heat capacity c. They are brought to a common temperature by a reversible process, using a Carnot engine, as shown in Fig. 7.7.

(a) Show that the final common temperature, T_z, for the bodies is

$$T_z = \sqrt{T_x T_y}.$$

(b) Show that the external work performed by the Carnot engine is given by

$$W_e = mc(\sqrt{T_x} - \sqrt{T_y})^2.$$

7.2.5. A stream of water at 20°C flows into a lake at a rate of 1 m^3 s^{-1}. The temperature of the lake, which is not disturbed significantly by the stream, is 10°C.

(a) What is the rate of increase of the entropy of this composite system due to this process, ignoring possible differences in chemical composition. (The density of water is 1000 kg m^{-3} and the specific heat is 4.18 kJ K^{-1} kg^{-1}.)

(b) Determine the rate at which external work could be performed if the same change of state were performed reversibly. Take the lake to be a suitable reservoir.

(c) The river is harnessed for hydro-electric power by erecting a dam and channelling the river flow through a turbine. The efficiency of the turbine-generator set is 85%. What height should the dam be in order to generate electrical power at the same rate as that obtained in (b)?

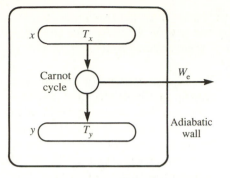

Fig. 7.7   Initial state of a system prior to a reversible heat transfer process.

7.2.6.   A composite system consists of two bodies, denoted by x and y, each of mass 1 kg, having specific heat capacities

$$c_x = aT, \qquad c_y = bT,$$

where T is the temperature. In each of the following processes the initial temperatures of x and y are T_1 and T_2, respectively, where $T_1 > T_2$. The thermal coefficient of volume expansion is negligible.

(a)   The bodies are brought into direct thermal contact while the system remains isolated. What is T_3, the final common temperature?

(b)   The bodies are brought reversibly to a common temperature, T_4, while interacting with a work reservoir. Determine T_4 and calculate how much work was performed on the system.

(c)   The bodies are brought reversibly to temperature T_3 (see part (a)) while interacting with a work reservoir and a heat reservoir at temperature T_0. Obtain the work performed on the system in this process.

7.2.7.   Two identical bodies, x and y, of mass m and specific heat capacity c are initially in thermal equilibrium at temperature T_1. A refrigerator operates between the two bodies, cooling x and heating y until x reaches a temperature T_2.

(a)   Given that volume expansion can be ignored and that c is constant, obtain a lower limit for the final temperature of y.

(b)   Show that the minimum work input to the system to complete the process is

$$W_{\min} = \frac{mc(T_1 - T_2)^2}{T_2}.$$

7.2.8. An object, of mass m, moving at speed v, collides inelastically with a stationary wall. The effective specific heat is c and the temperature of the environment is T_0. The initial temperature of the object is also T_0.

  (a) Obtain expressions for the change in the temperature and entropy of the object in terms of m, v, T_0, and c. (Assume there is no heat transfer between the object and the environment during the process of interest.)

  (b) Obtain an expression for the loss of availability, A, for the object as a result of the collision. Show that the loss of availability is less than the loss of kinetic energy. You may assume $\ln(1+x) \leqslant x.$

7.2.9. The initial state of $10\,\text{kg}$ of steam is $400\,°\text{C}$ at $50\,\text{MPa}$. The steam is expanded reversibly at constant temperature until the volume is $200\,\text{m}^3$, and then adiabatically and reversibly until the volume is $500\,\text{m}^3$. Given a reservoir at temperature T_r, it is found that this change of state can be executed without a *net* input of work. What is T_r? If no work is performed at *any* stage of the process, show that the process path must pass through the two-phase region.

7.3 Work processes in fluids

We considered previously only one type of reversible work process: volume change in a fluid system. In this section we shall summarize the results established already, as a point of reference, before we introduce other types of work process.

Simple fluids

In a reversible process the work done by external forces acting on a fluid, when the volume changes infinitesimally by dV, is given by

$$đW = -P\,dV. \tag{7.3.1}$$

If we have an isolated system, the total volume, V, must be a constant, but the pressure need not be fixed. For example, in a composite system we could remove an internal partition without doing work and this could result in a change of fluid pressure. For this reason the volume is the primary work variable for a fluid. Such variables may be known as *deformation parameters*.

  Now the first law, which applies to both reversible and irreversible state changes, is

$$dU = đQ + đW. \tag{7.3.2}$$

In the reversible limit, $đQ$ can also be expressed in terms of state variables:

$$\text{đ}Q = T\,dS \tag{7.3.3}$$

and hence we have the Gibbs equation, as in (4.1.13),

$$dU = T\,dS - P\,dV. \tag{7.3.4}$$

Since this is an equation between state variables, it applies whether or not the process of change is reversible. Thus (7.3.2) and (7.3.4) both apply to reversible and irreversible processes. But the process quantities, đQ and đW, are equal to the state quantities, $T\,dS$ and $-P\,dV$, respectively, only in the reversible limit. Equation (7.3.4) also implies that for a fixed amount of fluid, U is naturally a function of S and V and we can write

$$T = \left(\frac{\partial U}{\partial S}\right)_V, \tag{7.3.5}$$

$$P = -\left(\frac{\partial U}{\partial V}\right)_S. \tag{7.3.6}$$

The equation,

$$U = U(S, V) \tag{7.3.7}$$

is called a fundamental relation. The fundamental relation must be supplemented by our interpretation of the derivatives, (7.3.5) and (7.3.6), and the equations relating process and state quantities, (7.3.1) and (7.3.3). It is occasionally said that the Gibbs equation (7.3.4), is an expression of both the first and second law. On the other hand the classical laws are expressed in terms of processes, whereas the Gibbs equation is an expression of state properties. Thus one emphasizes the process view, the other the state representation.

Response functions

In addition to the simple derivatives for the pressure and temperature there are a number of characteristic response functions for fluids which can be expressed in derivative form. It is convenient to summarize them here.

1. The molar heat capacity at constant volume is

$$c_v = \frac{1}{N}\left(\frac{\text{đ}Q}{dT}\right)_{V,N} \tag{7.3.8}$$

and since the process may be executed quasistatically, đ$Q = T\,dS$,

$$c_v = \frac{T}{N}\left(\frac{\partial S}{\partial T}\right)_{V,N}. \tag{7.3.9}$$

2. The molar heat capacity at constant pressure, similarly to c_v, is

$$c_p = \frac{T}{N}\left(\frac{\partial S}{\partial T}\right)_{P,N}. \tag{7.3.10}$$

3. The coefficient of thermal expansion, also called the volume expansivity, is the fractional increase in volume with temperature at constant pressure:

$$\alpha = \frac{1}{V}\left(\frac{\partial V}{\partial T}\right)_{P,N}. \tag{7.3.11}$$

4. The isothermal compressibility is the fractional decrease in volume with increasing pressure in a constant temperature process:

$$\kappa_T = -\frac{1}{V}\left(\frac{\partial V}{\partial P}\right)_{T,N}. \tag{7.3.12}$$

5. The adiabatic compressibility is the fractional decrease in volume with increasing pressure in a reversible adiabatic process. This is also called the isentropic compressibility:

$$\kappa_S = -\frac{1}{V}\left(\frac{\partial V}{\partial P}\right)_{S,N}. \tag{7.3.13}$$

Note that $B = 1/\kappa$, called the bulk modulus, is a more convenient parameter than the compressibility in some systems (see Section 11.4).

It turns out that these properties are not independent. Since N is fixed, we may omit the subscript N in the partial derivatives without ambiguity. Consider the ratio

$$\frac{\kappa_T}{\kappa_S} = \frac{\left(\dfrac{\partial V}{\partial P}\right)_T}{\left(\dfrac{\partial V}{\partial P}\right)_S}.$$

From eqn (D.16) in Appendix D, we have

$$\left(\frac{\partial V}{\partial P}\right)_T = -\frac{\left(\dfrac{\partial T}{\partial P}\right)_V}{\left(\dfrac{\partial T}{\partial V}\right)_P} = -\frac{\left(\dfrac{\partial V}{\partial T}\right)_P}{\left(\dfrac{\partial P}{\partial T}\right)_V},$$

$$\frac{1}{\left(\dfrac{\partial V}{\partial P}\right)_S} = -\frac{\left(\dfrac{\partial S}{\partial V}\right)_P}{\left(\dfrac{\partial S}{\partial P}\right)_V}.$$

Using (D.28) and (7.3.10), the numerator is

$$\left(\frac{\partial V}{\partial T}\right)_P \left(\frac{\partial S}{\partial V}\right)_P = \left(\frac{\partial S}{\partial T}\right)_P = \frac{N}{T}c_p.$$

A similar relation holds in the denominator, but with V constant. Hence

$$\frac{\kappa_T}{\kappa_S} = \frac{c_p}{c_v}, \tag{7.3.14}$$

which is perhaps an unexpected connection when we consider that the procedures for measuring these parameters are quite unrelated.

Fluids coupled to work reservoirs

In many fluid processes the system is coupled to a work reservoir. However, when the interaction with the reservoir is reversible it is usually possible to separate the reservoir work inputs from others. Thus, although the system must remain physically linked to the work reservoir, thermodynamically we are able to separate the two. We have discussed this situation previously in Section 5.6 where we introduced the enthalpy function. Here we consider the enthalpy in a more general context.

Consider the coupled system shown in Fig. 7.8. This shows two bodies of fluid contained in cylinders, labelled 1 and 2. Each cylinder is maintained at a constant pressure by the weight of a piston which moves freely. Thus each cylinder is a fluid system coupled to a work reservoir, the elevated piston. The energy of the reservoir is the gavitational potential energy of

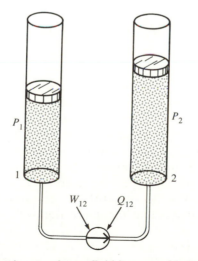

Fig. 7.8 Work and heat inputs when a fluid is pumped between two volume reservoirs at different pressures.

the piston. In any work or heat transfer process the work done by the reservoir must be included in the expression of the first law. This complication can be avoided, however, by using the enthalpy function.

Suppose we transfer a mass Δm of fluid from system 1 to system 2. During this process external work, W_{12} is done on, and heat Q_{12} is transferred to, the fluid. These heat and work processes need not be reversible, but we shall consider only processes in which the fluid interaction with the pistons is reversible. Then the first law for the composite system is

$$\Delta U = W_{12} + Q_{12} - P_1 \Delta V_1 - P_2 \Delta V_2. \tag{7.3.15}$$

Since the fluid in each cylinder is in equilibrium at a constant pressure, we can write $U_1 = m_1 u_1, U_2 = m_2 u_2$, where the specific internal energies u_1 and u_2 remain constant throughout the process and m_1 and m_2 are the masses in cylinders 1 and 2, respectively. Hence when Δm is transferred from cylinder 1 to cylinder 2

$$\Delta U = \Delta U_1 + \Delta U_2 = -u_1 \Delta m + u_2 \Delta m.$$

In addition, $\Delta V_1 = -\Delta m \, v_1$, $\Delta V_2 = \Delta m \, v_2$, where v_1 and v_2 are the specific volumes for the fluid for cylinders 1 and 2. Hence,

$$-u_1 + u_2 = w_{12} + q_{12} + P_1 v_1 - P_2 v_2$$

where $w_{12} = W_{12}/\Delta m$ and $q_{12} = Q_{12}/\Delta m$. This yields

$$h_2 - h_1 = w_{12} + q_{12}, \tag{7.3.16a}$$

or

$$H_2 - H_1 = W_{12} + Q_{12}, \tag{7.3.16b}$$

where $H = U + PV$ is the enthalpy of the fluid and $h = u + Pv$ is the specific enthalpy.

There is an important change of focus between (7.3.15) and (7.3.16a,b). Equation (7.3.15) applies to the composite system, the different parts of which are not in mutual equilibrium, being at different pressures. In (7.3.16a,b) the system is an element of the fluid in an equilibrium state. The state change is a discrete one in which the element is transferred from a state of equilibrium with the pressure reservoir 1 to another state of equilibrium with the pressure reservoir 2. Equations (7.3.15) and (7.3.16a,b) both hold whether or not the process of fluid transfer is reversible. All that is required is that the interaction with the pistons should be reversible.

When the transfer process $1 \rightarrow 2$ is reversible, the process quantities w_{12} and q_{12} may be expressed in terms of state quantities. From the Gibbs equation for energy,

$$dU = T\,dS - P\,dV,$$

and the definition of the enthalpy, we obtain the Gibbs equation for the enthalpy and specific enthalpy

$$dH = dU + d(PV) = T\,dS + V\,dP, \tag{7.3.17a}$$

$$dh = T\,ds + v\,dP. \tag{7.3.17b}$$

We can therefore identify the heat and work in the reversible limit,

$$đw = v\,dP, \tag{7.3.18a}$$

and

$$đq = T\,ds. \tag{7.3.18b}$$

These represent the work and heat per unit mass for an incremental reversible process of the type shown in Fig. 7.8. For a complete process in which the fluid is initially in equilibrium with a reservoir at pressure P_1 and finally in equilibrium with another at pressure P_2,

$$w_{12} = \int_{P_1}^{P_2} v\,dP, \tag{7.3.19a}$$

$$q_{12} = \int_{s_1}^{s_2} T\,ds, \tag{7.3.19b}$$

where the integral is over a suitable reversible path.

We have considered the fluid system in some detail because it illustrates how to separate the work terms when a system is coupled to a work reservoir. Note the distinction between the two forms of the reversible work on a fluid, $đW = -P\,dV$ and $đw = -v\,dP$. The first form represents the work done on a fixed mass of fluid when the volume is changed reversibly; the second is the work done per unit of fluid mass when an element of fluid is transferred reversibly from a fluid reservoir at pressure P to another at pressure $P + dP$. Some authors refer to this type of work as *rotating shaft work*, since it can represent the work performed on the fluid by a device, such as a pump driven by a rotating shaft. It is evident that the enthalpy has a similar role to the free energy, F, because it allows us to calculate the work input to the system in a particular process in terms of a function of state. The enthalpy, like the free energy, is said to be a thermodynamic potential.

Availability for fluid flow processes

We may extend the notion of availability to include processes involving steady flow. As in the case of a non-flow system considered in Section 7.2,

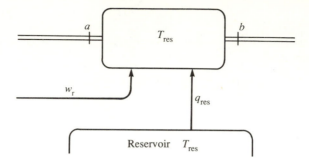

Fig. 7.9 Work and heat inputs in an isothermal fluid flow process in equilibrium with a heat reservoir.

the work input to a steady flow device is least when the relevant change of state, including heat transfer, occurs reversibly. Consider the control volume shown schematically in Fig. 7.9. If the change of state $a \rightarrow b$ occurs isothermally and reversibly then the change in specific entropy of the fluid in the control volume is

$$s_b - s_a = \frac{q_{res}}{T_{res}}$$

where q_{res} is the heat supplied reversibly across the control volume boundary at temperature T_{res} per kg of flow. For the moment we have taken T_{res} to be the temperature of both the heat reservoir and the control volume.

Then, from (7.3.16) we know that the work, w_r, performed on the fluid, per unit mass, is given by

$$w_r = \Delta h - q_{res},$$

where h is the specific enthalpy function for the fluid. Thus,

$$w_r = h_b - h_a - T_{res}(s_b - s_a). \tag{7.3.20}$$

Question

7.3.1. The environment temperature T_{res} is generally different from T, the process temperature, and in any case the process may not be isothermal. In this case the reversible work input is made up of two parts, w_e, the external work, required to operate the heat pump as shown in Fig. 7.10, and w_i, the internal work. This situation is just the same as that described in Section 7.2 for non-flow work.

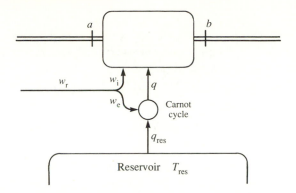

Fig. 7.10   A reversible fluid flow process in which the fluid has work and heat inputs. The fluid need not be in thermal equilibrium with the reservoir.

(a)   Show that in general,

$$q_{res} = T_{res}(s_b - s_a).$$

(b)   Show that the work input (shaft work) per kg of fluid in a reversible steady flow process is given by

$$w_r = w_i + w_e$$
$$= h_b - h_a - T_{res}(s_b - s_a).$$

We define the *specific steady flow availability function, b*, using the result from question 7.3.1:

$$b = h - T_{res}s, \tag{7.3.21}$$

Then the reversible shaft work input is given by

$$w_r = w_i + w_e = b_b - b_a. \tag{7.3.22}$$

Notice that the specific steady flow availability function, b, involves the thermodynamic state variables for the fluid as well as the temperature, T_{res}, for the heat reservoir. The interpretation of the non-flow availability function A given in section 7.2 also applies to b. That is, the value of b provides a measure of the potential for unit mass in a steady flowing fluid to perform work, given an environment characterized by the temperature T_{res}. This potential may only be realized in a reversible process, or, if the value of b changes between two points in a fluid flow:

(1)   the increase in b is the *minimum* external shaft work *input* necessary to achieve that change, per kg of fluid, by processes in the control volume;

(2)   the decrease in b is the *maximum* external shaft work *output* that could be done due to that change, per kg of fluid, by processes in the control volume.

Questions

7.3.2. Figure 7.11 represents a power station steam turbine. The steam
        flow rate is 200 kg s^{-1} and the entry and exit steam states, a and
        b, have the following parameters:

a: $T = 540°C$, $P = 16\,MPa$, $h = 3410.3\,kJ\,kg^{-1}$,
      $s = 6.4481\,kJ\,kg^{-1}\,K^{-1}$;

b: $T = 350°C$, $P = 4\,MPa$, $h = 3095.1\,kJ\,kg^{-1}$,
      $s = 6.5870\,kJ\,kg^{-1}\,K^{-1}$.

Take the temperature of the environmental reservoir to be 20°C.
Assume that changes in the kinetic and potential energy of the
steam at the entry and exit of the turbine can be neglected.

(a) What is the shaft power for the turbine as it stands, assuming
     there is no external heat input or output? (The turbine is
     adiabatic.)

(b) Determine the maximum possible shaft power, given the steam
     conditions at a and b and the flow rate.

(c) How much shaft power is lost due to irreversibility in the
     turbine?

7.3.3. For saturated water vapour at 40°C,

$$h = 2574.4\,kJ\,kg^{-1}, \qquad s = 8.2583\,kJ\,kg^{-1}\,K^{-1}.$$

For saturated liquid water at 40°C,

$$h = 167.45\,kJ\,kg^{-1}, \qquad s = 0.5721\,kJ\,kg^{-1}\,K^{-1}.$$

(a) Show that the changes in enthalpy and entropy given here are
     numerically consistent.

(b) Suppose a steady flow of steam, 200 kg s^{-1}, condenses at
     40°C in the condenser of a thermal power station. The entry

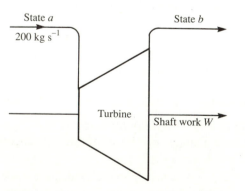

State a

200 kg s^{-1}

State b

Turbine

Shaft work W

Fig. 7.11 A simple steam turbine; a and b denote the steam states at the input and
output.

and exit states are dry saturated steam and saturated liquid
water, respectively. What is the rate of heat transfer across the
heat exchanger?

(c) What is the maximum shaft power which could be produced
by an ideal turbine which receives saturated steam at 40°C and
rejects the flow as saturated condensate at 40°C given an
environment temperature of 20°C? (You should use five
significant figures for the temperature.)

(d) Calculate the thermal efficiency for a Carnot engine operating
between reservoirs at 40°C and 20°C and compare this with
the work/heat transfer ratio obtained from (b) and (c) above.

(e) What would happen to the work potential you estimated in
(c) if the environment were at 40°C, or at least very close to
it. Recollect that, for a given heat input, the work output of
a Carnot engine approaches zero as the source–sink temperature difference vanishes.

Question 7.3.3 is an example of how the potential of a system to perform
work vanishes when it is in equilibrium with its environment. This state of
equilibrium is known as the *dead-state*. A useful dead-state for water is a
saturated state at the temperature of the environment. This state is relevant wherever a real water reservoir at that condition is available. Let
the values of h and s corresponding to the dead-state be h_0 and s_0, and
define

$$b_0 = h_0 - T_0 s_0. \qquad (7.3.23)$$

In the current example any mixture of saturated water and steam at T_0
may be taken to be the dead-state because changes in h_0 and s_0 at T_0 are
related by $\Delta h_0 = T_0 \Delta s_0$, and hence the value of b_0 has the same value for
all such saturated states.

We define the specific steady flow *exergy function* as

$$e = b - b_0. \qquad (7.3.24)$$

Evidently e is the maximum shaft work that can be produced, per kg of
fluid, in a process which results in the fluid being ultimately in thermal
equilibrium with the environment. For a process (Fig. 7.12) in which
external work is produced it is conventional to define the second law
efficiency as

$$\eta_e = \frac{w}{w_R}, \qquad (7.3.25)$$

where

$$w_R = b_a - b_b$$

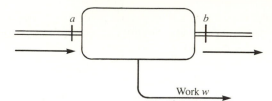

Fig. 7.12   External work performed in a steady fluid flow process.

is the reversible work output per kg and w is the actual work performed per kg. A number of other terms are in common use for η_e including: *rational efficiency, effectiveness, and exergetic efficiency.*

Finally, the expression for b may also include the flow kinetic and potential energy terms, when they are significant as in eqn (5.8.2):

$$b = h + \tfrac{1}{2}c^2 + gz - T_0 s, \tag{7.3.26}$$

where c is the mean axial velocity of the fluid and z is the elevation above some selected datum point. In this case we would normally expect to have $c_0 = 0$ for the dead-state. Then the dead-state availability is

$$b_0 = h_0 + gz_0 - T_0 s_0 \tag{7.3.27}$$

where z_0 is the elevation of the environment selected.

Question

7.3.4.   In an ideal steam turbine the specific entropy, s, for the steam at the entry and exit ports would be the same, but in a real turbine there is an increase in s due to the irreversibility. Ireversible processes include friction, mixing of steam at different states, and internal heat transfer within the turbine. As a result both the specific entropy and specific enthalpy of the exhaust steam are higher than they would be in an isentropic expansion process with the same terminal pressures.

This enthalpy increase is called the *reheat enthalpy*. It is common to regard the reheat enthalpy as a loss because the output power of the turbine is proportional to the difference in specific enthalpy between the entry and exit. Yet there remains an opportunity to use the reheat enthalpy further downstream in subsequent turbines, so it is difficult to establish a consistent standard of turbine 'perfection' using the isentropic model.

The following example illustrates the difference between the isentropic and exergetic models for rating the efficiency of a turbine. The relevant thermodynamic states for steam, x, y, and z, shown in Figs. 7.13–7.15, are given in the following table. The

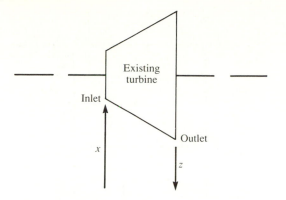

Fig. 7.13 Illustrating the states x and z in question 7.3.4.

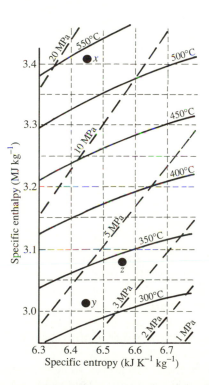

Fig. 7.14 Specific entropy–enthalpy diagram showing the states x, y, z referred to in question 7.3.4. The dashed lines are isobars; the solid lines are isotherms.

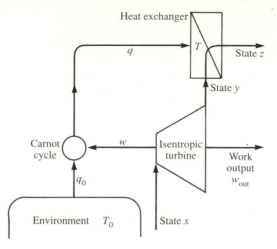

Fig. 7.15   A possible reversible process in which the steam inlet and outlet states x and z are the same as those for an existing turbine.

environmental temperature, T_0, is 12.29°C, and the steam flow rate is 209.26 kg s^{-1}. Ignore possible incidental heat and steam losses.

State	P (MPa)	T (°C)	h (kJ kg^{-1})	s (kJ kg^{-1} K^{-1})	e (kJ kg^{-1})
x	16.283	541.0	3409.94	6.4404	1571.63
y	4.172	321.58	3015.43	6.4404	1177.08
z	4.172	346.83	3083.51	6.5525	1213.19

(a)   Calculate the output shaft power of the turbine.

(b)   What is the shaft power of an isentropic turbine having the same entry condition and the same exit pressure as the existing turbine (isentropic ideal)?

(c)   Calculate the loss of power in the actual turbine relative to an isentropic turbine due to the reheat enthalpy being generated irreversibly.

(d)   What is the shaft power of a reversible turbine having the same entry and exit conditions as the existing turbine (exergetic ideal)?

(e)   Suppose the reheat enthalpy is introduced at the exit of the isentropic turbine by a reversible process (that is, using a heat pump) drawing heat from the environment as shown in Fig. 7.15. As an approximation, take the upper temperature, T, to be midway between that for T_y and T_z and use the

Carnot to formula (6.4.1) to calculate the mechanical power required.

(f) By coupling the isentropic ideal turbine to this reversible heat pump (as in Fig. 7.15) the steam condition at the outlet of this combined system can be arranged to be the same as that for the actual turbine. The net shaft power would be more than that for the actual turbine, but less than that for the hypothetical isentropic turbine. Calculate the output power in this case, using your results from (b) and (e). (This should be the same as that obtained in (d), within the approximation used in (e)).

Question 7.3.4 illustrates the difference between the isentropic and exergetic measures of turbine efficiency. The isentropic view is that the reheat loss, due to frictional dissipation, mixing, and other irreversible processes in the turbine, would be eliminated in an ideal device. Thus the specific entropy would be the same at the entry and exit.

The exergetic view is that the existence of reheat is not the real cause of loss. Rather the loss of work potential is caused by the reheat enthalpy being introduced to the steam flow irreversibly. The hypothetical ideal, in this picture, is a combined adiabatic reversible turbine and a reversible heat pump as in Fig. 7.15. It is evident that the exergetic efficiency for a turbine will always be greater than the corresponding isentropic efficiency.

7.4 Wires and surfaces

In regard to work processes, wires and surfaces represent one- and two-dimensional analogues of a fluid. First, consider a stretched wire of uniform cross-section, A, and homogeneous composition. Let τ be the tension and L the length. The work performed by external forces when the length is increased quasistatically by dL is

$$\mathrm{d}W = \tau\,\mathrm{d}L. \tag{7.4.1}$$

Here the quasistatic condition implies that the time over which the process occurs should be much longer than the time for a disturbance, such as a sound wave, to propagate the length of the wire. Given the process is reversible, we shall require that the wire undergoes elastic stretching only. If it were inelastic, a cyclic change in the length of the wire would result in a net amount of work being performed, even if the process were quasistatic. In such a process, which is said to be dissipative, there is no unique relation between the tension, τ, and the length, L. Equation (7.4.1) indicates that L is the deformation variable for the elastic wire.

For a given amount of wire, the first law and the Gibbs equation can be expressed as

$$dU = đQ + đW, \qquad (7.4.2)$$

and

$$dU = T\,dS + \tau\,dL, \qquad (7.4.3)$$

respectively. The comments made in Section 7.3 in relation to displacement work on a fluid apply here, with the replacements $P \to -\tau$ and $V \to L$. Thus U is a function of S and L and response functions similar to (7.3.8)–(7.3.13) may be defined. Corresponding to the bulk modulus $(B = 1/\kappa)$ it is usual to use Young's modulus, defined as

$$E_T = \frac{L}{A}\left(\frac{\partial \tau}{\partial L}\right)_T. \qquad (7.4.4)$$

Note, however, the isothermal Young's modulus cannot be used to relate τ and L in the Gibbs equation (7.4.3) since it is evaluated for a constant temperature process. Instead the relevant property is the isentropic Young's modulus,

$$E_S = \frac{L}{A}\left(\frac{\partial \tau}{\partial L}\right)_S. \qquad (7.4.5)$$

Corresponding to the enthalpy of a fluid, we may define a new function for the wire,

$$H = U - \tau L, \qquad (7.4.6)$$

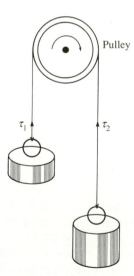

Fig. 7.16 A process involving an elastic wire analogous to the fluid pumping process shown in Fig. 7.8.

and corresponding to the specific enthalpy, we have

$$h = u - \tau l \tag{7.4.7}$$

where the lower case symbols denote specific quantities; thus l is the length per unit mass. An example of a process where the enthalpy function for a wire would be useful is illustrated in Fig. 7.16. This shows a wire supported by a pulley while the tension on each side is maintained by weights. Work is done on the system as the pulley is rotated; in this way elements of the wire are transferred from a state at tension τ_1 to another at tension τ_2. By using an argument similar to that presented in Section 7.3 it can be shown that

$$h_2 - h_1 = w_{12} + q_{12} \tag{7.4.8}$$

where w_{12} is the work performed by the pulley per unit mass of wire and q_{12} is the external heat input during the transfer process.

For a surface film, the reversible work performed in order to increase the surface area by dA is

$$\d W = \sigma \, dA, \tag{7.4.9}$$

where σ is the surface tension. The analogy between a film and a stretched wire is clear and the arguments presented for the elastic wire apply directly to the surface film with appropriate changes.

8
Electric and magnetic work processes

Several different types of work process may be identified when matter is subject to electric and magnetic fields. Consequently there are a number of expressions for the work performed. Each is useful in a particular physical situation. This chapter provides an introduction to these work processes and to the thermodynamic functions associated with them. While we shall refer to this material in subsequent chapters it may be passed over on a first reading without loss of continuity.

8.1   Dielectric media in an electric field

We shall consider the work processes for a dielectric medium in the presence of an electric field. S.I. units will be used. We assume that the medium does not exhibit any hysteresis in the processes of interest; in effect we require that quasistatic changes shall be reversible.

To be specific, we suppose that the electric field is produced by a conductor carrying a charge, q, as in Fig 8.1. Such a composite system can, if we choose, be isolated from its surroundings. This is important because the concept of an isolated system is a primitive notion which we use to define the energy of a system. Initially we shall assume that the position of the medium is fixed relative to the conductor.

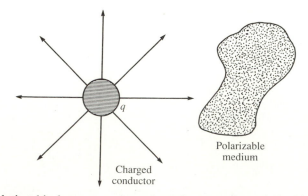

Fig. 8.1 Relationship between a charged conductor and a polarizable medium in an electric work process.

The charge as the constraint

The primary constraint which determines the energy of the composite system is the charge, q. For if the charge is constant, and no other work, heat, or mass transfer takes place, then the energy of the system will be constant. Suppose, on the other hand, we wished to maintain the potential of the conductor, ϕ, at a fixed value. Whilst the system remains isolated an internal process within the medium could cause the dielectric constant to change which could, in turn, affect the potential of the conductor. Thus ϕ can be held at a fixed value only if we choose to control the charge externally. External work would thereby be performed on the system. We therefore take q to be the primary work parameter for the electrostatic energy of the composite system. In this sense it is the deformation parameter for electric work, analogous to the volume of a fluid; ϕ, on the other hand, corresponds to $-P$.

We can illustrate the role of the charge by considering the work done against electrostatic forces when we add an element of charge, δq, brought quasistatically from infinity to the conductor. The work done on the composite system in this process is

$$\delta W_c = \phi \, \delta q. \tag{8.1.1}$$

Now the work done in this process can also be expressed in terms of the fields by a standard result, derived in Appendix C,

$$\phi \, \delta q = \int \mathbf{E} \cdot \delta \mathbf{D} \, \mathrm{d}V, \tag{8.1.2}$$

where $\mathbf{E}$ is the electric field and $\delta \mathbf{D}$ is the change in the electric displacement due to the additional charge δq. The volume integral extends over all space except the conductor. Hence,

$$\delta W_c = \int \mathbf{E} \cdot \delta \mathbf{D} \, \mathrm{d}V. \tag{8.1.3}$$

Since the work done in a reversible process depends only on the change of state, it is evident that $\mathbf{E}$ and $\mathbf{D}$ can also be used as state variables for the system, and that $\mathbf{D}$ has the role of the deformation parameter. Thus (8.1.3) represents the total work done on the system for any reversible process, whether or not the work is done by bringing up the additional charge, δq. For instance, the work might be done instead by changing the position of the dielectric medium.

We may write the first law for the composite system, denoted by the subscript c, when the system has a small reversible process,

$$\delta U_c = \delta Q + \int \mathbf{E} \cdot \delta \mathbf{D} \, \mathrm{d}V. \tag{8.1.4}$$

In general the conductor on which the source charge resides is not in thermal equilibrium with the medium. Hence $\delta Q \neq T \delta S_c$ because there is no unique temperature, T, for the system, so (8.1.4) cannot be put into the form of a Gibbs equation immediately. For this reason it is convenient to rewrite this equation in a way which involves state properties of the medium alone.

In order to specify the work done on the medium we shall adopt the following procedure. First calculate the work done on the composite system with the medium in place, then subtract the work required to set up the corresponding field which would exist in the vacuum in the absence of the medium. Certainly this procedure seems reasonable, but it has the status of a definition because the result depends on how we choose to define the applied field. We consider the following possibilities:

(1) the charge of the conductor is the same with and without the medium;

(2) the potential is the same.

In making this choice we must be guided by the problem in hand. Let us begin by using the source charge, q, as the primary constraint for the system. Thus we must take the applied fields E_0 and D_0 to be the electric field and displacement due to the same q on the same conductor in the absence of the medium. Due to the presence of the dielectric medium additional work must be done when $q \rightarrow q + \delta q$. This work, which we attribute to the medium, is given by

$$\delta W = \int E \cdot \delta D \, dV - \int E_0 \cdot \delta D_0 \, dV. \tag{8.1.5}$$

Notice that E and D are the actual fields, whereas we may regard E_0 and D_0 as the applied fields, being the fields in the absence of the medium.

We now transform this expression into an integral over the volume of the medium only. First we note that $D_0 = \varepsilon_0 E_0$, since these are vacuum fields, and then we utilize (8.1.2) and variations of that equation,

$$\delta W = \int (E \cdot \delta D - D_0 \cdot \delta E_0) \, dV = \phi \, \delta q - q_0 \, \delta\phi_0.$$

Since the charge of the conductor is the same for the applied field, $q = q_0$ and $\delta q = \delta q_0$, and hence,

$$\delta W = \phi \, \delta q_0 - q \, \delta\phi_0 = \int (E \cdot \delta D_0 - D \cdot \delta E_0) \, dV = \int (\varepsilon_0 E - D) \cdot \delta E_0 \, dV.$$

Now, P, the electric polarization for the medium, is defined in terms of the local fields by

$$D = \varepsilon_0 E + P. \tag{8.1.6}$$

It follows that

$$\delta W = - \int_{V_m} \mathbf{P} \cdot \delta \mathbf{E}_0 \, dV \qquad (8.1.7)$$

where V_m indicates that the integral extends over the volume of the medium alone. For an isotropic linear medium the polarization is related to the local electric field by

$$\mathbf{P} = \varepsilon_0 \chi_e \mathbf{E} \qquad (8.1.8)$$

where χ_e is the electric susceptibility of the medium.

To reiterate, δW is the additional work done on the free charge δq in bringing it to the conductor due to the medium being present. Per unit volume of the medium, the work is $-\mathbf{P} \cdot \delta \mathbf{E}_0$, where $\delta \mathbf{E}_0$ is the change in the applied field in the absence of the medium.

We can now write down the first law and the Gibbs equations for the medium. Let $U_c(q)$ be the energy of the composite system with the medium in place with charge q on the conductor and let $U_0(q)$ be the energy of the conductor having the same charge but in the absence of the medium. We define the energy of the medium, U_q,

$$U_q = U_c(q) - U_0(q). \qquad (8.1.9)$$

Clearly, for an isentropic charging process, $q \rightarrow q + \delta q$,

$$\delta U_q = \delta W = - \int_{V_m} \mathbf{P} \cdot \delta \mathbf{E}_0 \, dV \qquad (8.1.10)$$

and since the first law applies to the energy of the composite system it also holds for U_q. Thus,

$$\delta U_q = \delta Q + \delta W, \qquad (8.1.11)$$

where δQ and δW represent the heat and work inputs to the medium. Equation (8.1.11) holds whether or not the heat and work processes are reversible. By considering the reversible limit we obtain the Gibbs equation for U_q,

$$\delta U_q = T \delta S - \int_{V_m} \mathbf{P} \cdot \delta \mathbf{E}_0 \, dV. \qquad (8.1.12a)$$

Alternatively, denoting the energy and entropy per unit volume by lower case symbols,

$$\delta u_q = T \delta s - \mathbf{P} \cdot \delta \mathbf{E}_0. \qquad (8.1.12b)$$

In effect these equations use the local variable $\mathbf{E}_0$ as the deformation variable instead of the charge q. They have been obtained on the assumption that the medium and conductor have fixed positions, but if we hold the charge q fixed and move the medium relative to the conductor, δE_0 will represent the change in the applied field at each point in the medium due to translation. No work is done by electrical sources, since no free charge is moved. In this process δW will represent the mechanical work input to move the medium. Thus (8.1.7) can be used to determine the work done in a displacement process in which the applied field at each point in the medium changes by δE_0. Notice that δE_0 is a function of position in general.

The potential as the constraint

I have used the term *energy* for U_q because it satisfies the first law equation (8.1.11) for both reversible or irreversible processes, but it is better to regard U_q as a thermodynamic potential, being a function which can be used to relate heat and work quantities for part of a composite system. The form of the corresponding Gibbs equation arose from our decision to identify the applied field with q, rather than ϕ. But in appropriate situations we should take ϕ to be the fixed parameter.

To illustrate, suppose we introduce a dielectric body into a capacitor in which the potentials of the capacitor plates are maintained at fixed values by an external e.m.f. We might wish to determine the mechanical work done in moving the body. Here we have a composite system which is not isolated, since the external e.m.f. represents a work source. Because the electric potentials of the plates are fixed we shall take E_0 and D_0 to be the fields produced by the conductor (or conductors) in the absence of the medium, given that the potential is fixed. For simplicity we shall consider only one conductor.

The analysis is formally similar to the constant charge case considered above. The work done by the external e.m.f., attributable to the presence of the medium, is given by (8.1.5); but now the field changes are due to additional charges, δq, being brought up to the conductor in order to effect a change $\delta \phi$ in its potential. Hence $\phi = \phi_0$ and $\delta \phi = \delta \phi_0$ and we have

$$\delta W = \phi \, \delta q - q_0 \, \delta \phi_0 = \phi_0 \, \delta q - q_0 \, \delta \phi = \int (E_0 \cdot \delta D - \varepsilon_0 E_0 \, \delta E) \, dV.$$

Now the change in the polarization of the medium is

$$\delta P = \delta (D - \varepsilon_0 E), \tag{8.1.13}$$

and hence

$$\delta W = \int_{V_m} E_0 \cdot \delta P \, dV. \tag{8.1.14}$$

We define the energy of the medium by analogy with (8.1.9),

$$U_\phi = U_c(\phi) - U_0(\phi), \tag{8.1.15}$$

where ϕ indicates that the potential of the source conductor is the controlled parameter which defines the applied field. But while U_ϕ is a well-defined energy function, fixing the value of ϕ does not ensure that there is no electrical work done in a given process. Suppose, for instance, work W is performed on the medium in an adiabatic process at constant ϕ. In general $W \neq \Delta U_\phi$, because it will normally be necessary to perform electrical work also, changing q in order to hold ϕ constant. Instead, for this process we must have $W + W_e = \Delta U_\phi$, where W_e is the work performed by the external e.m.f. which maintains the value of ϕ. So, in general,

$$\Delta U_\phi = Q + W + W_e, \qquad (8.1.16)$$

for a process involving heat and work inputs, given that ϕ is the controlled variable. These processes need not all be reversible. The Gibbs equation for U_ϕ is obtained by considering a reversible process in which only electric work is done ($W = 0$). Then,

$$\delta U_\phi = T\delta S + \int_{V_m} \mathbf{E}_0 \cdot \delta \mathbf{P}\, \mathrm{d}V, \qquad (8.1.17a)$$

where the volume integral extends over the medium only, or, in terms of the energy and entropy per unit volume of the medium,

$$\delta u_\phi = T\delta s + \mathbf{E}_0 \cdot \delta \mathbf{P}. \qquad (8.1.17b)$$

If we perform mechanical work on the medium by moving it while ϕ is constant, work will be done both by the e.m.f. and by the external forces which move the medium. These can be separated by using the approach taken in Section 7.3 to determine the shaft work when a fluid is pumped between two volumes maintained at different pressures. Denote the initial and final positions of the medium by subscripts 1 and 2, respectively. The electric potential ϕ is constant throughout the process of change; we assume that the interaction between the conductor and the source e.m.f., which maintains ϕ, is reversible. However, the interaction between the mechanical forces and the medium need not be reversible. For this process the first law for the composite system comprising the medium and the conductor is $\Delta U_{\phi 12} = W_{m12} + W_{e12} + Q_{12}$, where W_{m12} is the mechanical work done on the medium and W_{e12} is the work performed by the e.m.f. Let δq be the additional charge required to keep ϕ constant. Then,

$$W_{e12} = \phi\, \delta q = \phi(q_2 - q_1) = \phi_0 q_2 - \phi q_0 - \phi_0 q_1 + \phi q_0$$

where q_0 is the charge of the conductor in the absence of the medium at the specified potential, $\phi_0 = \phi$. By applying (8.1.2), we have

$$\phi_0 q_2 - \phi q_0 = \int (\mathbf{E}_0 \cdot \mathbf{D}_2 - \mathbf{E}_2 \cdot \mathbf{D}_0)\, \mathrm{d}V = \int \mathbf{E}_0 \cdot (\mathbf{D}_2 - \varepsilon_0 \mathbf{E}_2)\, \mathrm{d}V = \int_{V_m} \mathbf{E}_0 \cdot \mathbf{P}_2\, \mathrm{d}V.$$

A similar expression is obtained for $-\phi_0 q_1 + \phi q_0$. Hence we obtain

$$W_{e12} = \int_{V_2} \mathbf{E}_0 \cdot \mathbf{P}_2 \, dV - \int_{V_1} \mathbf{E}_0 \cdot \mathbf{P}_1 \, dV, \tag{8.1.18}$$

where V_1 and V_2 define the medium at positions 1 and 2, respectively. Thus we obtain a first law equation:

$$Q_{12} + W_{m12} = \Delta U_{\phi 12} - \int_{V_2} \mathbf{E}_0 \cdot \mathbf{P}_2 \, dV + \int_{V_1} \mathbf{E}_0 \cdot \mathbf{P}_1 \, dV.$$

Following Section 7.3 we may identify another thermodynamic potential for the medium, designated as H_ϕ because it is analogous to the enthalpy of a fluid,

$$H_\phi = U_\phi - \int_{V_m} \mathbf{E}_0 \cdot \mathbf{P} \, dV, \tag{8.1.19a}$$

or, using lower case symbols to denote the energy and enthalpy per unit volume,

$$h_\phi = u_\phi - \mathbf{E}_0 \cdot \mathbf{P}. \tag{8.1.19b}$$

We see that the work done per unit volume by external mechanical forces when we move the medium in an adiabatic process is $w_{m12} = h_{\phi 2} - h_{\phi 1}$. If the process is not adiabatic,

$$w_{m12} + q_{12} = h_{\phi 2} - h_{\phi 1}, \tag{8.1.20}$$

where q_{12} is the input of heat per unit volume of the medium. This equation holds for both reversible and irreversible work and heat transfer provided the interaction with the source e.m.f. is reversible.

The Gibbs equation for h_ϕ is simply derived from (8.1.19b):

$$\delta h_\phi = \delta u_\phi - \delta(\mathbf{E}_0 \cdot \mathbf{P}), \tag{8.1.21}$$

and hence from (8.1.17b)

$$\delta h_\phi = T \, \delta s - \mathbf{P} \cdot \delta \mathbf{E}_0, \tag{8.1.22a}$$

or,

$$\delta H_\phi = T \, \delta S - \int_{V_m} \mathbf{P} \cdot \delta \mathbf{E}_0 \, dV, \tag{8.1.22b}$$

where $\delta \mathbf{E}_0$ is the change in the applied field at the medium due to its movement. Thus for a reversible process the mechanical work is

$$\delta w_m = -\mathbf{P} \cdot \delta \mathbf{E}_0, \tag{8.1.23a}$$

or

$$\delta W_m = -\int_{V_m} \mathbf{P} \cdot \delta \mathbf{E}_0 \, dV. \tag{8.1.23b}$$

Notice that the equation for a reversible mechanical work process is formally the same whether $\mathbf{E}_0$ is defined by constant potential or constant charge.

Example

A dielectric body is subject to an electric field due to a charged conductor on which the charge q is constant. If the medium undergoes an isentropic process in which the polarization changes $\mathbf{P} \rightarrow \mathbf{P} + \delta\mathbf{P}$, show that the work done on the medium can be expressed by

$$\delta H_q = \int_{V_{\mathrm{m}}} \mathbf{E}_0 \cdot \delta\mathbf{P}\, \mathrm{d}V. \tag{8.1.24}$$

Here H_q is the thermodynamic potential defined by

$$H_q = U_q + \int_{V_{\mathrm{m}}} \mathbf{E}_0 \cdot \mathbf{P}\, \mathrm{d}V, \tag{8.1.25}$$

and $\mathbf{E}_0$ is the field in the absence of the medium given that q is fixed. The medium remains stationary.

Solution

The energy of an isolated composite system, consisting of the charged conductor and the medium, is constant if q is fixed. We have defined a thermodynamic potential for the medium U_q, being the energy of the composite system minus the energy of the vacuum field corresponding to q. In the process of concern here the work term for U_q,

$$\delta U_q = \int_{\mathrm{medium}} \mathbf{P} \cdot \delta\mathbf{E}_0\, \mathrm{d}V,$$

will be zero. Nevertheless there is a transfer of energy between the field source and the medium and so this particular potential is not appropriate to represent this process.

If the potential changes $\phi \rightarrow \phi + \delta\phi$ whilst $q = q_0$ in the absence of the medium the potential energy of the conductor will change by $q_0 \delta\phi$. Hence $-q_0\delta\phi$ can be regarded as the energy transferred from the field source to the medium. Now,

$$q_0\,\delta\phi = \int \mathbf{D}_0 \cdot \delta\mathbf{E}\, \mathrm{d}V = \int \varepsilon_0 \mathbf{E}_0 \cdot \delta\mathbf{E}\, \mathrm{d}V = \int \mathbf{E}_0 \cdot (\delta\mathbf{D} - \delta\mathbf{P})\, \mathrm{d}V,$$

and

$$\int \mathbf{E}_0 \cdot \delta\mathbf{D}\, \mathrm{d}V = \phi_0\,\delta q = 0,$$

since $\delta q = 0$. Hence

$$q_0 \, \delta\phi = - \int_{V_m} \mathbf{E}_0 \cdot \delta\mathbf{P} \, \mathrm{d}V = - \int_{V_m} \delta(\mathbf{E}_0 \cdot \mathbf{P}) \, \mathrm{d}V.$$

Thus, if we define

$$H_q = U_q + \int_{V_m} \mathbf{E}_0 \cdot \mathbf{P} \, \mathrm{d}V, \qquad (8.1.26)$$

then given that q is constant, $\delta U_q = 0$, and

$$\delta H_q = \int_{V_m} \mathbf{E}_0 \cdot \delta\mathbf{P} \, \mathrm{d}V = -q_0 \, \delta\phi \qquad (8.1.27)$$

may be interpreted as the work done on the dielectric body by the field source, regarded as a work reservoir.

Example

A parallel plate capacitor is partly filled with a rectangular slab of dielectric material as shown in Fig. 8.2. The electric susceptibility, χ_e, is constant. Obtain an expression for the extra electrical work in charging the capacitor due to the material. Take the applied field to correspond to the same potential difference, V, between the plates of the capacitor. Ignore fringe effects at the edge of the capacitor. The surface area of a plate is A.

Solution

Let $\mathbf{E}_i$ be the electric field in the medium, and $\mathbf{E}_e$ the field in the void space. The electric displacement, $\mathbf{D} = \varepsilon_0 \mathbf{E} + \mathbf{P}$ (where $\mathbf{P} = \varepsilon_0 \chi_e \mathbf{E}$) is continuous across the surface of the dielectric. Hence we have

$$\mathbf{D} = \varepsilon_0 (1 + \chi_e) \mathbf{E}_i = \varepsilon_0 \mathbf{E}_e$$

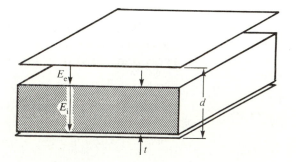

Fig. 8.2 Partly filled parallel plate capacitor. The plate separation is d and the thickness of the medium is t.

and thus

$$V = (d-t)E_e + tE_i = (d + (d-t)\chi_e)E_i.$$

Now $E_0 = V/d$ is the electric field in the absence of the medium, given that V is fixed. Thus

$$E_i = \frac{E_0 d}{d + (d-t)\chi_e}.$$

We can calculate the electrical work in two ways.

(1) $W = \int_{V_m} \int E_0 \, dP \, dV = At \int E_0 \frac{\varepsilon_0 \chi_e d}{d + (d-t)\chi_e} \, dE_0$

$\qquad = \tfrac{1}{2} C_0 V^2 \frac{t\chi_e}{d + (d-t)\chi_e}$

where $C_0 = \varepsilon_0 A/d$ is the capacitance of the empty capacitor.

(2) Alternatively, $W = \tfrac{1}{2}CV^2 - \tfrac{1}{2}C_0 V^2$. Now $C = \dfrac{q}{V} = \dfrac{\sigma_f A}{V}$ where $V = E_0 d$ and σ_f is the density of free surface charge on the capacitor plate. Then,

$$\sigma_f = D = \varepsilon_0 (1 + \chi_e)E_i = \frac{\varepsilon_0 d (1 + \chi_e)E_0}{d + (d-t)\chi_e}.$$

Hence

$$C = \frac{\varepsilon_0 A}{d} \frac{d(1 + \chi_e)}{d + (d-t)\chi_e} = C_0 \frac{d(1 + \chi_e)}{d + (d-t)\chi_e},$$

from which we get the expression for W obtained above.

Example

A rectangular slab of dielectric fits neatly between the plates of a parallel plate capacitor. The plate separation is d, and the surface is a square, the length of the edge being y. Find the extra electrical work required to charge the capacitor due to the medium being present when the slab protrudes a distance z into the gap (Fig. 8.3). Ignore fringe fields. The charging process is adiabatic. Take q to be the reference parameter.

Solution

Since q is the controlled variable, the electrical work due to the medium is

$$W = \int dU_q = -\int_{V_m} \int \mathbf{P} \cdot d\mathbf{E}_0 \, dV$$

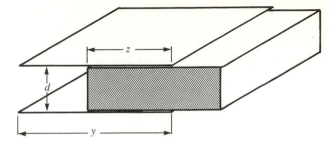

Fig. 8.3 Dielectric slab located partly within a parallel plate capacitor.

where $\mathbf{E}_0$ is the electric field in the absence of the medium when the charge of the capacitor is q. First we need to obtain a relation between $\mathbf{E}_0$ and $\mathbf{P}$. Now the electric field in the medium $\mathbf{E}_i$ is the same as in the void space, $\mathbf{E}_e$, and the surface charge on the capacitor plates, q, will distribute itself accordingly. Let σ_i be the free surface charge density on the capacitor plate adjacent to the dielectric slab; let σ_e be the surface charge in the void space. Then $\varepsilon_0(1 + \chi_e)E_i = D_i = \sigma_i$ and $\varepsilon_0 E_e = D_e = \sigma_e$. Hence, since $E_i = E_e$,

$$q = y(y - z)\sigma_e + yz\sigma_i = \varepsilon_0 y(y + \chi_e z)E_i.$$

But $q = \varepsilon_0 y^2 E_0$, and thus

$$E_i = \frac{yE_0}{y + \chi_e z}.$$

Hence, using $P = \varepsilon_0 \chi_e E_i$ we have

$$W = -yzd \int \frac{\varepsilon_0 \chi_e y E_0}{y + \chi_e z}\, \mathrm{d}E_0 = -\frac{q^2 d}{2\varepsilon_0 y^2} \frac{\chi_e z}{y + \chi_e z}$$

$$= -\frac{q^2}{2C_0} \frac{\chi_e z}{y + \chi_e z}$$

where C_0 is the capacitance of the empty capacitor.

Questions

8.1.1. A dielectric body undergoes an adiabatic reversible process in which the polarization of the medium changes $\mathbf{P} \to \mathbf{P} + \delta\mathbf{P}$, where $\mathbf{P}$ is a function of position. The body is subject to an electric field due to a charged conductor. During the process the dielectric is stationary relative to the conductor. Given that the potential of the conductor ϕ is constant, show that the work performed by the external e.m.f is

$$\delta U_\phi = \int_{V_m} \mathbf{E}_0 \cdot \delta \mathbf{P} \, \mathrm{d}V$$

where $\mathbf{E}_0$ is the electric field due to the charged conductor in a vacuum, and δU_ϕ is evaluated at constant entropy.

8.1.2. Let E_i be the electric field in a dielectric medium located in a parallel plate capacitor. The applied fields corresponding to constant q and V are denoted E_{0q} and E_{0V}, respectively.

(a) For the system in Fig. 8.2 show that

$$E_{0q} = \frac{d(1 + \chi_e)}{d + (d - t)\chi_e} E_{0V} \tag{8.1.28}$$

and

$$E_i = \frac{E_{0q}}{1 + \chi_e}. \tag{8.1.29}$$

(b) For the system in Fig. 8.3 show that

$$E_{0q} = \frac{y + z\chi_e}{y} E_{0V} \tag{8.1.30}$$

and

$$E_i = E_{0V}. \tag{8.1.31}$$

(c) In the limit of a small dielectric body in the capacitor, show that $E_{0q} = E_{0V}$ in both cases.

8.1.3. A parallel plate capacitor is filled with a dielectric medium, the electric susceptibility, χ_e, of which is constant. The capacitance of the empty capacitor is $C_0 = \dfrac{\varepsilon_0 A}{d}$.

(a) If the capacitor charge, q, is fixed, show that the mechanical work required to move the medium into the capacitor reversibly is

$$W_{mq} = -\frac{q^2}{2C_0} \frac{\chi_e}{1 + \chi_e}.$$

(b) If the potential difference, V, of the capacitor plates is constant show that the mechanical work required to move the medium into the capacitor is

$$W_{mV} = -\frac{\chi_e}{2} C_0 V^2,$$

and that the work done by the e.m.f. which maintains V at a constant value is

$$W_e = \chi_e C_0 V^2.$$

8.1.4. A parallel plate capacitor, of plate area A and plate separation d, contains a slab of dielectric material of thickness t and surface area A.

(a) Given that the applied field corresponds to the same charge q, show that the electric work required to polarize the medium is

$$W_q = - \frac{q^2}{2C_0} \frac{t\chi_e}{d(1 + \chi_e)},$$

where C_0 is the capacitance of the empty capacitor.

(b) If V is constant, show that the mechanical work required to insert the medium into the capacitor is

$$W_{mV} = - \tfrac{1}{2} C_0 V^2 \frac{t\chi_e}{d + (d - t)\chi_e}.$$

(c) The mechanical work required to insert the medium into the capacitor, given that q is constant, is $W_{mq} = W_q$. Why?

(d) Show that the actual charge and potential difference are related by

$$q = C_0 V \frac{d(1 + \chi_e)}{d + (d - t)\chi_e}$$

and hence that $W_{mq} \neq W_{mv}$, except in the limit $t \ll d$. Do you expect this?

8.1.5. A well-fitting dielectric slab is inserted between the plates of a parallel plate capacitor. The capacitor plate is square, the length of the side being y, and the slab is inserted a distance z as shown in Fig. 8.3. The capacitance of the empty capacitor is C_0.

(a) Suppose q is constant. Show that the mechanical work done on the medium when inserted a distance z is

$$W_{mq} = - \frac{q^2}{2C_0} \frac{\chi_e z}{y + \chi_e z}.$$

(b) Given that V is maintained at a constant value by an external e.m.f., show that

$$W_{mv} = - \tfrac{1}{2} C_0 V^2 \frac{\chi_e z}{y}.$$

(c) Show that the charge and potential difference (measured at the same time) are related by

$$q = C_0 V \left(1 + \frac{\chi_e z}{y} \right) \tag{8.1.32}$$

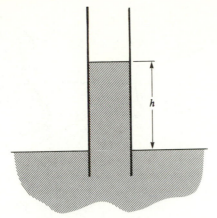

Fig. 8.4 Parallel plate capacitor partly immersed in a dielectric fluid. The plates are oriented vertically.

and hence that $W_{mq} \neq W_{mv}$, except in the limit $\chi_e z \ll y$.

(d)   Show that the force with which the medium is drawn into the capacitor is given by

$$F_q = \frac{q^2}{2C_0} \frac{\chi_e y}{(y + \chi_e z)^2}$$

given that q is specified, and

$$F_v = \tfrac{1}{2} C_0 V^2 \frac{\chi_e}{y}$$

in terms of V. Use (8.1.32) to show that $F_q = F_v$.

(e)   A parallel plate capacitor is partly immersed in a dielectric liquid of density ρ, as presented in Fig. 8.4. Show that the height to which the fluid is drawn into the plates is given by

$$h = \frac{\varepsilon_0 \chi_e V^2}{2d^2 \rho g},$$

ignoring the surface tension of the liquid. Here d is the plate separation and g is the acceleration due to gravity.

8.2   Magnetic work

The analysis of work processes involving magnetizable media subject to a magnetic field is analogous to that for electric work processes, but there are important differences. Again we shall use S.I. units and we assume that quasistatic changes in the medium are reversible. We exclude media which exhibit magnetic hysteresis.

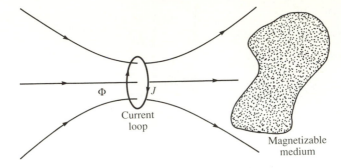

Fig. 8.5 Relationship between a magnetic field source (the current loop) and a magnetizable medium in a magnetic work process.

The first task is to identify a system in which the processes of interest occur whilst being thermodynamically isolated. This allows us to identify the primary work parameter, or deformation variable, for the system. We might, for instance, consider a permanent magnet interacting with a magnetizable body, but processes involving a permanent magnet are not sufficiently transparent to offer a clear thermodynamic view. Instead we shall take a closed conducting loop carrying a current J to be the field source, as in Fig. 8.5.

In order for the system to have a stable state (without J being zero) the conductor must have no electrical resistance. On the other hand, for simplicity we assume that the cross-sectional area of the conductor is negligible compared with the area enclosed by the loop. Let Φ denote the flux linkage (or flux-turns) for the circuit. Suppose the medium undergoes a change of state while the system remains isolated. In general the magnetic properties of the medium will change. The question to be considered is: will J change, or will Φ change, or both? The answer is that Φ will remain constant for an isolated system, and the current will respond accordingly. For if Φ were to change due to a change in the state of the medium, then by Lenz's law there would be an induced e.m.f. which would effect a change in J so as to annul the change in Φ. Hence Φ rather than J is the conserved property for this system. Thus the deformation variable corresponding to q in an electrostatic system is Φ, whereas J is the analogue of the electrostatic potential ϕ.

Suppose the current carrying circuit is linked to a current source and we arrange that Φ increases, by adjusting J if necessary. Let $\Phi \to \Phi + \delta\Phi$ in a time interval δt. The corresponding induced e.m.f. in the current loop is $\delta\Phi/\delta t$ and hence the current source will do work at a rate $J\,\delta\Phi/\delta t$. Given that this change occurs quasistatically, we need not consider radiative fields. Then the total work performed on the composite system is given by a relation analogous to (8.1.1),

$$\delta W_{\mathrm c} = J\,\delta\Phi. \qquad (8.2.1)$$

In Appendix C it is shown that the right-hand side can also be expressed in terms of the magnetic intensity, **H**, and the magnetic induction **B**,

$$J\,\delta\Phi = \int \mathbf{H}\cdot\delta\mathbf{B}\,\mathrm dV \qquad (8.2.2)$$

where the volume integral extends over all space, including the conductor, since the fields are generally non-zero within the conductor. Thus,

$$\delta W_{\mathrm c} = \int \mathbf{H}\cdot\delta\mathbf{B}\,\mathrm dV. \qquad (8.2.3)$$

The flux as the constraint

We shall follow the procedures used for electrostatic work in Section 8.1, and begin by taking Φ to be the primary constraint for the composite system. Here the applied fields, $\mathbf{H}_0$ and $\mathbf{B}_0$, are those obtained in the absence of the medium given that Φ is the same. We take the work done on the medium when $\Phi \to \Phi + \delta\Phi$ to be

$$\delta W = \int (\mathbf{H}\cdot\delta\mathbf{B} - \mathbf{H}_0\cdot\delta\mathbf{B}_0)\,\mathrm dV, \qquad (8.2.4)$$

and hence from (8.2.2), and related results (Appendix C) we have

$$\delta W = J\,\delta\Phi - \Phi_0\,\delta J_0,$$

where we have used $\mathbf{B}_0 = \mu_0\mathbf{H}_0$ and $\mathbf{H}_0\cdot\delta\mathbf{B}_0 = \mathbf{B}_0\cdot\delta\mathbf{H}_0$. Now since $\Phi = \Phi_0$ and $\delta\Phi = \delta\Phi_0$,

$$\delta W = J\,\delta\Phi_0 - \Phi\,\delta J_0 = \int (\mathbf{H}\cdot\delta\mathbf{B}_0 - \mathbf{B}\cdot\delta\mathbf{H}_0)\,\mathrm dV = \int \left(\mathbf{H} - \frac{1}{\mu_0}\mathbf{B}\right)\cdot\delta\mathbf{B}_0\,\mathrm dV.$$

The magnetization vector of the medium, **M**, is given by

$$\mathbf{B} = \mu_0(\mathbf{H} + \mathbf{M}), \qquad (8.2.5)$$

and hence,

$$\delta W = -\int_{V_{\mathrm m}} \mathbf{M}\cdot\delta\mathbf{B}_0\,\mathrm dV, \qquad (8.2.6)$$

which is analogous to (8.1.7); here $V_{\mathrm m}$, the volume of the medium, defines the boundary of the volume integral. For an isotropic linear medium the magnetization is related to the local magnetic intensity **H**,

$$\mathbf{M} = \chi_{\mathrm m}\mathbf{H}, \qquad (8.2.7)$$

where $\chi_{\mathrm m}$ is the magnetic susceptibility. The relationship between **H**, $\mathbf{H}_0$, **B**, and $\mathbf{B}_0$ must be considered in each particular situation.

The energy of the medium, U_Φ, is defined as follows. Let $U_c(\Phi)$ be the energy of the composite system with the medium in place. Let $U_0(\Phi)$ be the energy of the system in the absence of the medium given the same flux, Φ, and define

$$U_\Phi = U_c(\Phi) - U_0(\Phi). \tag{8.2.8}$$

Clearly, for an isentropic process in which $\Phi \to \Phi + \delta\Phi$,

$$\delta U_\Phi = W = - \int_{V_m} \mathbf{M} \cdot \delta\mathbf{B}_0 \, dV. \tag{8.2.9}$$

In addition U_Φ must satisfy the first law,

$$\delta U_\Phi = \delta W + \delta Q. \tag{8.2.10}$$

This equation, like (8.1.11), holds for both reversible and irreversible processes. The Gibbs equation for U_Φ is obtained from (8.2.10) in the reversible limit:

$$\delta U_\Phi = T \delta S - \int_{V_m} \mathbf{M} \cdot \delta\mathbf{B}_0 \, dV, \tag{8.2.11a}$$

where S is the entropy of the medium. Alternatively, in terms of the energy and entropy per unit volume of the medium,

$$\delta u_\Phi = T \delta s - \mathbf{M} \cdot \delta\mathbf{B}_0. \tag{8.2.11b}$$

If Φ is constant no work is performed by the external current source. Under this restriction a change in U_Φ due to translation of the medium is the external mechanical work done on the medium. Thus the mechanical work, δW_m, due to an infinitesimal reversible adiabatic process at constant Φ is given by

$$\delta W_m = - \int_{V_m} \mathbf{M} \cdot \delta\mathbf{B}_0 \, dV. \tag{8.2.12}$$

Here $\delta\mathbf{B}_0$ is the change in the applied field at a point in the medium due to translation.

The current as the constraint

The medium and the current carrying coil comprise a composite system. Obviously the medium cannot be treated as an isolated system whilst maintaining the magnetic interaction. If we choose to specify the current, J, rather than Φ, then we must identify the applied fields, $\mathbf{H}_0$ and $\mathbf{B}_0$, with the fields in the absence of the medium for the same current in the conducting circuit. This definition of $\mathbf{H}_0$ and $\mathbf{B}_0$ is normally adopted implicitly. It is analogous to the definition of $\mathbf{D}_0$ and $\mathbf{E}_0$ in which we took the potential of the conductor to be specified. In both situations the systems must be linked to external work reservoirs. In the magnetic context the conducting circuit must contain a current source.

We take the medium to be stationary. As was the case when we considered the flux as the constraint, the extra work done by the current source, due to the medium, when $J \rightarrow J + \delta J$ is

$$\delta W = J \delta \Phi - \Phi_0 \delta J_0 = J_0 \delta \Phi - \Phi_0 \delta J,$$

since $J = J_0$ and $\delta J = \delta J_0$. Hence, using (8.2.2) and related results,

$$\delta W = \int (\mathbf{H}_0 \cdot \delta \mathbf{B} - \mathbf{B}_0 \cdot \delta \mathbf{H}) \, dV = \int \mathbf{B}_0 \cdot \left(\frac{1}{\mu_0} \delta \mathbf{B} - \delta \mathbf{H} \right) dV,$$

and therefore,

$$\delta W = \int_{V_m} \mathbf{B}_0 \cdot \delta \mathbf{M} \, dV. \tag{8.2.13}$$

This work is done by the current source. In addition one may show that if J remains constant whilst the magnetization of the medium changes, $\mathbf{M} \rightarrow \mathbf{M} + \delta \mathbf{M}$ due to some reversible internal process, then (8.2.13) represents the work done on the system by the current source.

The relevant energy function for the medium, U_J, is defined as follows. Let $U_c(J)$ be the energy of the composite system with the medium in place given a source current J. Let $U_0(J)$ be the corresponding energy in the absence of the medium. Then

$$U_J = U_c(J) - U_0(J). \tag{8.2.14}$$

In one sense U_J is the energy of the medium corresponding to J as the control variable. But U_J does not satisfy a simple first law equation. Instead U_J satisfies a first law relation like (8.1.16),

$$\Delta U_J = Q + W + W_e, \tag{8.2.15}$$

for a process in which work W and heat Q are transferred to the system. Here W_e is the work performed by the external current source in order to maintain J at a given value. In addition, by taking $W = 0$, and W_e given by (8.2.13), we obtain the Gibbs equation for U_J:

$$\delta U_J = T \delta S + \int_{V_m} \mathbf{B}_0 \cdot \delta \mathbf{M} \, dV. \tag{8.2.16}$$

The volume-specific form, corresponding to (8.2.11b) follows directly.

By adapting the argument used in considering the potential as the constraint in Section 8.1 we may also determine the work done by mechanical forces as the medium is moved when J is fixed. Given that the initial and final states are denoted 1 and 2, respectively, the work of the current source is

$$W_{e12} = \int_{V_2} \mathbf{B}_0 \cdot \mathbf{M}_2 \, dV - \int_{V_1} \mathbf{B}_0 \cdot \mathbf{M}_1 \, dV, \qquad (8.2.17)$$

which is analogous to (8.1.18). The mechanical work done is expressed in terms of a magnetic enthalpy-like potential,

$$H_J = U_J - \int_{V_m} \mathbf{B}_0 \cdot \mathbf{M} \, dV, \qquad (8.2.18a)$$

or in terms of the volume-specific quantities

$$h_J = u_J - \mathbf{B}_0 \cdot \mathbf{M}. \qquad (8.2.18b)$$

The first law equation for a process in which a body is moved from position 1 to position 2 is given by

$$w_{12} + q_{12} = h_{J2} - h_{J1}, \qquad (8.2.19)$$

where w_{12} and q_{12} are the work and heat input per unit volume in this process. Note that the work w_{12} is delivered by sources other than the current source. In addition, not all these heat and work processes need to be reversible; however the magnetic interaction must be reversible. The corresponding form for the Gibbs equation is

$$\delta h_J = T \, \delta s - \mathbf{M} \cdot \delta \mathbf{B}_0, \qquad (8.2.20a)$$

or

$$\delta H_J = T \, \delta S - \int_{V_m} \mathbf{M} \cdot \delta \mathbf{B}_0 \, dV. \qquad (8.2.20b)$$

If the process is reversible the mechanical work performed in moving the medium is

$$\delta W_m = - \int_{V_m} \mathbf{M} \cdot \delta \mathbf{B}_0 \, dV. \qquad (8.2.21)$$

While this equation formally is the same as (8.2.12), the physical constraint on the system is different. This difference is illustrated in the examples.

Example

A magnetizable medium is subject to a magnetic field, the source of which is a current carrying circuit consisting of a lossless conducting loop. Whilst the medium remains stationary its magnetization undergoes a reversible change, $\mathbf{M} \to \mathbf{M} + \delta \mathbf{M}$. Show that the work done on the medium can be expressed in terms of the potential

$$H_\Phi = U_\Phi + \int_{V_m} \mathbf{B}_0 \cdot \mathbf{M} \, dV \qquad (8.2.22)$$

by

$$\delta W = \delta H_\Phi = \int_{V_m} \mathbf{B}_0 \cdot \delta\mathbf{M} \, dV, \qquad (8.2.23)$$

where $\mathbf{B}_0$ is the magnetic field in the absence of the medium given that Φ is constant.

Solution

Since Φ is constant in this process $\delta U_\Phi = 0$, so while U_Φ is useful for determining the work done on the medium in establishing Φ it does not allow us to represent the transfer of energy between the current loop and the medium. But in the process of interest the current will change, $J \to J + \delta J$, and there will be a corresponding change in the energy of the current loop. Now $\Phi \delta J$ is the change in the energy of the current loop if it had occurred in the absence of the medium, given the same current change and the same flux. Consider the function, $U_\Phi - \Phi J$. In an adiabatic reversible process, $\mathbf{M} \to \mathbf{M} + \delta\mathbf{M}$, $\delta(U_\Phi - \Phi J)$ may be treated as the work done on the medium when Φ is constant. In that situation,

$$\delta(\Phi J) = \Phi \, \delta J = \Phi_0 \, \delta J = \int \mathbf{B}_0 \cdot \delta\mathbf{H} \, dV = \int \mu_0 \mathbf{H}_0 \cdot \left(\frac{\delta\mathbf{B}}{\mu_0} - \delta\mathbf{M}\right) dV.$$

But

$$\int_{V_m} \mathbf{H}_0 \cdot \delta\mathbf{B} \, dV = J \, \delta\Phi = 0,$$

since Φ is constant. Thus for a reversible adiabatic process in which $\mathbf{M} \to \mathbf{M} + \delta\mathbf{M}$ at constant Φ, we may regard

$$\delta H_\Phi = \int_{V_m} \mathbf{B}_0 \cdot \delta\mathbf{M} \, dV$$

as the work done on the medium, where H_Φ is defined by (8.2.22)

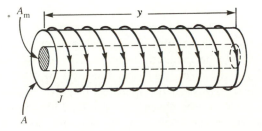

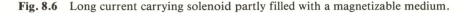

Fig. 8.6   Long current carrying solenoid partly filled with a magnetizable medium.

Example

A long solenoid contains a cylindrical magnetizable medium. The cross-sections of the solenoid and medium are A and A_m, respectively (Fig. 8.6). Derive an expression for the additional work done by an external current source in establishing a current J in the solenoid, due to the presence of the medium. The solenoid has n turns per metre, and the length of the solenoid and medium is y. The magnetic susceptibility of the medium, χ_m, is constant.

Solution

Since the solenoid is long $\mathbf{H}$, $\mathbf{B}$, and $\mathbf{M}$ will be axial and $\mathbf{H}$ will be continuous where it is tangential to the boundary of the medium. Since the current J is the control variable, the additional work is

$$W = \int_{V_m} \int \mathbf{B}_0 \cdot d\mathbf{M}\, dV = \int_{V_m} \int B_0\, dM\, dV,$$

where $B_0 = \mu_0 H_0 = \mu_0 nJ$ and $M = \chi_m H_0 = \chi_m nJ$. Hence,

$$W = A_m y \int \mu_0 nJ \chi_m n\, dJ = \mu_0 \chi_m y A_m \tfrac{1}{2} n^2 J^2.$$

Equally, we could calculate W in terms of the inductance of the coil, L:

$$W = \tfrac{1}{2} L J^2 - \tfrac{1}{2} L_0 J^2,$$

where L_0 is the inductance in the absence of the medium. Now,

$$L = yn \frac{d\Phi}{dJ}$$

and

$$\Phi = \mu A_m (1 + \chi_m) nJ + \mu_0 (A - A_m) nJ = \mu_0 (A + A_m \chi_m) nJ.$$

Thus $L = \mu_0 y n^2 (A + A_m \chi_m)$ and the result for W obtained above follows.

Example

A cylindrical magnetizable medium extends a distance z into a long solenoid, the solenoid and the medium having the same cross-sectional area A, as shown in Fig. 8.7. The magnetic susceptibility of the medium, χ_m, is constant. Determine the electrical work required to establish a magnetic field having Φ_t flux turns with the medium in place relative to the work required to establish Φ_t in the absence of the medium. Assume $z \gg \sqrt{A}$ in which case the demagnetizing field (see, for instance, Robinson (1973)) can be ignored. The density of turns on the solenoid is n per metre and the length of the solenoid is y.

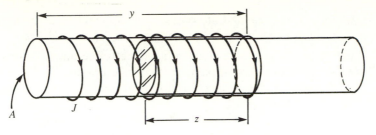

Fig. 8.7 Cylindrical magnetizable medium extending partly into a long current carrying solenoid.

Solution

The work done is given by

$$W = - \int_{V_m} \int \mathbf{M} \cdot d\mathbf{B}_0 \, dV,$$

where $\mathbf{B}_0$ is the magnetic field in the absence of the medium, given by $\Phi_t = nyAB_0$. Since we can ignore the demagnetizing field, the magnetic field intensity in the medium is $H = nJ$ and hence $M = \chi_m nJ$, where J is the solenoid current when the medium is present. Then the field in the medium is $B_i = \mu_0 nJ(1 + \chi_m)$, and the field in the void space of the solenoid is $B_e = \mu_0 nJ$. Hence,

$$\Phi_t = An(y - z)B_e + AnzB_i = \mu_0 An^2 J(y - z + z(1 + \chi_m)).$$

Thus we have

$$M = \frac{\Phi_t \chi_m}{\mu_0 An(y + \chi_m z)} = B_0 \frac{y\chi_m}{\mu_0(y + \chi_m z)},$$

yielding

$$W = - \frac{zyA\chi_m}{\mu_0(y + \chi_m z)} \int B_0 \, dB_0 = - \frac{zyA\chi_m B_0^2}{2\mu_0(y + \chi_m z)} = - \frac{\Phi_t^2}{2L_0} \frac{\chi_m z}{y + \chi_m z}.$$

Questions

8.2.1. A magnetizable body undergoes an isentropic process in which the magnetization changes $\mathbf{M} \to \mathbf{M} + \delta\mathbf{M}$. The medium is subject to a magnetic field due to a current, J, maintained in a lossless loop by a current source. The geometry of the system is fixed. Show that the work performed by the current source is given by

$$\delta U_J = \int_{V_m} \mathbf{B}_0 \cdot \delta\mathbf{M} \, dV, \qquad (8.2.24)$$

where δU_J is evaluated at constant entropy and $\mathbf{B}_0$ is the magnetic field due to the current J in the absence of the medium.

8.2.2. Let B_i be the magnetic field in a magnetizable medium located in a long solenoid. The applied fields, corresponding to fixed values of Φ and J in the absence of the medium, are denoted $B_{0\Phi}$ and B_{0J}, respectively.

(a) For the system geometry shown in Fig. 8.6 assume the demagnetizing field is negligible and show that

$$B_{0\Phi} = \frac{A + A_m \chi_m}{A} B_{0J} \qquad (8.2.25)$$

and

$$B_i = (1 + \chi_m) B_{0J}. \qquad (8.2.26)$$

Note that in the limit of a very small body when $A_m \chi_m \ll A$, $B_{0\Phi} = B_{0J}$. In addition, when $A_m = A$, $B_{0\Phi} = B_i$.

(b) For the system in Fig. 8.7 (assuming $z \gg \sqrt{A}$ in which case the demagnetizing field can be ignored) show that

$$B_{0\Phi} = \frac{y + z\chi_m}{y} B_{0J} \qquad (8.2.27)$$

and

$$B_i = (1 + \chi_m) B_{0J}. \qquad (8.2.28)$$

Note that in the limit $z\chi_m \ll y$, $B_{0\Phi} = B_{0J}$, provided the condition $z \gg \sqrt{A}$ holds.

(c) For a thin flat disc, as in Fig. 8.8, the demagnetizing field is axial and its magnitude (Robinson 1973) is approximately $-M$. The magnetic intensity in the medium is then

$$H_i = H - M,$$

where $H = nJ$. For this limiting case show that

$$B_{0\Phi} = B_{0J} = B_i = \mu_0 nJ. \qquad (8.2.29)$$

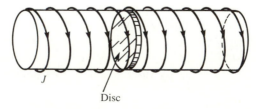

J

Disc

Fig. 8.8 Thin flat magnetizable disc located within a long current carrying solenoid.

8.2.3. A long solenoid of circular cross-section is filled with a magnetiz-
        able medium having constant magnetic susceptibility, χ_m. The
        inductance of the empty solenoid is L_0.
    (a) Given that the flux-turns, Φ_t, is constant, show that the
         mechanical work required to place the medium into the sole-
         noid in a reversible process is

$$W_{m\Phi} = -\frac{\Phi_t^2}{2L_0}\frac{\chi_m}{1+\chi_m}.$$

    (b) If the current J in the solenoid windings is maintained at a
         constant value by an external current source, show that the
         mechanical work required to place the medium into the
         capacitor is

$$W_{mJ} = -\tfrac{1}{2}L_0J^2\chi_m,$$

and that in this process the work done by the external current
source is

$$W_e = L_0J^2\chi_m.$$

8.2.4. A partly filled solenoid consists of a long cylindrical solenoid of
        cross-sectional area A, containing a cylinder of magnetizable
        material of the same length y, but having cross-sectional area A_m,
        as in Fig. 8.6. The magnetic susceptibility, χ_m, is constant and
        there are n turns per metre.
    (a) In establishing a magnetic field, having Φ_t flux-turns in the
         solenoid, show that the additional work due to the medium
         is given by

$$W_\Phi = -\frac{\Phi_t^2}{2L_0}\frac{A_m\chi_m}{A+A_m\chi_m},$$

where L_0 is the inductance of the empty solenoid.

    (b) If the current J is constant, show that the mechanical work
         required to insert the medium into the solenoid is

$$W_{mJ} = -\tfrac{1}{2}L_0J^2\frac{A_m\chi_m}{A}.$$

    (c) Explain why the mechanical work required to insert the
         medium, given that Φ_t is constant, is $W_{m\Phi} = W_\Phi$.
    (d) Show that the flux-turns Φ_t and J are related by

$$\Phi_t = L_0J\left(1+\frac{A_m\chi_m}{A}\right),$$

and hence, that $W_{mJ} \neq W_{m\Phi}$ except in the limit $A_m\chi_m \ll A$.
Explain why this is to be expected.

8.2.5   A magnetizable rod of cross-sectional area A extends a distance z
into a solenoid of the same cross-section (Fig. 8.7). The solenoid
has n turns per metre and the total length is y. Assume $z/\sqrt{A} \gg 1$
so that the demagnetizing field can be neglected. The magnetic
susceptibility of the medium, χ_m, is constant and the process of
inserting the rod is reversible.

(a)   Given that the flux-turns Φ_t for the solenoid is constant,
show that the mechanical work done on the rod is given by

$$W_{m\Phi} = -\frac{\Phi_t^2}{2L_0}\frac{\chi_m z}{y + \chi_m z},$$

where L_0 is the inductance of the empty solenoid.

(b)   Given that J is kept at a constant value by an external current
source, show that the mechanical work done is

$$W_{mJ} = -\tfrac{1}{2}L_0 J^2 \frac{\chi_m z}{y}.$$

(c)   If Φ_t and J are measured at the same time, show that

$$\Phi_t = L_0 J\left(1 + \frac{\chi_m z}{y}\right),$$

and hence that $W_{m\Phi} \neq W_{mJ}$ unless $\chi_m z \ll y$. Is this result
reasonable?

(d)   Use the expression for W_{mJ} to show that the force on the rod
into the solenoid, given that J is specified, is

$$F_J = \tfrac{1}{2}L_0 J^2 \frac{\chi_m}{y}.$$

Use the expression for $W_{m\Phi}$ to determine the force F_Φ given
that Φ_t is specified, and show that $F_J = F_\Phi$.

Part III
Gibbsian Thermodynamics

9

The Gibbs formulation

9.1 The state viewpoint

When a fluid system changes state we can express dU using the first law in terms of the process quantities, heat and work,

$$dU = đQ + đW, \qquad (9.1.1)$$

or we can use the Gibbs equation, which is expressed in terms of state quantities,

$$dU = T\,dS - P\,dV. \qquad (9.1.2)$$

Both equations apply to reversible and irreversible processes, but the terms appearing in the equations are equivalent only for reversible processes.

When Gibbs introduced this equation (Gibbs 1873) he emphasized the state viewpoint by drawing attention to its solution, $U = U(S, V)$. He called this the *fundamental relation*. Thus in Gibbsian thermodynamics the immediate focus of attention moves from the system and its processes, to the states as represented by the fundamental relation and its properties. Of course the solution, $U = U(S, V)$, is useful only when the amount of matter of a given type is fixed. Gibbs showed that this limitation may be avoided by treating the amount of matter as an independent thermodynamic variable as well. In this way the Gibbs formulation offers a natural way for dealing with processes in which the quantity of matter changes. Such processes, which are important in chemistry, physics, and astrophysics, will be considered in Chapter 12.

9.2 The fundamental relations

Simple systems

We shall initially consider only simple systems, which we identify as follows. First, a simple system is homogeneous, having no internal walls or partitions. Second, we assume that its equilibrium state is identified by the value of U, the amount of each of its chemical components, expressed as mol numbers, $N_1, N_2, \ldots, N_r$ and the deformation variables for the system. For instance, the deformation variable of a simple fluid is its volume, V. Hence the equilibrium state of a simple fluid is assumed to be

completely identified by the variables $(U, V, N_1, \ldots, N_r)$. In Chapters 7
and 8 deformation variables were introduced for a surface, an elastic wire,
and for dielectric and magnetizable media. In the following we shall
illustrate simple systems using a fluid, but with suitable changes the results
may also be used for other simple systems. Normally all that is necessary
is to introduce the appropriate deformation variables.

We know the entropy of a system in equilibrium depends on the thermodynamic state alone. Given a value for the entropy when the system is
in one particular state, the entropy can be calculated in any equilibrium
state, using eqn (6.7.3). Since the state of a simple system is identified by
$U, V, N_1, \ldots, N_r$, the entropy can be expressed in the form

$$S = S(U, V, N_1, \ldots, N_r). \tag{9.2.1}$$

Following Gibbs and others, we shall call this a fundamental relation
(according to Kubo (1976) Planck called this form the *canonical equation
of state*). We shall also assume in the following that S is a well-behaved
function of its arguments. As we show below, such equations allow us to
calculate all other thermodynamic functions for the system directly. In this
sense the fundamental relation provides a complete representation of the
thermodynamic properties of the system.

Energy and entropy representations

We know from the reversible process theorem (see question 7.1.1) that S
must be a monotonic increasing function of U when $V, N_1, \ldots, N_r$ are
fixed. Hence the coordinates $(S, V, N_1, \ldots, N_r)$ may also be used to identify the equilibrium state of a simple system. Equation (9.2.1) may therefore
be re-expressed in the equivalent form

$$U = U(S, V, N_1, \ldots, N_r). \tag{9.2.2}$$

Equations (9.2.1) and (9.2.2) are known, respectively, as the entropy and
energy representations of the fundamental relation.

Evidently Maxwell was particularly taken by the U–S–V state representation of Gibbs. As a token of his esteem he created a solid model of the
U–S–V surface for water and presented a plaster cast of it to Gibbs. Figure
9.1 illustrates this relation. Observe that S is a monotonic increasing function of U, as we expect. The simplicity of this representation can be
appreciated by comparing this figure with the diagrams in Appendix B
showing the intensive variables. Now the intensive variables, P and T, are
given by the gradient of the U–S–V surface where it intersects certain
planes. For if we compare the derivative of the energy equation (9.2.2) at
constant mole number,

$$dU = \left(\frac{\partial U}{\partial S}\right)_{V, N_1, N_2, \ldots} dS + \left(\frac{\partial U}{\partial V}\right)_{S, N_1, N_2, \ldots} dV, \tag{9.2.3}$$

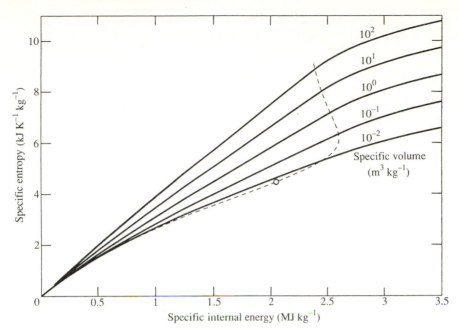

Fig. 9.1 Isochores for water illustrated using S–U coordinates. The dashed line indicates the boundary of the two-phase region and the small circle represents the critical point. To the left of the dashed line the system is a mixture of liquid water and steam.

with the Gibbs equation (9.1.2), we have

$$T = \left(\frac{\partial U}{\partial S}\right)_{V,N_1,N_2,\ldots}, \tag{9.2.4}$$

$$P = -\left(\frac{\partial U}{\partial V}\right)_{S,N_1,N_2,\ldots}. \tag{9.2.5}$$

Equivalently, from eqn (D.17), we can obtain T in the entropy representation:

$$\frac{1}{T} = \left(\frac{\partial S}{\partial U}\right)_{V,N_1,N_2,\ldots}. \tag{9.2.6}$$

Homogeneous property

We have assumed in Section 6.7 that the entropy is an additive property, as are U, V, and N_i $(i = 1, \ldots, r)$. Thus when a composite system has λ subsystems, its entropy, S_c, is

$$S_c = S_1 + S_2 + \cdots + S_\lambda, \tag{9.2.7}$$

where S_i is the entropy of the ith subsystem. Now suppose the subsystems are identical, each being a simple system whose entropy is given by the fundamental relation

$$S = S(U, V, N_1, \ldots, N_r).$$

The composite system will also be a simple system, its state being identified by the extensive parameters $(\lambda U, \lambda V, \lambda N_1, \ldots, \lambda N_r)$. Hence,

$$S_c = S(\lambda U, \lambda V, \lambda N_1, \ldots, \lambda N_r), \tag{9.2.8}$$

and since $S_c = \lambda S$,

$$\lambda S = S(\lambda U, \lambda V, \lambda N_1, \ldots, \lambda N_r). \tag{9.2.9}$$

A function of this type is said to be a *homogeneous function of first order*: S is a first-order homogeneous function of the extensive variables, U, V, and N_i, in a simple fluid system. Similarly in the energy representation U is a first-order homogeneous function of the extensive variables, S, V, and N_i.

The third law

Consider a chemical reaction such as

$$2N_2O_5 \leftrightarrow 4NO_2 + O_2. \tag{9.2.10}$$

In order to evaluate the change in the entropy of a system as a result of this process we need to be able to relate the entropy of the different molecular species, N_2O_5, NO_2, and O_2. This means we shall need to integrate dQ/T over a reversible process which links N_2O_5 in its initial thermodynamic state with NO_2 and O_2 in the final state.

In general the task of finding such a reversible process is difficult. However, this difficulty can be avoided by using the following empirical result, which is usually referred to as the third law. Like the other laws of thermodynamics it has been expressed in many forms and here we adopt a particular expression, the Planck statement:

In the limit as $T \to 0$, the entropy of a system in equilibrium approaches zero.

The implications of this statement are far-reaching (see, for example, Wilks (1961) and Tisza (1966)). In the present context it means that we may consistently fix the constant of integration in the definition of the entropy of a simple system. Consequently, eqn (6.7.7), which asserts that the entropy is an additive variable for a composite system, is justified. In particular, we can establish an absolute value for the entropy of each of the chemical species, N_2O_5, NO_2, and O_2.

First we set the entropy to be zero in the limit $T \to 0$. Then the molar entropy of each species can be evaluated at any temperature, T, and pressure, P, by using relations of the form

$$s = \int_0^T \frac{c_p \, dT'}{T'},$$
(9.2.11)

where c_p is the molar heat capacity at constant pressure, P. In practice, transformations such as phase changes must also be included in this calculation. Thus we may determine the change in the entropy of a system due to a chemical reaction from a knowledge of the molar heat capacity of the participating chemical components.

Summary

For a simple fluid system:

(1)   the entropy is a continuous differentiable function of the extensive parameters, $(U, V, N_1, \ldots, N_r)$;

(2)   the entropy is a monotonic increasing function of U, given $V, N_1, \ldots, N_r$ are constant;

(3)   the entropy of a composite system is additive over its subsystems;

(4)   the energy and entropy fundamental relations are homogeneous functions of first order;

(5)   the entropy vanishes in the limit $T \to 0$, T being given by (9.2.4) or (9.2.6).

For other simple systems the variable V should be replaced by the deformation variable or variables appropriate to that system.

9.3   Application of the fundamental relation

A fundamental relation for a particular system may be established from empirical measurements. For instance, the graphical representation of the thermodynamic properties of water in Appendix B was obtained using such equations. But for systems which can be analysed from first principles, statistical mechanics methods may be used to determine the fundamental relation. The equations listed below were obtained in this way.

Illustrative fundamental relations

In the following equations $\tilde{N}_A$ is Avogadro's number, h is Planck's constant, m is the mass of an atom or molecule of the gas, R is the molar ideal gas constant, c is the velocity of light in vacuum, and k ($= R/\tilde{N}_A$) is the Boltzmann constant. N denotes the number of moles of atoms or molecules. Some equations do not satisfy the third law, but they are nevertheless valid provided we do not try to use them in the limit $T \to 0$. It is assumed that the mean molecular speed is small compared to c, except in examples (11) and (12).

1. A single-component ideal gas of structureless particles (that is, single atoms of spin zero), excluding the limit $T \rightarrow 0$:

$$S = NR \left\{ \frac{5}{2} - \ln \sigma + \ln \left(\frac{U^{3/2} V}{N^{5/2}} \right) \right\}, \qquad (9.3.1)$$

where (see (9.3.10a) for particles with spin)

$$\sigma = \tilde{N}_A \left(\frac{3 \tilde{N}_A h^2}{4 \pi m} \right)^{3/2}. \qquad (9.3.1a)$$

2. A mixture of ideal gases of structureless particles, excluding the limit $T \rightarrow 0$:

$$S = NR \left\{ \frac{5}{2} + \ln \left(\frac{U^{3/2} V}{N^{5/2}} \right) - \frac{N_1}{N} \ln \sigma_1 - \frac{N_2}{N} \ln \sigma_2 - \cdots \right.$$

$$\left. - \frac{N_1}{N} \ln \left(\frac{N_1}{N} \right) - \frac{N_2}{N} \ln \left(\frac{N_2}{N} \right) - \cdots \right\}. \qquad (9.3.2)$$

Here,

$$N = N_1 + N_2 + \cdots, \qquad (9.3.2a)$$

and

$$\sigma_i = \tilde{N}_A \left(\frac{3 \tilde{N}_A h^2}{4 \pi m_i} \right)^{3/2}, \qquad i = 1, 2, \ldots, \qquad (9.3.2b)$$

and $m_1, m_2, \ldots$, represent the mass per atom for the species $1, 2, \ldots$, of the mixture (unit: kg).

3. A photon gas (black-body radiation):

$$U = \left(\frac{3}{4} \right)^{4/3} \left(\frac{c}{4\sigma} \right)^{1/3} \frac{S^{4/3}}{V^{1/3}}, \qquad (9.3.3)$$

where σ is the Stefan–Boltzmann constant,

$$\sigma = \frac{2 \pi^5 k^4}{15 h^3 c^2}. \qquad (9.3.3a)$$

4. A system consisting of N mol of weakly interacting linear harmonic oscillators:

$$S = NR \ln \left(1 + \frac{U}{N E_v} \right) + \frac{RU}{E_v} \ln \left(1 + \frac{N E_v}{U} \right), \qquad (9.3.4)$$

where

$$E_v = \tilde{N}_A h f_o, \tag{9.3.4a}$$

and f_o is the classical frequency for one oscillator.

5. Molecular rotational motion in a diatomic gas, excluding the limit $T \to 0$:

$$S = NR \left\{ \ln \left[\frac{U}{NE_r} \right] + 1 \right\}. \tag{9.3.5}$$

Here,

$$E_r = \tilde{N}_A \varepsilon_o, \tag{9.3.5a}$$

where $\varepsilon_o = h^2/2I$ and I is the moment of inertia of a molecule.

6. A system of particles, each having only two accessible quantum states, energy 0 and ε:

$$S = \left(\frac{U}{E} - N \right) R \ln \left(1 - \frac{U}{NE} \right) - \frac{U}{E} R \ln \left(\frac{U}{NE} \right). \tag{9.3.6}$$

Here, $E = \tilde{N}_A \varepsilon$ and (9.3.6) is valid for $0 < U < NE/2$. (This equation does not include the translational motion of the particles.)

7. A gas of spin-$\frac{1}{2}$ particles confined to a volume V at temperatures near $T = 0$:

$$U = \frac{3}{5} N \tilde{N}_A \left(\frac{3N\tilde{N}_A}{\pi V} \right)^{2/3} \frac{h^2}{8m} \left\{ 1 + \frac{5}{3} \left(\frac{S}{\pi NR} \right)^2 \right\}. \tag{9.3.7}$$

This is valid for $T \ll T_f$, where T_f is the Fermi temperature,

$$T_f = \left(\frac{3\tilde{N}_A N}{\pi V} \right)^{2/3} \frac{h^2}{8mk}. \tag{9.3.7a}$$

8. A gas of spin-zero particles in a volume V at temperatures close to $T = 0$:

$$U = \frac{h^3}{7.46 \pi mk} V^{-2/3} S^{5/3}. \tag{9.3.8}$$

This is valid for $T < T_c$, where T_c is the Bose–Einstein condensation temperature,

$$T_c = \left(\frac{\tilde{N}_A N}{V} \right)^{2/3} \frac{h^2}{11.9 mk}. \tag{9.3.8a}$$

9. A gas of spin-$\frac{1}{2}$ particles confined to a two-dimensional area, A, near $T = 0$:

$$U = \frac{N^2 \tilde{N}_A^2}{\pi A} \frac{h^2}{8m} \left\{ 1 + 3 \left(\frac{S}{\pi NR} \right)^2 \right\}. \tag{9.3.9}$$

This is valid for $T \ll T_f$, where,

$$T_f = \frac{\tilde{N}_A N h^2}{4\pi m A k}. \tag{9.3.9a}$$

10. A gas of particles with spin-s in r dimensions confined to a hypercube of volume $V = a^r$, excluding the limit $T \to 0$. Here $g = 2s + 1$:

$$U = N \left(\frac{N\sigma}{V}\right)^{2/r} \exp\left\{\frac{2S}{rNR} - \frac{r+2}{r}\right\}, \tag{9.3.10}$$

where

$$\sigma = \frac{\tilde{N}_A}{g} \left(\frac{r\tilde{N}_A h^2}{4\pi m}\right)^{r/2}. \tag{9.3.10a}$$

11. A gas of relativistic particles with spin s, the mean kinetic energy of a particle being large compared to mc^2, where m is the rest mass of a particle. In this fundamental relation U includes the rest mass energy of the particles and $g = 2s + 1$:

$$U = 3N \left(\frac{\sigma N}{V}\right)^{1/3} \exp\left(\frac{S}{3NR} - \frac{4}{3}\right), \tag{9.3.11}$$

where

$$\sigma = \frac{h^3 c^3 \tilde{N}_A^4}{8\pi g}. \tag{9.3.11a}$$

This is applicable when $T \gg T_f$, where

$$T_f = \frac{hc}{k} \left(\frac{3\tilde{N}_A N}{4\pi g V}\right)^{1/3}. \tag{9.3.11b}$$

12. A gas of relativistic spin-$\frac{1}{2}$ particles, the mean kinetic energy of the particles being large compared to mc^2, where m is the rest mass of a particle. As in (11), U includes the rest mass energy of the particles:

$$U = \frac{3}{2} \left(\frac{\sigma N^4}{V}\right)^{1/3} \left\{1 + \frac{2}{3} \left(\frac{S}{\pi NR}\right)^2\right\}. \tag{9.3.12}$$

Here σ is given by (9.3.11a), taking $g = 2$, and (9.3.12) is applicable when $T \ll T_f$, where T_f is given by (9.3.11b).

Interpretation: Gibbs paradox

The form of the fundamental relation for a mixture of ideal gases raises some interesting and important issues first identified by Gibbs. The problem concerns the change in the entropy of a system due to the mixing of two gases. Consider the system shown in Fig. 9.2. Initially we have N mol each of species 1 and species 2 occupying separate volumes, V, the

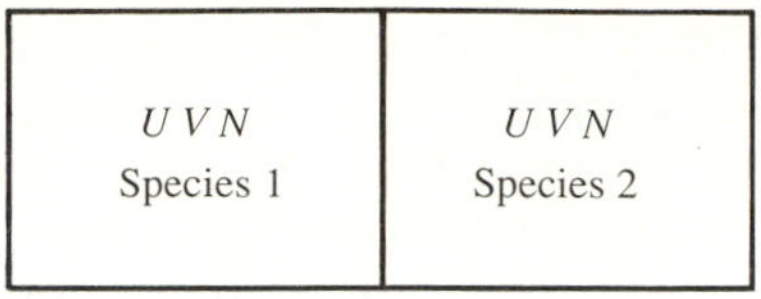

UVN Species 1	UVN Species 2

Fig. 9.2 Initial state of a composite system consisting of equal amounts of different monatomic ideal gases, 1 and 2, in equal volumes and having the same energy.

subsystems each having internal energy U. The entropy of the composite system is

$$S_c = S_1 + S_2,$$

where (eqn (9.3.1))

$$S_i = NR \left\{ \frac{5}{2} - \ln \sigma_i + \ln \left(\frac{U^{3/2} V}{N^{5/2}} \right) \right\}, \qquad (i = 1, 2)$$

and σ_i is given by (9.3.2b).

Let $U_c = 2U$, $V_c = 2V$, $N_c = 2N$. While the partition remains in place,

$$S_c = N_c R \left\{ \frac{5}{2} - \frac{1}{2} \ln (\sigma_1 \sigma_2) + \ln \left(\frac{U_c^{3/2} V_c}{N_c^{5/2}} \right) \right\}.$$

When we remove the partition there are two possibilities: the gases are either indistinguishable or distinguishable. In the first case no thermodynamic process occurs on removing the partition and hence the entropy of the system does not change. But if the species 1 and 2 are distinguishable, we must use (9.3.2) to determine the entropy of the mixed system, S_m. We then find

$$S_m = S_c + N_c R \ln 2.$$

Thus, depending on whether or not the gases are distinguishable, the entropy of the mixed system is larger than that of the unmixed system by $N_c R \ln 2$. This quantity is called the *entropy of mixing*.

From a purely macroscopic point of view it is surprising that no parameter exists to indicate a degree of distinguishability. For instance, suppose we have the ability to differentiate between atoms using a continuously variable parameter, such as their 'colour'. In that situation red and blue atoms would be distinguishable. Suppose also that we are able to perform a controlled colour change process, gradually turning the red ones blue. Then the question would arise, if the atoms are identical in all other respects, at what point in this process would the mixing term disappear?

This question, called the *Gibbs paradox*, presupposes that it is possible to have a continuous parameter, such as colour, which can distinguish atomic species. But atoms are either distinguishable or indistinguishable because the structure of atoms is ultimately quantized at the microscopic level. There can be no continuous colour-like identification variable. In this example, therefore, the entropy of mixing is either zero or $N_c R \ln 2$. Thus the relations of macroscopic thermodynamics exhibit discrete characteristics which derive from the distinguishability of systems in a quantum sense.

An interesting illustration of distinguishability is provided by homonuclear molecules, such as H_2, for which the nuclei must be in one of two possible symmetry states, called ortho-hydrogen and para-hydrogen. Because the allowed quantum states of a homonuclear molecule are determined by symmetry, the molar heat capacity and other thermodynamic properties for ortho-hydrogen and para-hydrogen differ. At room temperature normal hydrogen is a mixture of 75% ortho-hydrogen and 25% para-hydrogen. However, the fraction of para-hydrogen is enhanced at low temperatures, a transformation which can be hastened by the presence of catalysts. As a consequence the heat capacity of gaseous H_2 at temperatures below room temperature depends on the method of preparation, because this affects the composition of the system.

This effect is illustrated in Fig. 9.3 which shows the contributions to the molar heat capacities of ortho-hydrogen and para-hydrogen due to molecular rotation. It shows that while the atomic properties which distinguish

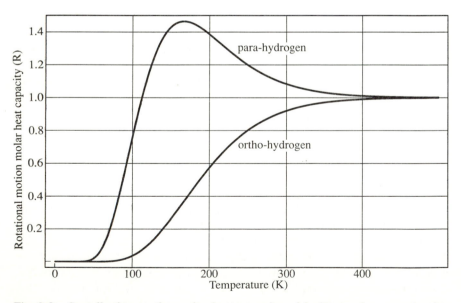

Fig. 9.3 Contribution to the molar heat capacity of hydrogen due to molecular rotation. Ortho-hydrogen and para-hydrogen are distinguished by the symmetry of the nuclear spin quantum state.

the two systems are quite subtle, their influence on thermodynamic properties may be important. This effect illustrates the difficulty of identifying the thermodynamic variables of a system from a purely macroscopic viewpoint. The Gibbs paradox signals the limitations of that viewpoint.

Questions

9.3.1.   (a)   Obtain (9.3.1) from (9.3.10).

(b)   Use the fundamental relation (9.3.1) for a monatomic ideal gas to establish the equations of state:

$$PV = NRT, \qquad U = \tfrac{3}{2}NRT.$$

(c)   Establish the corresponding equations for examples (8), (10), and (12).

9.3.2.   Confirm that equations (9.3.2), (9.3.3), (9.3.4), (9.3.5), (9.3.6), (9.3.9), and (9.3.11) each satisfy the following requirements for a fundamental relation.

(a) S is a monotonic increasing function of U (or, equivalently, U is a monotonic increasing function of S).

(b) S is a first-order homogeneous function of the extensive variables relevant to the system, such as U, V, and N (or U is a first-order homogeneous function of extensive variables, such as S, V, and N).

(c)   Unless the equation excludes the limit $T \to 0$ from its range of application, then $S \to 0$ as $T \to 0$, where T is given by (9.2.4) or (9.2.6).

9.3.3.   Show that the fundamental relation for a single-component monatomic ideal gas (eqn (9.3.1)) can be expressed as

$$U = U_0 \left(\frac{V_0}{V} \right)^{2/3} \left(\frac{N}{N_0} \right)^{5/3} \exp \left\{ \frac{2}{3R} \left(\frac{S}{N} - \frac{S_0}{N_0} \right) \right\}, \qquad (9.3.13)$$

where the subscript 0 denotes a reference state for the gas such that

$$U_0 = U(S_0, V_0, N_0). \qquad (9.3.14a)$$

Let $U_0 = \tfrac{3}{2} N_0 \tilde{N}_A k T_0$, where $k = R/\tilde{N}_A$ is the Boltzmann constant, and let $\tilde{v}_0 = V_0/N_0 \tilde{N}_A$ be the volume per atom. Show that

$$S_0 = N_0 R \left\{ \frac{5}{2} + \ln \left(\frac{\tilde{v}_0}{\lambda^3} \right) \right\}, \qquad (9.3.14b)$$

where

$$\lambda = \left(\frac{h^2}{2\pi m k T_0} \right)^{1/2}. \qquad (9.3.15)$$

In statistics λ is called the *thermal wavelength* of the atom. It may be shown that λ is $\pi^{-1/2}$ times the de Broglie wavelength for an atom having kinetic energy kT_0.

9.3.4. Use (9.3.2) to obtain the fundamental relation for a mixture of ideal gases, labelled $i = 1, 2, \ldots, r$, in the energy representation:

$$U = N \exp\left(\frac{2S}{3NR} - \frac{5}{3}\right) \prod_{i=1}^{r} \left(\frac{N_i \sigma_i}{V}\right)^{\frac{2N_i}{3N}}. \quad (9.3.16)$$

9.3.5. Give the reasons why the following equations are unsuitable as fundamental relations for a simple fluid system. Take α and β to be positive constants.

(a)
$$U = \alpha \left(\frac{N^3 S}{V}\right)^{1/4}.$$

(b)
$$S = \alpha \left(\frac{NU}{V^2}\right) - \beta \left(\frac{N^2 V}{U}\right)^{1/2}.$$

(c)
$$U = \alpha \left(\frac{S^2}{N}\right) \exp\left(\frac{-\beta S V}{N^2}\right).$$

(d)
$$S = \alpha N \left(\frac{U}{V}\right)^{1/2} \ln\left(1 + \frac{\beta N}{U}\right).$$

9.3.6. Consider a solid incompressible body which exhibits zero thermal expansion. For the states of interest the specific heat capacity, c, can be taken to be constant. Show that the fundamental relation can be expressed in the form

$$S = Mc \ln\left(\frac{U}{Mc}\right), \quad (9.3.17)$$

where M is the mass. The effect of the restriction on c is that the fundamental relation does not satisfy the third law. However, provided we do not attempt to use the equation in the limit $T \to 0$, this need not introduce difficulties.

9.3.7. Use (9.3.2) to show that the pressure of a mixture of ideal gases, P, can be expressed as

$$P = \sum_i P_i, \quad (9.3.18)$$

where P_i, called the *partial pressure* of component i, is given by

$$P_i = \frac{N_i R T}{V}. \quad (9.3.19)$$

9.3.8. A system consists of two equal volumes V, separated by a common wall, each containing monatomic ideal gases. Side A contains a mixture of N_1 mol of species 1 and N_2 mol of species 2. Side B contains N_1 mol of species 1 only. Initially sides A and B have the same temperature. Show that when the common wall is removed the entropy increase due to mixing is the same as the entropy increase in a free-expansion (Joule) process in which N_2 mol of ideal gas expands from a volume V to a volume $2V$.

9.3.9.   A system contains r different ideal gases. Initially all components are located in separate subvolumes but they have the same temperature and pressure. The gases are then allowed to mix, the total volume remaining fixed. Use eqn (9.3.2) to show that the increase in entropy in this process is

$$\Delta S = -NR \sum_{i=1}^{r} x_i \ln (x_i), \qquad (9.3.20)$$

where $x_i = N_i/N$ and $N = \Sigma N_i$.

9.3.10.   Show that the pressure of a photon gas, described by eqn (9.3.3), depends on the temperature only, not on the volume.

9.3.11.   As $T \to 0$ the specific heat capacity for a certain solid body has the limiting form

$$c = aT^{\lambda},$$

where a and λ are constants. Demonstrate that $\lambda > 0$ using the third law.

9.4   Maximum entropy principle

Introduction

Consider a composite system in which the subsystems are not in equilibrium with each other, in the sense that they do not have the same temperature or pressure. If the system is held in a stable state by an internal constraint, it is said to be in a state of constrained equilibrium. Now if the constraint is released the system will normally evolve to a new equilibrium state. Generally this is just one of many states permitted by the new constraints. Here we shall show that the new equilibrium state is that state for which the entropy is maximized. This is the *maximum entropy principle*.

To illustrate, take two solid bodies, A and B, having masses of 1 kg and 2 kg, respectively. The initial temperatures of the bodies, which are separated by an adiabatic wall, are 900 K and 300 K, respectively. For simplicity take the specific heat capacity to be $c = 1\,\text{kJ}\,\text{kg}^{-1}\text{K}^{-1}$ for both A and B and ignore volume changes. Then the entropy for each body, given by eqn (9.3.17), is

$$S = Mc \ln \left(\frac{U}{Mc} \right) \tag{9.4.1}$$

where M is the mass of the body and $U = McT$.

The initial state of the composite system is fixed by the equations of constraint:

$$U_A = 900 \, \text{kJ}, \tag{9.4.2a}$$

$$U_B = 600 \, \text{kJ}. \tag{9.4.2b}$$

Suppose the two bodies are allowed to exchange heat, by removing the adiabatic wall. The system now has a new energy equation of constraint,

$$U_A + U_B = 1500 \, \text{kJ}, \tag{9.4.3}$$

which is less restrictive than (9.4.2a, b). However, this equation does not uniquely identify the new state of the system since there is a manifold of possible states represented by the pairs (U_A, U_B) which satisfy (9.4.3). Yet in the absence of the adiabatic wall, only one of these states will be an equilibrium state. How can we identify that state?

First, we can use the zeroth law according to which the equilibrium state is identified by $T_A = T_B$. One can easily show that the equilibrium temperature is 500 K in this case, the corresponding energies being $U_A = 500 \, \text{kJ}$ and $U_B = 1000 \, \text{kJ}$.

But the equilibrium state can also be obtained by finding the allowed state for which the entropy of the composite system, S_c, is a maximum. This condition is illustrated in Figs 9.4 and 9.5. Here we show S_c, the entropy of the composite system as a function of U_A and U_B. The figures have been obtained using (9.4.3) and

$$S_c = S_A + S_B = \ln (U_A) + 2 \ln (U_B/2). \tag{9.4.4}$$

We see that S_c is maximized when $U_A = 500 \, \text{kJ}$ and $U_B = 1000 \, \text{kJ}$. This illustrates how the maximum entropy condition identifies the same equilibrium state obtained from the zeroth law.

Formal expression

Suppose an isolated composite system, Σ_c, is initially maintained in a state of thermodynamic equilibrium by internal constraints. The constraints, which restrict the interactions between the subsystems of Σ_c, are formally expressed in terms of the extensive variables, eqn (9.4.2) being an example. Now suppose the state of Σ_c is allowed to change by relaxing a constraint condition, so that one of the constraint variables, X, can be treated as a free parameter. In the example above X would be either U_A or U_B. As a result the system will evolve to a new state in which X will change from an initial value, X_i, to a new equilibrium value, X_e.

Since Σ_c is isolated the change of state will be uninhibited, like a free-

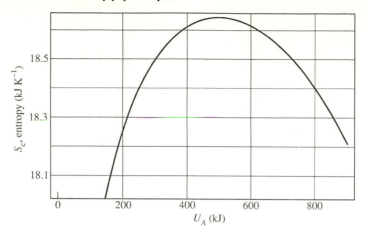

Fig. 9.4 Entropy of the composite system consisting of the two bodies A and B as a function of U_A, given $U_A + U_B = 1500\,\text{kJ}$.

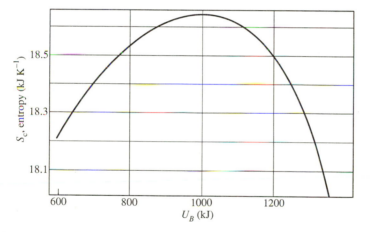

Fig. 9.5 Entropy of the composite system consisting of the two bodies A and B as a function of U_B, given $U_A + U_B = 1500\,\text{kJ}$.

expansion process. Hence the process which follows such a relaxation of constraints will be non-quasistatic and therefore by the second law, irreversible. Consequently, the entropy of Σ_c will increase. But if $X_i = X_e$, there will be a null process, in which case the entropy will remain constant. Thus the change of state $X_i \rightarrow X_e$ will be attended by an entropy increase whenever $X_i \neq X_e$. Writing the entropy of Σ_c, S_c, as a function of the free variable X alone, we have: the value of X which maximizes $S_c(X)$, subject to fixed constraints, is the equilibrium value, X_e. Thus we have the *maximum entropy principle*:

$$S_c(X) < S_c(X_e) \qquad \text{when } X \neq X_e. \qquad (9.4.5)$$

The maximum entropy principle says nothing directly about S_c *during* the process. Rather it says that, given the set of equilibrium states allowed by the new constraint condition, the equilibrium state is the allowed state which maximizes S_c. During the process the subsystems of Σ_c are generally not in equilibrium. The entropy of non-equilibrium systems is not always well defined.

Simple systems

To illustrate the use of the maximum entropy principle, consider a composite system, Σ_c, comprising several simple systems, Σ_i. The entropy of Σ_c can be written as

$$S_c = \sum S_i, \qquad (9.4.6)$$

where

$$S_i = S_i(U_i, V_i, N_{i1}, \ldots, N_{ir}). \qquad (9.4.7)$$

We may express the maximum entropy principle as follows. Suppose particular variables, denoted X_1 and X_2, are permitted to change whilst the remaining variables are fixed. Then the equilibrium values of X_1 and X_2 are such that

$$dS_c = 0, \qquad (9.4.8)$$

$$d^2S_c < 0, \qquad (9.4.9)$$

subject to the restriction $dX_j = 0$, $j \neq 1, 2$. Here dS_c and d^2S_c are the first and second differentials of S_c for arbitrary increments dX_1, dX_2 (see Appendix D):

$$dS_c = \left(\frac{\partial S_c}{\partial X_1}\right) dX_1 + \left(\frac{\partial S_c}{\partial X_2}\right) dX_2 \qquad (9.4.10)$$

$$d^2S_c = \left(\frac{\partial^2 S_c}{\partial X_1^2}\right) (dX_1)^2 + 2\left(\frac{\partial^2 S_c}{\partial X_1 \partial X_2}\right) dX_1 \, dX_2 + \left(\frac{\partial^2 S_c}{\partial X_2^2}\right) (dX_2)^2. \qquad (9.4.11)$$

By solving (9.4.8) and (9.4.9) the entropy fundamental relation yields the equilibrium value for the unconstrained variables in a composite system.

In principle the entropy fundamental relation expresses all the equilibrium thermodynamic properties of the system. However, it does so in a rather inflexible way since the independent variables are extensive; the equilibrium condition is really useful only in an isolated system. Thus, the entropy representation is not always a convenient form for the fundamental relation since we are often interested in systems which are not isolated. The systems might, for instance, be in contact with a heat reservoir or a work source. In Chapter 10 we shall show how other forms of the fundamental relation can be constructed for use in these situations.

9.5 Formal relations

The Gibbs equation has already been introduced in connection with the entropy of an ideal gas (Section 4.1) and with different work processes (Sections 7.3, 7.4, 8.1, and 8.2). Here we reconsider the Gibbs equation from a different perspective in which we take the fundamental relation as the starting point.

Energy representation

We start with the energy fundamental relation for a simple fluid system:

$$U = U(S, V, N_1, \ldots, N_r),$$

and consider a small change in state. The first-order differential is

$$dU = T\,dS - P\,dV + \sum_{j=1}^{r} \mu_j\,dN_j. \tag{9.5.1}$$

This is the general form of the Gibbs equation for the energy of a simple fluid system. Here T and P are familiar:

$$T = T(S, V, N_1, \ldots, N_r) = \left(\frac{\partial U}{\partial S}\right)_{V, N_i}, \tag{9.5.2}$$

$$P = P(S, V, N_1, \ldots, N_r) = -\left(\frac{\partial U}{\partial V}\right)_{S, N_i}. \tag{9.5.3}$$

In addition, we have new intensive parameters, $\mu_1, \mu_2, \ldots$, defined as

$$\mu_j = \mu_j(S, V, N_1, \ldots, N_r) = \left(\frac{\partial U}{\partial N_j}\right)_{S, V, N_i, i \neq j}. \tag{9.5.4}$$

μ_j is called the *chemical potential* for component j.

Equations (9.5.2)–(9.5.4) express the intensive parameters as functions of the independent variables in the energy representation, $S, V, N_1, \ldots, N_r$. They are intensive in the sense that T, P, and μ_j are homogeneous functions of zero order in the extensive variables. This can be shown as follows. Let Σ_c be a composite system comprising λ identical subsystems. Let T_c be the temperature, U_c the energy, and S_c the entropy of Σ_c. For one of the subsystems suppose that

$$T = T(S, V, N_1, \ldots, N_r), \qquad U = U(S, V, N_1, \ldots, N_r).$$

Then since the composite system is in equilibrium,

$$T_c = T(\lambda S, \lambda V, \lambda N_1, \ldots, \lambda N_r),$$

and

$$U_c = U(\lambda S, \lambda V, \lambda N_1, \ldots, \lambda N_r) = \lambda U,$$

$$S_c = \lambda S.$$

Now,

$$T_c = \frac{\partial U_c}{\partial S_c} = \frac{\partial(\lambda U)}{\partial(\lambda S)} = \frac{\partial U}{\partial S} = T,$$

and therefore

$$T(\lambda S, \lambda V, \lambda N_1, \ldots, \lambda N_r) = T(S, V, N_1, \ldots, N_r), \tag{9.5.5}$$

showing that T is a homogeneous function of zero order.

While the Gibbs equation (9.5.1) is a relation between thermodynamic variables, we have seen previously that the terms of the equation represent process quantities when a change of state is reversible. To summarize:

1. The first term is the contribution to the change in internal energy due to reversible heat transfer at constant volume and mole numbers:

$$(đQ)_{rev} = T\,dS. \tag{9.5.6}$$

2. The second term is the contribution to the change in energy due to a reversible volume change at constant mole numbers. The condition that the change of state should be reversible implies $dS = 0$:

$$(đW)_{rev\,displ} = -P\,dV. \tag{9.5.7}$$

3. The third term we interpret as the energy contribution due to matter transfer at constant volume and entropy. The chemical work done on the system due to a reversible increment dN_i in component i, under these conditions is:

$$(đW_i)_{rev\,chemical} = \mu_i\,dN_i. \tag{9.5.8}$$

Local equilibrium

For systems which are not homogeneous the fundamental relation cannot be written as a first-order homogeneous function of the extensive variables. For instance, it may happen that the pressure of a fluid varies with position, as in the atmosphere. Such a system is not in a state of thermodynamic equilibrium although its properties may be constant in time.

In such cases it is usually possible to divide the system into subsystems, each being small enough that it is in a state of thermodynamic equilibrium with well-defined values for the temperature, pressure, and chemical potentials. In this situation the system is said to be in a state of *local equilibrium*.

The fundamental relation and the Gibbs equation are then most usefully expressed in terms of specific variables, such as the energy per unit mass. These quantities may also be called *densities*.

For example, in a fluid we might identify a small volume δV in which the variations of the pressure and temperature can be ignored. Let δM be the mass of fluid in the volume δV, and let δU and δN_j be the corresponding energy and mole numbers. Then the energy, volume, entropy, and mole densities are defined by

$$u = \frac{\delta U}{\delta M}, \qquad v = \frac{\delta V}{\delta M}, \qquad s = \frac{\delta S}{\delta M}, \qquad n_j = \frac{\delta N_j}{\delta M}, \qquad (9.5.9)$$

and the fundamental relation has the form

$$u = u(s, v, n_1, n_2, \ldots n_r). \qquad (9.5.10)$$

The corresponding Gibbs equation is

$$du = T\,ds - P\,dv + \sum_{j=1}^{r} \mu_j\,dn_j. \qquad (9.5.11)$$

Notice, however, that the increments, dn_j, are not independent in this representation since δM is fixed. Hence,

$$dM = \sum_{j=1}^{r} M_{gj}\,dn_j = 0, \qquad (9.5.12)$$

where M_{gj} is the molecular weight of the component j (in $kg\,mol^{-1}$).

In chemical applications it is useful to set up a local form of the fundamental relation using mole-specific instead of mass-specific variables. In that case the mole-specific densities are $\delta U/\delta N$, $\delta S/\delta N$, ..., where δN is the mole content of the volume δV. The corresponding fundamental relation and Gibbs equation are formally the same as (9.5.10) and (9.5.11). However, corresponding to (9.5.12), the permitted variations in the mole fractions must sum to zero.

In dealing with electric and magnetic interactions with polarizable media a volume-specific form of the fundamental relation is useful (Sections 8.1 and 8.2). In order to avoid complexity in the notation we shall use lower case symbols to denote the volume-specific variables, since we can normally distinguish these from mass and molar densities by the context. To illustrate, consider a medium subject to a magnetic field, $\mathbf{B}_0$. The Gibbs equation has two forms, (8.2.11a, b) and (8.2.16), corresponding to two ways for defining the applied field in the absence of the system, $\mathbf{B}_0$. We shall consider the case in which $\mathbf{B}_0$ is defined by the current, J, in a field coil and for completeness we will include the chemical terms. Hence, from (8.2.16),

$$\delta u_J = T\,\delta s + \mathbf{B}_0 \cdot \delta \mathbf{M} + \sum_{j=1}^{r} \mu_j\,\delta n_j. \tag{9.5.13}$$

The corresponding fundamental relation has the form

$$u_J = u_J(s, \mathbf{M}, n_1, n_2, \ldots, n_r), \tag{9.5.14}$$

and $\mathbf{B}_0$ is given by

$$B_{0i} = \left(\frac{\partial u_J}{\partial M_i} \right)_{s, n_j}, \qquad i = x, y, z. \tag{9.5.15}$$

Both $\mathbf{M}$ and $\mathbf{B}_0$ are independent of the size of the system, but whereas $\mathbf{B}_0$ is a property of the environment of the system, $\mathbf{M}$ normally depends on the concentration of matter in the system. Hence we shall regard $\mathbf{B}_0$ as the magnetic analogue of an intensive parameter whilst $\mathbf{M}$ is a density, or volume-specific quantity, rather than an intensive quantity. In many systems u_J is a first-order homogeneous function of $s, \mathbf{M}, n_1, n_2, \ldots, n_r$, so they represent magnetic analogues of simple fluid systems.

In particular circumstances the magnetic work term can be expressed as a product of the form $P\,\delta X$, where X is an extensive state variable. For instance, suppose the field $\mathbf{B}_0$ and the magnetization $\mathbf{M}$ are collinear and constant over the volume of the system. We may now drop the vector notation. The magnetic moment for the medium is

$$I = \int_{V_m} M\,\mathrm{d}V = MV, \tag{9.5.16}$$

and the Gibbs equation for U_J, which can now be written in terms of extensive independent variables, will normally include the volume displacement work as an additional term:

$$\delta U_J = T\,\delta S - P\,\delta V + B_0\,\delta I + \sum_{j=1}^{r} \mu_j\,\delta N_j. \tag{9.5.17}$$

The corresponding fundamental relation has the form

$$U_J = U_J(S, V, I, N_1, N_2, \ldots, N_r). \tag{9.5.18}$$

In this representation, the fundamental relation for a system subject to a magnetic field is a first-order homogeneous function of extensive variables. While there is an obvious analogy with a simple fluid system, there are also important differences. In particular, the volume is a constraint for an isolated fluid system, but the magnetic moment I is not a constraint for a body in a magnetic field (see Section 8.2). The magnetic constraint is best expressed in the U_ϕ representation, in which the corresponding field, B_0, is identified by the flux-turns of the source coil. Corresponding considerations apply to a dielectric subject to an electric field.

Entropy representation

The entropy representation yields a different form of the Gibbs equation. Starting with

$$S = S(U, V, N_1, \ldots, N_r), \tag{9.5.19}$$

for a simple system, we have

$$dS = \left(\frac{\partial S}{\partial U}\right)_{V, N_i} dU + \left(\frac{\partial S}{\partial V}\right)_{U, N_i} dV + \sum_{j=1}^{r} \left(\frac{\partial S}{\partial N_j}\right)_{U, V, N_k, k \neq j} dN_j \tag{9.5.20}$$

$$= \beta\, dU + \gamma\, dV + \sum_{j=1}^{r} \varphi_j\, dN_j, \tag{9.5.21}$$

where the quantities β, γ, and φ_j are the intensive parameters for the entropy fundamental relation. These can be expressed in terms of the energy representation intensive parameters by using the quotient rule, Appendix D, eqn (D,16):

$$\gamma = \left(\frac{\partial S}{\partial V}\right)_{U, N_j} = -\frac{\left(\dfrac{\partial U}{\partial V}\right)_{S, N_j}}{\left(\dfrac{\partial U}{\partial S}\right)_{V, N_j}} = \frac{P}{T}. \tag{9.5.22}$$

Similarly, we can show

$$\beta = \frac{1}{T}, \tag{9.5.23}$$

$$\varphi_j = -\frac{\mu_j}{T}. \tag{9.5.24}$$

Note that the entropy intensive parameters, β, γ, and φ_j, are explicit functions of $(U, V, N_1, \ldots, N_r)$. On the other hand, the intensive variables of the energy representation, P, T, and μ_j, are explicit functions of $(S, V, N_1, \ldots N_r)$. For this reason we have used different symbols for the entropic intensive parameters, instead of $1/T$, P/T, and $-\mu_j/T$.

The Euler theorem

Consider the energy fundamental relation for a simple fluid system:

$$U = U(S, V, N_1, \ldots, N_r). \tag{9.5.25}$$

The first-order homogeneity property is expressed by

$$\lambda U = U(\lambda S, \lambda V, \lambda N_1, \ldots, \lambda N_r). \tag{9.5.26}$$

Now differentiate both sides with respect to λ, keeping S, V, $N_1, \ldots, N_r$ constant. Since U is also constant the left-hand side becomes

$$\frac{\partial(\lambda U)}{\partial \lambda} = U. \tag{9.5.27}$$

On the right-hand side of (9.5.26), we take U to be a function of the variables $\lambda S, \lambda V, \ldots, \lambda N_r$ and for brevity we denote $U(\lambda S, \lambda V, \ldots, \lambda N_r)$ by $U(\lambda)$. Then from Appendix D, eqn (D.26),

$$\frac{\partial U(\lambda)}{\partial \lambda} = \frac{\partial U(\lambda)}{\partial(\lambda S)} \cdot \frac{\partial(\lambda S)}{\partial \lambda} + \frac{\partial U(\lambda)}{\partial(\lambda V)} \frac{\partial(\lambda V)}{\partial \lambda} + \cdots. \tag{9.5.28}$$

Now $S, V, N_1, \ldots, N_r$ are constant in the derivatives with respect to λ. Hence

$$\frac{\partial(\lambda S)}{\partial \lambda} = S. \tag{9.5.29}$$

On the other hand, in the derivative with respect to (λS) the variables $\lambda V, \ldots, \lambda N_r$ are fixed. Thus,

$$\frac{\partial}{\partial(\lambda S)} U(\lambda S, \lambda V, \ldots, \lambda N_r) = T. \tag{9.5.30}$$

Thus (9.5.28) yields

$$U = TS - PV + \sum_{j=1}^{r} \mu_j N_j. \tag{9.5.31}$$

This is the *Euler form* of the energy fundamental relation for a simple fluid system. A similar resalt can be obtained in the entropy representation,

$$S = \beta U + \gamma V + \sum_{j=1}^{r} \varphi_j N_j, \tag{9.5.32}$$

where $\beta, \gamma, \varphi_1, \ldots, \varphi_r$ denote the entropy representation intensive variables. When the fundamental relation of a system subject to a magnetic field is expressed in a first-order homogeneous form, as in (9.5.14) (see question 9.5.6) or (9.5.18), we obtain, respectively,

$$u_J = Ts + \mathbf{B}_0 . \mathbf{M} + \sum_{j=1}^{r} \mu_j n_j, \tag{9.5.33}$$

$$U_J = TS - PV + B_0 I + \sum_{j=1}^{r} \mu_j N_j. \tag{9.5.34}$$

Gibbs–Duhem relation

If we compare the Euler form, eqn (9.5.31), with the Gibbs equation (9.5.1), the two equations are not obviously consistent. The derivative of (9.5.31) is

$$dU = T\,dS + S\,dT - P\,dV - V\,dP + \sum_{j=1}^{r} \mu_j\,dN_j + \sum_{j=1}^{r} N_j\,d\mu_j, \quad (9.5.35)$$

whereas the corresponding Gibbs equation is

$$dU = T\,dS - P\,dV + \sum_{j=1}^{r} \mu_j\,dN_j. \quad (9.5.36)$$

Hence we must have

$$S\,dT - V\,dP + \sum_{j=1}^{r} N_j\,d\mu_j = 0. \quad (9.5.37)$$

This is a new condition on the intensive variables, called the *Gibbs–Duhem relation*. Similar expressions may be obtained from (9.5.32)–(9.5.34). In the particular case of a single-component simple system the Gibbs–Duhem relation shows that the chemical potential, μ, is fixed entirely by the other intensive variables, P and T:

$$S\,dT - V\,dP + N\,d\mu = 0. \quad (9.5.38)$$

This equation yields partial differential expressions of the form

$$\left(\frac{\partial P}{\partial T}\right)_\mu = \frac{S}{V}, \quad (9.5.39)$$

and other relations which involve partial derivatives of μ. We shall consider a number of applications of these relations in Chapter 12.

Questions

9.5.1. Use eqn (9.3.2) to show that the chemical potential for component i of a mixture of monatomic ideal gases can be expressed as

$$\mu_i = -RT\ln\left(\frac{V(1.5RT)^{3/2}}{N_i\sigma_i}\right) \quad (9.5.40)$$

$$= -\tilde{N}_A kT\ln\left\{\frac{V}{N_i}\left(\frac{2\pi m_i kT}{h^2}\right)^{3/2}\right\} \quad (9.5.41)$$

where σ_i is given by (9.3.2b), $\tilde{N}_i = N_i\tilde{N}_A$, and $\tilde{N}_A$ is Avogadro's number.

9.5.2. The fundamental relation for a gas of spin-$\frac{1}{2}$ particles near $T=0$ is given by (9.3.7). Establish the following expression for the chemical potential:

$$\mu = RT_f\left\{1 - \frac{1}{3}\left(\frac{\pi T}{2T_f}\right)^2\right\}, \quad (9.5.42)$$

where T_f is the Fermi temperature (eqn (9.3.7a)).

9.5.3.   A certain paramagnetic system is subject to an applied magnetic field, B, identified by the current in the source circuit (Section 8.2). The corresponding fundamental relation is

$$u_J = nu_0 + \frac{nc_0}{T_0} - \left\{ \frac{n^2 c_0^2}{T_0^2} + 2nc_0(ns_0 - s) - \frac{\mu_0 c_0}{c_c} M^2 \right\}^{1/2}, \quad (9.5.43)$$

where $n = N/V$, $u_J = U_J/V$, $s = S/V$, and M is the magnetization. Here c_0, c_c, u_0, s_0, and T_0 are constants; μ_0 is the permeability for a vacuum. The system is dilute so that we may take $B/\mu_0 = H$, the magnetic intensity within the system.

(a)   Confirm that u_J is a first-order homogeneous function of s, n, and M; show that the equation must fail when $T \to 0$.

(b)   Show that the magnetization is given by

$$M = \frac{nc_c B}{\mu_0 T}. \quad (9.5.44)$$

(This relation between M and B is the *Curie law*.)

(c)   For this system show that u_J is a function of T and n alone:

$$u_J = nu_0 + nc_0 \left(\frac{1}{T_0} - \frac{1}{T} \right). \quad (9.5.45)$$

9.5.4.   (a)   The magnetic enthalpy function which corresponds to u_J is given by (8.2.18b),

$$h_J = u_J - MB. \quad (9.5.46)$$

Obtain h_J for the system described in question 9.5.3:

$$h_J = nu_0 + \frac{nc_0}{T_0} - \left\{ nc_0 + \frac{nc_c B^2}{\mu_0} \right\}^{1/2} \left\{ \frac{nc_0}{T_0^2} + 2(ns_0 - s) \right\}^{1/2}. \quad (9.5.47)$$

(b)   Demonstrate that h_J given in (9.5.47) is not a first-order homogeneous function of s, B, and n, but instead,

$$h_J(\lambda s, B, \lambda n) = \lambda h_J. \quad (9.5.48)$$

9.5.5.   (a)   Show that the Euler form for the fundamental relation of a magnetic system expressed in the h_J representation, (9.5.46), is

$$h_J = Ts + \sum_{i=1}^{r} \mu_i \, dn_i \quad (9.5.49)$$

where h_J, s, and n_i are the enthalpy, entropy, and mole numbers per unit volume.

(b)   Confirm that (9.5.47) satisfies this expression.

(c)   Establish the Gibbs–Duhem relation corresponding to h_J.

9.5.6.   For a uniform magnetic system U_J is a first-order homogeneous function of S, V, N, and I where $I = MV$. Yet in question 9.5.3 we found that $u_J = U_J/V$ is a first-order homogeneous function of $s = S/V$, M, and $n = N/V$ for the system represented by the fundamental relation (9.5.43). Show that this result holds for other systems only if we can take the pressure, P, to be vanishingly small. (Hint: use the Euler form.) Illustrate by showing that s for a monotomic ideal gas (see eqn (9.3.1)) is not a homogeneous first-order function of $u = U/V$ and $n = N/V$, although S is a homogeneous first-order function of U, V, and N.

9.5.7.   A simple fluid system has r components. Let $N = \Sigma N_j$, $u = U/N$, $s = S/N$, $v = V/N$, and $n_j = N_j/N$. Show that the Gibbs equation can be written as

$$du = T\,ds - P\,dv + \sum_{j=1}^{r-1} (\mu_j - \mu_r)\,dn_j. \qquad (9.5.50)$$

9.6   The conditions for equilibrium

In Section 9.4 we considered the equilibrium condition for a composite system whose parts could exchange heat. We showed that the equilibrium state could be obtained by either of two conditions:

(1)   the subsystems should have the same temperature;

(2)   the entropy of the composite system should be a maximum.

Here we shall illustrate the connection between these two criteria.

Suppose a composite system, Σ_c, has two simple subsystems, Σ_A and Σ_B, which are separated initially by an adiabatic wall. At some time the wall is replaced by a thermally conducting wall. In the process which follows the volume and molar composition of Σ_A and Σ_B remain unchanged. Since Σ_c remains otherwise isolated, its energy is

$$U_c = U_A + U_B = \text{const}, \qquad (9.6.1)$$

where U_A and U_B represent the energy of Σ_A and Σ_B, respectively.

Let S_c be the entropy of Σ_c in equilibrium,

$$S_c = S_A + S_B. \qquad (9.6.2)$$

Now consider a displaced state of the system, infinitesimally different from the equilibrium state. In order to establish the entropy of Σ_c in this state, it must be in a state of constrained equilibrium and we shall assume that the adiabatic wall is temporarily restored accordingly. We can express the entropy for Σ_c in the displaced state as

$$S_c + dS_c = S_A + dS_A + S_B + dS_B,$$

where

$$dS_A = \frac{\partial S_A}{\partial U_A} dU_A + \frac{\partial S_A}{\partial V_A} dV_A + \cdots, \qquad (9.6.3)$$

and similarly for dS_B. Now the constraints on Σ_A are

$$dV_A = 0, \qquad dN_A = 0 \qquad \text{and} \qquad dU_A + dU_B = 0.$$

Hence,

$$dS_A = \frac{\partial S_A}{\partial U_A} dU_A, \qquad (9.6.4a)$$

$$dS_B = \frac{\partial S_B}{\partial U_B} dU_B = -\frac{\partial S_B}{\partial U_B} dU_A. \qquad (9.6.4b)$$

Thus, for an infinitesimal variation in U_A about the equilibrium value, the entropy change is

$$dS_c = \left(\frac{\partial S_A}{\partial U_A} - \frac{\partial S_B}{\partial U_B} \right) dU_A. \qquad (9.6.5)$$

But since S_c is a maximum when Σ_c is in equilibrium at constant U_c, $dS_c = 0$. Thus,

$$\frac{\partial S_A}{\partial U_A} = \frac{\partial S_B}{\partial U_B}, \qquad (9.6.6)$$

and hence

$$T_A = T_B. \qquad (9.6.7)$$

Thus the maximum entropy principle yields the same equilibrium condition we previously obtained from the zeroth law (Section 5.3). So have we really achieved anything new? The answer is definitely yes. The maximum entropy principle is a much more general equilibrium condition than the zeroth law, being applicable to volume and matter exchange, as well as heat.

Questions

9.6.1. Figure 9.6 shows a heat conducting ('diathermal') piston separating two chambers, A and B. Both contain a single-component simple fluid. Given that the piston is free to slide, use the maximum entropy principle to show that the condition for equilibrium is

$$P_A = P_B, \qquad T_A = T_B.$$

9.6.2. In the composite system shown in Fig. 9.7 each side contains two components, a and b. Sides A and B are separated by a fixed

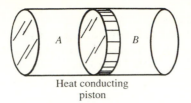

Heat conducting
piston

Fig. 9.6   Heat conducting piston separating two fluid systems, A and B. The piston is free to slide within the cylinder.

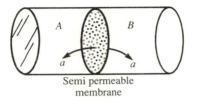

Semi permeable
membrane

Fig. 9.7   Fixed semipermeable membrane separating two fluid systems, each consisting of two mixed components, a and b. The membrane is permeable to component a, but is impermeable to b.

membrane which is permeable to component a, but not to b. The membrane is also a conductor of heat.

(a)   Show, using the maximum entropy condition, that the equilibrium state is identified by the equations

$$T_A = T_B, \qquad \mu_{aA} = \mu_{aB},$$

where T is the temperature and μ_a is the chemical potential for component a.

(b)   Given that the systems are monatomic ideal gases, show that in equilibrium the partial pressure of component a is the same, for sides A and B, but that the total pressure need not be the same. (Use eqns (9.3.9) and (9.5.40).)

9.7   The minimum energy principle

An example

In certain processes the maximum entropy principle is unable to identify the equilibrium condition. Here we illustrate this difficulty and show how it can be circumvented by using the energy, rather than the entropy, to express the condition for equilibrium.

We reconsider the system shown in Fig. 9.6 but we now assume that the piston is adiabatic. In addition we suppose that all effects related to viscosity can be suppressed. Under these restrictions the system will have no dissipative mechanisms so it will undergo only reversible processes. Consequently, if the piston is displaced from equilibrium and released, it will simply oscillate back and forth in undamped motion. Since the entropy of the composite system will be the same for all positions of the piston the entropy maximum condition is inapplicable.

Let us see what happens here in detail. First we suppose the piston has an equilibrium position for which the volume of side A is V_A. Consider a nearby state for the system, the volume of side A being $V_A + dV_A$. Then,

$$dS_A = \frac{dU_A}{T_A} + \frac{P_A}{T_A} dV_A, \tag{9.7.1}$$

and similarly for side B. But dU_A and dV_A are not independent in the absence of heat conduction, matter transfer, and other processes. Instead, $dU_A = -P_A dV_A$ and hence $dS_A = 0$. Similarly, $dS_B = 0$. Thus for the composite system, $dS_c = 0$ for an arbitrary displacement, dV_A.

The maximum entropy condition can therefore tell us nothing about the equilibrium state of this system. Physically this means that, when a constraint is released, this system cannot proceed by itself from one equilibrium state to another. But we would expect that the system can be put into an equilibrium state. For instance, we might simply take hold of the piston and gradually allow it to move into its equilibrium position. Evidently, outside intervention is required.

Suppose then that we modify the system to ensure that it does have an equilibrium state, by attaching a damping mechanism to the piston as illustrated in Fig. 9.8. The damping vane, located in a viscous fluid, is

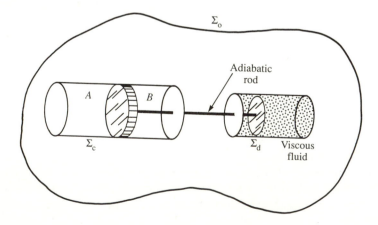

Fig. 9.8 Mechanism for damping the free oscillations of a piston which can undergo only isentropic processes within the system Σ_c.

attached to the piston by an adiabatic connecting rod. We may identify the following systems: Σ_c, the original composite system; Σ_d, the damping system which is coupled to Σ_c by an adiabatic rod; and Σ_o, the overall system. We shall denote the energy and entropy of Σ_c by U_c and S_c, respectively, and use appropriate subscripts for Σ_d and Σ_o.

When the piston is released from its initial position the process is reversible and adiabatic within Σ_c. Hence S_c is constant. Now Σ_o is an isolated system. Thus the equilibrium state is that which maximizes S_o given that U_o is constant. Since $S_o = S_c + S_d$, maximizing S_o implies that S_d is maximized, and because U_d is a monotonic increasing function of S_d, U_d must also be maximized. Finally, since $U_o = U_c + U_d = \text{const}$, U_c must be minimized.

We have now a new equilibrium condition, the *minimum energy principle*:

> At equilibrium U_c is minimized subject to S_c being constant.

This result has been established for a particular system, but the argument holds quite generally. For the particular case shown in Fig. 9.8,

$$dS_A = 0 \qquad \text{and} \qquad dS_B = 0 \tag{9.7.2}$$

are the process constraints, together with

$$dV_A + dV_B = 0. \tag{9.7.3}$$

Then the equilibrium state is that for which

$$dU_c = 0, \qquad d^2U_c > 0. \tag{9.7.4}$$

Now,

$$dU_c = dU_A + dU_B, \tag{9.7.5}$$

where,

$$dU_A = -P_A\, dV_A \qquad \text{and} \qquad dU_B = -P_B\, dV_B. \tag{9.7.6}$$

Thus for a small deviation about the equilibrium state,

$$dU_c = (-P_A + P_B)\, dV_A = 0, \tag{9.7.7}$$

and hence,

$$P_A = P_B, \tag{9.7.8}$$

as we would expect for mechanical equilibrium.

Generalizing

In the above example we have seen that a composite system Σ_c, which can undergo only reversible processes, cannot change from one equilibrium state to another whilst it is isolated. On the other hand, if Σ_c is permitted to interact adiabatically with a dissipative system (one which can execute

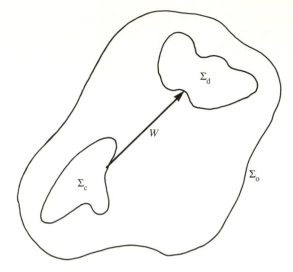

Fig. 9.9 Relationship between Σ_c, Σ_d, and Σ_o. Σ_c is a composite system which undergoes an isentropic process. Σ_d is a damping system which can have only a work interaction with Σ_c.

irreversible processes) then Σ_c may change to a new equilibrium state. That state is identified by the minimum energy condition, which is expressed in terms of the state properties of Σ_c only. Here we shall see how the energy minimum condition can be derived from the maximum entropy principle directly. (See also Section 12.2.)

We consider a composite system, Σ_c, having constant entropy S_c. Σ_c is linked adiabatically to another system, Σ_d, which may have an irreversible change of state as a result of this interaction. We assume that the thermodynamic state of Σ_d is represented by one thermodynamic coordinate only, the energy, U_d. Figure 9.9 illustrates the relationship between Σ_c and Σ_d. We denote the overall system by Σ_o.

Since Σ_o is isolated we can apply the maximum entropy principle to find the equilibrium value of some thermodynamic parameter, X. Thus,

$$\frac{\partial S_o}{\partial X} = 0; \qquad \frac{\partial^2 S_o}{\partial X^2} < 0, \tag{9.7.9}$$

subject to the restrictions,

$$S_o = S_d + S_c, \tag{9.7.10}$$

$$S_c = \text{const}, \tag{9.7.11}$$

and

$$U_o = U_d + U_c = \text{const}. \tag{9.7.12}$$

We assume the fundamental relations for the systems Σ_c and Σ_d can be expressed as

$$U_c = U_c(S_c, X), \tag{9.7.13}$$

$$S_d = S_d(U_d). \tag{9.7.14}$$

Now, from (9.7.10),

$$\left(\frac{\partial S_o}{\partial X}\right)_{S_c, U_o} = \left(\frac{\partial S_d}{\partial X}\right)_{S_c, U_o}$$

Hence, applying the quotient rule, eqn (D.16), to the right-hand-side,

$$\left(\frac{\partial S_o}{\partial X}\right)_{S_c, U_o} = -\frac{\left(\frac{\partial U_o}{\partial X}\right)_{S_d, S_c}}{\left(\frac{\partial U_o}{\partial S_d}\right)_{X, S_c}} = -\frac{\left(\frac{\partial U_c}{\partial X}\right)_{S_c}}{\frac{dU_d}{dS_d}} = -\frac{1}{T_d}\left(\frac{\partial U_c}{\partial X}\right)_{S_c}. \tag{9.7.15}$$

Here we have used the following: U_d is constant when S_d is fixed; U_c is constant when X and S_c are fixed; $dU_d/dS_d = T_d$ is the temperature of Σ_d. Thus, from (9.7.9) we get

$$\left(\frac{\partial U_c}{\partial X}\right)_{S_c} = 0. \tag{9.7.16}$$

In addition, using (9.7.15),

$$\left(\frac{\partial^2 S_o}{\partial X^2}\right)_{S_c, U_o} = -\frac{\partial}{\partial X}\left(\frac{1}{T_d}\right) \cdot \left(\frac{\partial U_c}{\partial X}\right)_{S_c} - \frac{1}{T_d}\left(\frac{\partial^2 U_c}{\partial X^2}\right)_{S_c}.$$

Here the first term on the right-hand side vanishes, by (9.7.16), and hence,

$$\left(\frac{\partial^2 S_o}{\partial X^2}\right)_{S_c, U_o} = -\frac{1}{T_d}\left(\frac{\partial^2 U_c}{\partial X^2}\right)_{S_c}.$$

Therefore, from (9.7.9), we have

$$\left(\frac{\partial^2 U_c}{\partial X^2}\right)_{S_c} > 0. \tag{9.7.17}$$

Equations (9.7.16) and (9.7.17) represent the minimum energy principle: the energy of Σ_c is minimized with respect to changes in the unconstrained variable at equilibrium, given that the entropy of Σ_c is constant.

Example

A system consisting of N mol of an ideal monatomic gas is confined to a cylinder by a piston of mass m_1, whose weight is supported by the pressure of the gas (rather like the system shown in Fig. 3.14.) An additional mass Δm is placed on the piston, bringing the total mass to m_2. Determine V_2/V_1 and T_2/T_1 (where the subscripts correspond to the masses m_1 and m_2). Neglect heat transfer to the piston and cylinder walls. The process of reaching equilibrium after Δm is added is executed in two ways:

(a) the total energy of the piston plus gas remains constant after Δm is released;

(b) the process is isentropic.

Solution (a)

The fundamental relation (9.3.1) can be expressed as

$$\frac{S}{NR} - \frac{S_0}{N_0 R} = \frac{3}{2} \ln\left\{\frac{U}{U_0}\right\} + \ln\left\{\frac{V}{V_0}\right\} - \frac{5}{2} \ln\left\{\frac{N}{N_0}\right\}, \qquad (9.7.18)$$

where S_0, U_0, V_0, and N_0 are constants (see question 9.3.3). Consider a piston of mass m_1, which comes to equilibrium such that the total energy, U_T, is constant:

$$U_T = U + m_1 gh = U + \frac{m_1 g V_1}{A} = \text{const.} \qquad (9.7.19)$$

Here A is the surface area of the piston and V_1 is the volume enclosed in the cylinder. At equilibrium $dS = 0$ for virtual changes of state, subject to $dU_T = 0$ and $dN = 0$. Thus in equilibrium we have from (9.7.18),

$$\frac{dS}{NR} = \frac{3}{2}\frac{dU}{U} + \frac{dV_1}{V_1} = 0, \qquad (9.7.20)$$

together with the constraint equation (9.7.19), which may be expressed

$$dU_T = dU + \frac{m_1 g\, dV_1}{A} = 0. \qquad (9.7.21)$$

Combining (9.7.20) and (9.7.21) we have

$$U = \frac{3}{2}\frac{m_1 g V_1}{A}, \qquad (9.7.22)$$

and, expressing V_1 in terms of U_T, using (9.7.19),

$$V_1 = \frac{2A U_T}{5 m_1 g}. \qquad (9.7.23)$$

Now add a mass Δm to m_1, without moving the piston, and release the system. The new total mass is $m_2 = m_1 + \Delta m$ and the new total energy is (from 9.7.19),

$$U_T' = U_T + \Delta m\, g V_1 / A. \qquad (9.7.24)$$

By expressing U_T in terms of V_1 (eqn (9.7.23)) we find

$$U_T' = \frac{g V_1}{2A} (3m_1 + 2m_2).$$ (9.7.25)

But from (9.7.23),

$$V_2 = \frac{2A U_T'}{5 m_2 g},$$ (9.7.26)

and hence

$$\frac{V_2}{V_1} = \frac{3m_1 + 2m_2}{5m_2}.$$ (9.7.27)

The corresponding ratio of the temperatures is obtained from

$$\frac{T_2}{T_1} = \frac{P_2 V_2}{P_1 V_1},$$ (9.7.28)

where $P_1 = m_1 g/A$, and similarly for P_2. Hence,

$$\frac{T_2}{T_1} = \frac{m_2 V_2}{m_1 V_1} = \frac{3m_1 + 2m_2}{5m_1}.$$ (9.7.29)

Notice that when $m_2 \gg m_1$, $V_2/V_1 \to 0.4$ and $T_2/T_1 \to \infty$. The potential energy lost by the piston as it compresses the gas irreversibly is taken up by the gas as internal energy.

Solution (b)

If the process takes place reversibly and adiabatically, the potential energy lost by the piston in reaching equilibrium must be dissipated elsewhere. In this case U_T is minimized subject to S being constant. Thus,

$$dU_T = dU + \frac{mg \, dV}{A} = 0.$$ (9.7.30)

Here dU is obtained from (9.7.18) taking $dS = 0$, $dN = 0$:

$$dU = U_0 V_0^{2/3} \left(\frac{N}{N_0} \right)^{5/3} \exp \left\{ \frac{2}{3} \left(\frac{S}{NR} - \frac{S_0}{N_0 R} \right) \right\} \left(\frac{-2 \, dV}{3 V^{5/3}} \right),$$ (9.7.31)

and hence

$$V = \left(\frac{2A}{3mg} \right)^{3/5} N \exp \left\{ \frac{2}{5} \left(\frac{S}{NR} - \frac{S_0}{N_0 R} \right) \right\}.$$ (9.7.32)

Since S is constant, the ratio of the equilibrium volumes for two masses m_1 and m_2 is

$$\frac{V_2}{V_1} = \left(\frac{m_1}{m_2} \right)^{3/5}.$$ (9.7.33)

The temperature ratio is obtained from (9.7.28) as in part (a):

$$\frac{T_2}{T_1} = \left(\frac{m_2}{m_1}\right)^{2/5}. \tag{9.7.34}$$

These results illustrate the difference between processes in which S is maximized at constant U, and those in which U is minimized at constant S.

Questions

9.7.1. The fundamental relation for a relativistic gas (eqn (9.3.11)) can be expressed as

$$\left(\frac{U}{U_0}\right)^3 \left(\frac{V}{V_0}\right) \left(\frac{N_0}{N}\right)^4 = \exp\left(\frac{S}{NR} - \frac{S_0}{N_0 R}\right).$$

Suppose the gas is confined to a cylinder by a piston of mass m_1 and area A, the weight of the piston $m_1 g$ being supported by the pressure of the gas. The walls and piston are adiabatic. An additional weight, Δm, is then added, initially keeping the volume fixed. Determine V_2/V_1 and T_2/T_1, in terms of m_1 and $m_2 = m_1 + \Delta m$. Assume that after Δm is added the subsequent process takes place (a) at fixed total energy and (b) at constant entropy.

9.7.2. Use the minimum energy principle to show that the equilibrium temperatures of two systems which can exchange energy must be the same. Show that the chemical potentials for particles of the same species must also be the same when these can be exchanged.

10
The potentials

10.1 Introduction

For an isolated system the equilibrium condition is expressed in a natural way by the maximum entropy principle. Alternatively, when a system interacts adiabatically with its surroundings the minimum energy principle provides an equivalent condition for equilibrium.

But often we are interested in systems which are linked with their surroundings in other ways. For example, suppose a chemical reaction takes place in the laboratory at atmospheric pressure. The system is then coupled to the atmosphere which represents a volume reservoir at constant pressure. In such cases we may apply the maximum entropy principle (or the minimum energy principle) to the composite system comprising the system of interest and the reservoir. However, the extensive variable of the reservoir is involved in this calculation. This is an unnecessary complication because only the intensive parameter for the reservoir is relevant. We shall now show how the minimum energy principle can be re-expressed in such cases using new functions, called the thermodynamic potentials.

We have already introduced some of the thermodynamic potentials, such as the enthalpy and the Helmholtz free energy for a fluid system, and the energy and enthalpy for media subject to electric or magnetic fields. In this chapter we shall develop the properties and applications of the thermodynamic potentials formally.

10.2 The Helmholtz free energy

In Section 7.2 we considered the Helmholtz free energy for a system undergoing a reversible isothermal process, $a \rightarrow b$, while in thermal equilibrium with a reservoir Σ_r at temperature T_r. We showed the work input $W = F_b - F_a$, where $F = U - TS$ is the free energy evaluated at temperature T_r.

In the following, Σ_c is a composite system which consists of two or more simple systems. We shall denote the combined system, consisting of Σ_c and Σ_r, by Σ_o, as in Fig. 10.1. The energy of Σ_c, Σ_r, and Σ_o will be denoted by U_c, U_r, and U_o, respectively; other variables will be indicated similarly.

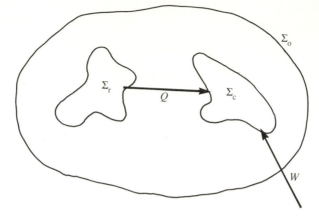

Fig. 10.1 Heat and work inputs to a system Σ_c which undergoes a reversible process whilst in equilibrium with a heat reservoir, Σ_r.

Let X be a parameter of Σ_c which is free to change. We know that the equilibrium value of X is such that

(1) S_o is a maximum when U_o is constant (maximum entropy) or

(2) U_o is a minimum when S_o is constant (minimum energy).

Both these criteria are inconvenient in practice because the extensive properties of the reservoir, Σ_r, are explicitly involved, yet only the intensive property, T_r, the temperature of the reservoir, is physically defined. We shall now show how F_c, the Helmholtz free energy for Σ_c, can be used to express the equilibrium criterion in a way which avoids this difficulty.

The equilibrium condition

Suppose Σ_c is initially in state a while in equilibrium with Σ_r. Thus all the subsystems of Σ_c are at the temperature T_r. While maintaining this temperature, Σ_c moves to state b as a result of a change in the state variable X. In this process Σ_c exchanges heat, Q, with Σ_r and work, W, with the surroundings, as indicated in Fig. 10.1. Within the overall system Σ_o the process $a \rightarrow b$ is reversible. Thus S_o is constant, and hence the minimum energy theorem applies.

For the process $a \rightarrow b$ the change in U_o is given by

$$\Delta U_o = \Delta U_c + \Delta U_r \qquad (10.2.1)$$

and since the processes of Σ_o are reversible,

$$\Delta U_r = T_r \Delta S_r. \qquad (10.2.2)$$

Also since $\Delta S_o = 0$,

$$\Delta S_c = -\Delta S_r, \tag{10.2.3}$$

and hence

$$(\Delta U_o)_{S_o} = \Delta U_c - T_r \Delta S_c, \tag{10.2.4}$$

or since,

$$F_c = U_c - T_c S_c, \tag{10.2.5}$$

$$(\Delta U_o)_{S_o} = (\Delta F_c)_{T_c = T_r}. \tag{10.2.6}$$

Equation (10.2.6) is a state relationship which holds for a finite process. It follows that the partial derivatives of U_o at constant S_o may be replaced by the corresponding derivatives of F_c evaluated at constant $T_c = T_r$. In particular,

$$\left(\frac{\partial U_o}{\partial X}\right)_{S_o} = \left(\frac{\partial F_c}{\partial X}\right)_{T_c = T_r}, \tag{10.2.7}$$

and

$$\left(\frac{\partial^2 U_o}{\partial X^2}\right)_{S_o} = \left(\frac{\partial^2 F_c}{\partial X^2}\right)_{T_c = T_r}. \tag{10.2.8}$$

Now the minimum energy principle for the overall system, Σ_o, (Fig. 10.1) is

$$\left(\frac{\partial U_o}{\partial X}\right)_{S_o} = 0, \qquad \left(\frac{\partial^2 U_o}{\partial X^2}\right)_{S_o} > 0. \tag{10.2.9}$$

Hence the equilibrium value of the free variable, X, may also be obtained by solving the equations expressing the minimization of the Helmholtz free energy:

$$\left(\frac{\partial F_c}{\partial X}\right)_{T_c = T_r} = 0, \qquad \left(\frac{\partial^2 F_c}{\partial X^2}\right)_{T_c = T_r} > 0. \tag{10.2.10}$$

This equation expresses the equilibrium condition for Σ_o in a way which involves only the properties of Σ_c and T_r, the only relevant thermodynamic property of Σ_r. Further, since the external work done on Σ_c is given by

$$W = \Delta U_o, \tag{10.2.11}$$

the external work done on Σ_c in the reversible change of state is

$$W = (\Delta F_c)_{T = T_r}. \tag{10.2.12}$$

Hence, from the reversible process theorem (Section 7.1) we see that ΔF_c is the minimum work input necessary to effect the change of state given that Σ_c exchanges heat with Σ_r only. Equivalently, $-\Delta F_c$ is the maximum work output of Σ_c in this process.

Thus we can say that F represents the *thermodynamic potential* for external work in a process linking two states of a system having the same temperature. It is normal to refer to F as the *free energy* of the system, or the *Helmholtz free energy* if the context is ambiguous. These terms are reasonable given eqn (10.2.12). The terms *Helmholtz function* and *Helmholtz potential* may also be used.

The Legendre transform

The Helmholtz function $F = U - TS$ is an example of a *Legendre transform* of U. We shall illustrate the meaning of this relationship by showing that F is naturally a function of $T, V, N_1, \ldots, N_r$ in the same sense that U is naturally a function of $S, V, N_1, \ldots N_r$. This means that all the thermodynamic properties expressed by the relation $U = U(S, V, N_1, \ldots, N_r)$ are also expressed by $F = F(T, V, N_1, \ldots, N_r)$. In effect the Helmholtz function provides another representation of the fundamental relation.

From the definition, $F = U - TS$, we have

$$dF = dU - d(TS) = dU - T\,dS - S\,dT. \tag{10.2.13}$$

Hence, from the Gibbs equation for dU, (9.5.1),

$$dF = -S\,dT - P\,dV + \sum_{i=1}^{r} \mu_i\,dN_i. \tag{10.2.14}$$

Thus the Legendre transform, F, expresses the fundamental relation in a form with T as the independent variable instead of S:

$$F = F(T, V, N_1, \ldots, N_r). \tag{10.2.15}$$

In addition the Euler form for F follows from the Euler form for U, eqn (9.5.31):

$$F = U - TS = -PV + \sum_{i=1}^{r} \mu_i N_i, \tag{10.2.16}$$

and following (9.5.2)–(9.5.4) we have

$$S = S(T, V, N_1, \ldots, N_r) = -\left(\frac{\partial F}{\partial T}\right)_{V, N_i} \tag{10.2.17}$$

$$P = P(T, V, N_1, \ldots, N_r) = -\left(\frac{\partial F}{\partial V}\right)_{T, N_i} \tag{10.2.18}$$

$$\mu_j = \mu_j(T, V, N_1, \ldots, N_r) = \left(\frac{\partial F}{\partial N_j}\right)_{T, V, N_i, i \neq j}. \tag{10.2.19}$$

Obviously this procedure can also be invoked for the other extensive variables, either individually or together. In this way the basic fundamental relations for U and S give rise to a number of different Legendre transforms; these are described in Section 10.3

We shall now illustrate how the original fundamental relation may be reconstructed from the Legendre transform. For simplicity, consider a fixed amount of fluid in a fixed volume. In this situation we may write $U = U(S, V, N_1, \ldots, N_r) = U(S)$, so that

$$dU = T\,dS, \tag{10.2.20}$$

where

$$T = \frac{dU}{dS} = T(S). \tag{10.2.21}$$

We are unable to reconstruct the original fundamental relation from the equation $T = T(S)$ because information is lost when we take the derivative. After all, $T = T(S)$ is consistent with any function of the type $U = U(S) + \text{const}$. So if we want a fundamental relation in which T is a variable we cannot use (10.2.21).

Consider the Legendre transform $F = U - TS$. The geometric construction shown in Fig. 10.2, shows that F is the intercept of the tangent to the curve $U = U(S)$ and the line $S = 0$. Hence $F = F(T)$ expresses a relation between the gradient of the tangent line and its intercept on the U-axis. In effect the function $F = F(T)$ allows the original curve to be reconstructed from the envelope of the family of tangents. It is evident that this procedure can be used only when T is a monotonic increasing or decreasing function

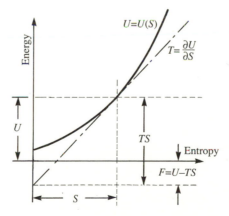

Fig. 10.2   Geometric relationship between the function $U = U(S)$ and its Legendre transform $F = F(T)$. F is the U-intercept of the tangent line at the point (S, U).

of S. In Chapter 12 we shall see that simple systems in equilibrium satisfy this condition because T must then increase monotonically with S.

To obtain the original fundamental relation $U = U(S)$ from its Legendre transform, $F = F(T)$, we first determine S as a function of T using $S = -\partial F/\partial T = S(T)$. We then invert this function to obtain T as a function of S. Since $U = F + TS$, we may now eliminate T as the independent variable and reconstruct the original relation $U = U(S)$ again. In this way the properties of the original function are preserved by the Legendre transform.

Example

The fundamental relation for a photon gas, given by (9.3.3), is

$$U = \left(\frac{3}{4}\right)^{4/3} \left(\frac{c}{4\sigma}\right)^{1/3} S^{4/3} V^{-1/3} \tag{10.2.22}$$

where σ is the Stefan–Boltzmann constant, given by eqn (9.3.3a).

(a) Obtain $F(T, V)$.

(b) Show that $U = U(S, V)$ can be obtained from F by taking the inverse Legendre transform.

(c) Obtain an expression for the pressure, P, as a function of S and V and as a function of T and V.

Solution (a)

We seek to eliminate S from the equation $F = U - TS$ and make T the independent variable. First we have

$$T = \left(\frac{\partial U}{\partial S}\right)_V = \left(\frac{3}{4}\right)^{1/3} \left(\frac{c}{4\sigma}\right)^{1/3} S^{1/3} V^{-1/3}, \tag{10.2.23}$$

and hence

$$S = \frac{4}{3}\left(\frac{4\sigma}{c}\right) V T^3, \tag{10.2.24}$$

$$TS = \frac{4}{3}\left(\frac{4\sigma}{c}\right) V T^4, \tag{10.2.25}$$

$$U = \left(\frac{4\sigma}{c}\right) V T^4. \tag{10.2.26}$$

Thus the Helmholtz potential is

$$F = -\frac{1}{3}\left(\frac{4\sigma}{c}\right) V T^4. \tag{10.2.27}$$

Solution (b)

Notice that the expressions for S and U as functions of T and V are not fundamental relations because, taken alone, they cannot be used in general to reconstruct the relation $U = U(S, V)$. But the equation $F = F(T, V)$ is a fundamental relation, as we can illustrate by taking the inverse Legendre transform. Begin by calculating S using

$$S = - \left(\frac{\partial F}{\partial T} \right)_V = \frac{4}{3} \left(\frac{4\sigma}{c} \right) V T^3. \tag{10.2.28}$$

Hence we can express T as a function of S, yielding (10.2.23) again. Then by eliminating T from the equation, $U = F + TS$, we finally recover the original fundamental relation, (10.2.22).

Solution (c)

We can calculate P in two ways. First, as a function of S and V:

$$P(S, V) = - \left(\frac{\partial U}{\partial V} \right)_S = \frac{1}{3} \left(\frac{3}{4} \right)^{4/3} \left(\frac{c}{4\sigma} \right)^{1/3} S^{4/3} V^{-4/3}. \tag{10.2.29}$$

Notice that $-(\partial U/\partial V)_T \neq P$; instead from (10.2.26) we get

$$- \left(\frac{\partial U}{\partial V} \right)_T = - \left(\frac{4\sigma}{c} \right) T^4 \neq P.$$

This illustrates the importance of clearly indicating the fixed variables when taking partial derivatives, unless they are really obvious. However, we can obtain P as a function of T and V by using (10.2.24) to eliminate S. Alternatively, we can simply take the derivative of F with respect to V using (10.2.27):

$$P(T, V) = - \left(\frac{\partial F}{\partial V} \right)_T = \frac{1}{3} \left(\frac{4\sigma}{c} \right) T^4. \tag{10.2.30}$$

In this form we see that P is a function of T alone.

Example

The free energy for a crystal in which the ions have just two quantum states is readily obtained in statistical mechanics (Riedi 1988). Given that the ions have states of energy 0 and ε it is found that

$$F = -NRT \ln \left(1 + e^{-E/RT} \right), \tag{10.2.31}$$

where $E = \tilde{N}_A \varepsilon$ is constant.

(a)   Obtain the fundamental relation in the entropy representation.

(b)   Express the molar heat capacity as a function of T.

Solution (a)

Since E is constant,

$$S = -\left(\frac{\partial F}{\partial T}\right)_E = NR \ln\left(1 + e^{-E/RT}\right) + \frac{NE/T}{e^{E/RT} + 1}, \qquad (10.2.32)$$

and we have

$$U = F + TS = \frac{NE}{e^{E/RT} + 1}. \qquad (10.2.33)$$

Since we are seeking the entropy fundamental relation, we now invert this relation to express T as a function of U. We can then eliminate T from (10.2.32). First we have

$$e^{E/RT} + 1 = \frac{NE}{U},$$

$$e^{-E/RT} + 1 = \left(1 - \frac{U}{NE}\right)^{-1},$$

and

$$\frac{NE}{T} = NR \ln\left(\frac{NE}{U} - 1\right).$$

Thus on substituting in (10.2.32), and rearranging the terms,

$$S = \left(\frac{U}{E} - N\right) R \ln\left(1 - \frac{U}{NE}\right) - \frac{UR}{E} \ln\left(\frac{U}{NE}\right), \qquad (10.2.34)$$

which agrees with (9.3.6).

Solution (b)

Given E is fixed, the molar heat capacity is

$$c_E = \frac{T}{N}\left(\frac{\partial S}{\partial T}\right)_E \qquad (10.2.35)$$

and substituting from (10.2.32) and rearranging, we find

$$c_E = R\frac{(E/RT)^2}{\{\exp(E/2RT) + \exp(-E/2RT)\}^2}. \qquad (10.2.36)$$

This function, which is illustrated in Fig. 10.3, exhibits a peak at $T \approx 0.4E/R$, known as the *Schottky anomaly*. The presence of a peak such as this is a characteristic of a system in which the atoms have a few low-lying closely spaced quantum levels. Where a Schottky peak occurs at low temperatures it may completely dominate all other contributions to the molar heat capacity of a solid (see question 10.2.5).

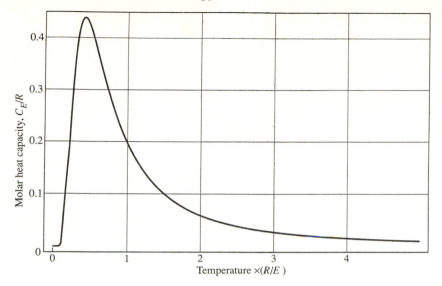

Fig. 10.3   Molar heat capacity of a system in which the particles have just two quantum states, illustrating the Schottky anomaly at $T \approx 0.4E/R$.

When T is significantly greater than E/R, the denominator is approximately 4 and then

$$c_E = R\left(\frac{E}{2RT}\right)^2. \tag{10.2.37}$$

However, according to (10.2.36), $c_E \to 0$ as $T \to 0$, as we expect from the third law.

Questions

10.2.1. The fundamental relation for a gas of spin-zero particles at low temperature is given in the energy representation by (9.3.8):

$$U = AV^{-2/3}S^{5/3}, \tag{10.2.38}$$

where A is a positive constant. Obtain an expression for the free energy and show that the pressure is given by

$$P = \frac{2}{5}\left(\frac{3}{5A}\right)^{3/2} T^{5/2}.$$

10.2.2. Show that

$$U = F - T\left(\frac{\partial F}{\partial T}\right)_{V, N_i}. \tag{10.2.39}$$

This is one of several equations called *Gibbs–Helmholtz equations*.

10.2.3. (a) Obtain expressions for the Helmholtz free energy of a single-component ideal gas and for a mixture of ideal gases, starting from (9.3.1) and (9.3.2). Let $F_i(V, T, N_i)$ be the free energy for species i when it occupies the volume V alone at temperature T. Show that the free energy for the mixture can be written as

$$F(V, T, N_1, N_2, \ldots, N_r) = \sum_i F_i(V, T, N_i)$$

(10.2.40)

where

$$F_i(V, T, N_i) = -N_i RT \left\{ 1 + \frac{3}{2} \ln\left(\frac{3RT}{2}\right) - \ln\left(\frac{N_i \sigma_i}{V}\right) \right\},$$

(10.2.41)

and σ_i is given by (9.3.2b).

(b) Use (10.2.41) to show that the chemical potential for component i is given by (9.5.40).

10.2.4. The Debye theory for a crystalline solid yields an expression for the contribution to the internal energy due to the lattice vibrations (Riedi 1988). By adopting suitable approximations the Helmholtz free energy is found to be

$$F = -NR \frac{\pi^4 T^4}{5T_D^3}.$$

(10.2.42)

Here T_D is a constant called the *Debye temperature* and the approximation is applicable provided $T \ll T_D$. Representative values of T_D in metals are (Rosenberg 1963): lead (108 K), copper (348 K), and silver (225 K). For diamond T_D is unusually high, being approximately 2000 K. Assume $T \ll T_D$ in the following.

(a) Show that the fundamental relation in the energy representation is

$$U = \frac{3}{4} T_D \left(\frac{5S^4}{4\pi^4 NR}\right)^{1/3}.$$

(10.2.43)

(b) Demonstrate that the molar heat capacity at constant volume due to lattice vibrations in a solid is given by

$$c_{vL} = \frac{1}{N}\left(\frac{\partial U}{\partial T}\right)_{V,N} = \frac{12\pi^4}{5} R \left(\frac{T}{T_D}\right)^3.$$

(10.2.44)

This equation shows that the molar heat capacity due to the lattice vibrations of solids at low temperatures should approach zero as T^3. Other contributions to the molar heat capacity may also be important under these conditions. (See question 10.2.5.)

10.2.5.   The fundamental relation for an electron gas at low temperature, (9.3.7), can be written

$$U = \frac{3}{5} A N^{5/3} V^{-2/3} \left(1 + \frac{5}{3} \left(\frac{S}{\pi NR} \right)^2 \right), \qquad (10.2.45)$$

where A is a constant given by

$$A = \frac{\tilde{N}_A h^2}{8m} \left(\frac{3\tilde{N}_A}{\pi} \right)^{2/3}. \qquad (10.2.46)$$

Here m is the electron mass, $\tilde{N}_A$ is Avogadro's number, and h is Planck's constant.

(a)   Show that the free energy is

$$F = \frac{3}{5} A N^{5/3} V^{-2/3} - \frac{\pi^2 R^2}{4A} N^{1/3} V^{2/3} T^2. \qquad (10.2.47)$$

(b)   Hence show that

$$P = \frac{2}{5} \frac{NRT_f}{V} \left\{ 1 + \frac{5\pi^2}{12} \left(\frac{T}{T_f} \right)^2 \right\}, \qquad (10.2.48)$$

where T_f is the Fermi temperature, given by (9.3.7a). (These equations are applicable to the conduction electrons in a metal. Since the Fermi temperature is typically 40×10^3 K, the pressure of a degenerate electron gas at room temperature is only weakly dependent on T.)

(c)   Show that the molar heat capacity at constant volume, c_{ve}, due to the free electrons of a metal is

$$c_{ve} = \frac{\pi^2 R}{2} \left(\frac{T}{T_f} \right), \qquad (10.2.49)$$

and compare c_{ve} with the molar heat capacity of a monatomic ideal gas at room temperature. Take $T_f = 40,000$ K.

(d)   Compare the contributions to the molar heat capacity of a metal due to: (i) lattice vibrations, c_{vL} (question 10.2.4); and (ii) free electrons, c_{ve}. Assume there is one free electron for each metal atom. Take $T_D = 200$ K, $T_f = 40,000$ K, and show that

$$\frac{c_{vL}}{c_{ve}} = 0.237 \, T^2. \qquad (10.2.50a)$$

Thus while the electron gas makes a negligible contribution to the heat capacity of a metal at room temperature, typically it is the dominant contribution when $T < 1$ K.

(e) A metal exhibits a Schottky anomaly for which the molar heat capacity, c_E, is given by (10.2.36). Suppose $E/R = 0.01$ K. In addition there is one free electron per atom and $T_f = 40{,}000$ K. Show that, provided $T \gg 0.01$ K,

$$\frac{c_E}{c_{ve}} = \frac{2}{\pi^2 T^3}. \tag{10.2.50b}$$

Show that the Schottky anomaly is by far the most dominant contribution to the molar heat capacity as the temperature approaches 0.01 K.

10.2.6. The Helmholtz free energy for a particular system is

$$F = -A N^{2/3} V^{1/3} T^2, \tag{10.2.51}$$

where A is a positive constant.

(a) The system undergoes a reversible expansion process having initial and final volumes V and $8V$, whilst in equilibrium with a heat reservoir at temperature T. Show that the work input, W, and heat input, Q, are given by

$$W = -A N^{2/3} V^{1/3} T^2, \tag{10.2.52}$$

$$Q = 2A N^{2/3} V^{1/3} T^2, \tag{10.2.53}$$

and confirm that $\Delta U = Q + W$.

(b) In a heating process at constant volume, V, the initial and final temperatures are T and $2T$, respectively. Show that

$$Q = 3A N^{2/3} V^{1/3} T^2. \tag{10.2.54}$$

(c) In a reversible adiabatic expansion the initial temperature and volume are T and V; the final volume is $8V$. Show that the final temperature is $T/2$ and that the work done on the system is

$$W = -\tfrac{1}{2} A N^{2/3} V^{1/3} T^2. \tag{10.2.55}$$

10.2.7. An approximate expression for the Helmholtz free energy, F_r, due to the rotational motion of the molecules of a diatomic gas is

$$F_r = -NRT \ln \left(\frac{RT}{E_r} + \frac{1}{3} + \frac{E_r}{15RT} \right), \tag{10.2.56}$$

where E_r is a constant given by (9.3.5a). Obtain expressions for U and S as functions of T and N. In the limit $RT/E_r \gg 1$ show that the fundamental relation in the entropy representation is given by (9.3.5).

10.2.8. Use the fundamental relation for a set of weakly coupled harmonic oscillators, eqn (9.3.4), to establish the Helmholtz free energy, F_v, for this system:

$$F_v = NRT \ln \left\{ 1 - \exp\left(-\frac{E_v}{RT} \right) \right\}. \qquad (10.2.57)$$

10.2.9. The Helmholtz function for a single-component ideal gas of diatomic molecules can be written as the sum of the separate contributions due to translational motion, F_t, rotational motion, F_r, and vibrational motion, F_v, of the molecules:

$$F = F_t + F_r + F_v. \qquad (10.2.58)$$

Here F_t, F_r, and F_v are given by (10.2.41), (10.2.56), and (10.2.57), respectively.
(a) Typically at room temperature $RT/E_r \gg 1$ whereas $RT/E_v \ll 1$. For this situation show that

$$F = -NRT \left[\ln \left\{ \left(\frac{T}{T_0} \right)^{5/2} \frac{N_0 V}{N V_0} \right\} + \phi_r \right]. \qquad (10.2.59)$$

Here T_0, V_0, and N_0 are constants and

$$\phi_r = 1 + \ln \left\{ \left(\frac{3}{2} \right)^{3/2} \frac{(RT_0)^{5/2} V_0}{\sigma E_r N_0} \right\}, \qquad (10.2.60)$$

where σ is defined by (9.3.1a).
Show that the molar heat capacity at constant volume is $5R/2$.
(b) At sufficiently elevated temperatures, $RT/E_r \gg 1$ and $RT/E_v \gg 1$. Establish that

$$F = -NRT \left[\ln \left\{ \left(\frac{T}{T_0} \right)^{7/2} \frac{N_0 V}{N V_0} \right\} + \phi_{rv} \right], \qquad (10.2.61)$$

where T_0, V_0, and N_0 are constants and

$$\phi_{rv} = 1 + \ln \left\{ \left(\frac{3}{2} \right)^{3/2} \frac{(RT_0)^{7/2} V_0}{\sigma E_r E_v N_0} \right\}. \qquad (10.2.62)$$

Show that the molar heat capacity at constant volume is $7R/2$.

10.3 Properties of the potentials

The formal and physical properties of other thermodynamic potentials may be established by the same arguments as those used in Section 10.2. Here we summarize these relationships for the principal potentials.

Enthalpy, H

Define the enthalpy by

$$H = U + PV. \tag{10.3.1}$$

The corresponding Gibbs equation is

$$dH = dU + P\,dV + V\,dP$$
$$= T\,dS + V\,dP + \sum_{i=1}^{r} \mu_i\,dN_i, \tag{10.3.2}$$

and hence, as a function of its proper variables,

$$H = H(S, P, N_1, \ldots, N_r). \tag{10.3.3}$$

In addition we have the derivative relations,

$$T = \left(\frac{\partial H}{\partial S}\right)_{P,N_i}, \qquad V = \left(\frac{\partial H}{\partial P}\right)_{S,N_i}, \qquad \mu_i = \left(\frac{\partial H}{\partial N_i}\right)_{S,P,N_j,j \neq i}. \tag{10.3.4}$$

We see that the enthalpy is the potential in which P replaces V as the independent variable. From Section 7.3, and by analogy with the Helmholtz potential, we obtain the following physical applications for H. The equilibrium state for a system maintained at constant pressure, P_r, by a reservoir whilst separated from its surroundings by adiabatic walls is expressed in terms of the enthalpy. The reservoir exchanges volume with the system, rather than energy, as in the Helmholtz case. Thus one may show that H is a minimum at equilibrium, subject to $P = P_r$, $S = $ const. We may regard the enthalpy as the potential for external work in a process at constant entropy when the system interacts with a constant pressure reservoir.

Like U, H also satisfies a form of the first law, as we showed in Section 7.3. Only the potentials in which S is a natural variable have first law process equations of this type. Thus when a system is transferred from a state of equilibrium with a constant pressure reservoir at pressure P_1 to another at pressure P_2,

$$H_2 - H_1 = W_{12} + Q_{12}, \tag{10.3.5}$$

where Q_{12} is the external heat input and W_{12} the external work input. (W_{12} does not include the work done by the reservoirs.) This relation holds for both reversible and irreversible processes, provided the interaction with the reservoirs takes place quasistatically. Thus for a process carried out at constant pressure in which $W_{12} = 0$, H may be regarded as the potential for heat transfer. For this reason H is sometimes called the *heat function*.

Gibbs function, G

The Gibbs function is the double Legendre transform of U,

$$G = U - TS + PV. \tag{10.3.6}$$

The corresponding Gibbs equation is

$$dG = dU - T\,dS - S\,dT + P\,dV + V\,dP$$

$$= -S\,dT + V\,dP + \sum_{i=1}^{r} \mu_i\,dN_i, \tag{10.3.7}$$

and hence in terms of its natural variables,

$$G = G(T, P, N_1, \ldots, N_r). \tag{10.3.8}$$

The derivative relations are

$$S = -\left(\frac{\partial G}{\partial T}\right)_{P, N_i}, \qquad V = \left(\frac{\partial G}{\partial P}\right)_{T, N_i}, \qquad \mu_i = \left(\frac{\partial G}{\partial N_i}\right)_{T, P, N_j, j \neq i}. \tag{10.3.9}$$

By adapting the arguments used in Section 10.2, one may show that the Gibbs function is the potential for work in a reversible process where, P and T are established by external reservoirs. For this reason G is also referred to as the *Gibbs free energy*.

In addition, the equilibrium state of an unconstrained parameter is that which minimizes G subject to P and T being fixed by the reservoirs, and to other constraints on the extensive variables. This potential is especially useful in connection with chemical processes at constant P and T (Section 12.6). Also when a system is moved reversibly from a state of equilibrium with a constant pressure reservoir at pressure P_1 to another at pressure P_2, whilst remaining in equilibrium with a heat reservoir, the external work done on the system is $W = \Delta G$.

Grand potential, Ψ_i

The grand potential is the relevant thermodynamic function when a system of defined volume is in equilibrium with both a heat reservoir and a particle reservoir. The particle reservoir is characterized by its chemical potential. The name grand potential (or grand canonical potential) arises from its association with the grand canonical ensemble in statistical mechanics. Let i denote a component of a simple system which is exchanged with the reservoir, and let

$$\Psi_i = U - TS - N_i\mu_i. \tag{10.3.10}$$

Then

$$d\Psi_i = -S\,dT - P\,dV - N_i\,d\mu_i + \sum_{j \neq i}\mu_j\,dN_j \tag{10.3.11}$$

and, in terms of its natural variables,

$$\Psi_i = \Psi_i(T, V, \mu_i, N_j), \tag{10.3.12}$$

where $j = 1, \ldots, r$ with the restriction that $j \neq i$. In addition,

$$S = -\left(\frac{\partial\Psi_i}{\partial T}\right)_{V, \mu_i, N_j, j \neq i}, \qquad P = -\left(\frac{\partial\Psi_i}{\partial V}\right)_{T, \mu_i, N_j, j \neq i} \tag{10.3.13a}$$

and

$$N_i = -\left(\frac{\partial\Psi_i}{\partial\mu_i}\right)_{T, V, N_j, j \neq i}, \qquad \mu_j = \left(\frac{\partial\Psi_i}{\partial N_j}\right)_{T, V, \mu_i, N_k, k \neq i, k \neq j}. \tag{10.3.13b}$$

Consider a system having a defined volume in thermal equilibrium with a heat reservoir and in equilibrium with respect to the exchange of particles i in a particle reservoir.

The equilibrium value of an unconstrained parameter, X, is such that Ψ_i is minimized subject to the restrictions $T = T_r$ and $\mu_i = \mu_{ri}$. In addition, in a reversible process, the end states of which have fixed values for T and μ_i, then $W = \Delta\Psi_i$. Thus Ψ_i represents the potential for external work in such a process.

Magnetic media

Because a body subject to a magnetic field cannot be treated as an isolated system, the definition of its magnetic energy requires some care. This matter has been considered in Section 8.2 where we identified thermodynamic potentials to represent different work processes. Two potentials, U_Φ and U_J, have the character of energy functions. Two others, H_Φ and H_J, which are related to U_Φ and U_J by the Legendre transforms,

$$H_\Phi = U_\Phi + \int \mathbf{M} \cdot \mathbf{B}_0 \, dV \tag{10.3.14}$$

and

$$H_J = U_J - \int \mathbf{B}_0 \cdot \mathbf{M} \, dV, \tag{10.3.15}$$

were referred to as the magnetic enthalpy functions.

We shall consider first the case in which the current, J, of the source coil is the control parameter, since it is normal to identify the field $\mathbf{B}_0$ in the absence of the medium by J. Equation (8.2.15) expresses the first law for a process in which work W and heat Q are input to the system of interest:

$$\Delta U_J = Q + W + W_e. \tag{10.3.16}$$

Here W_e is the associated work done by the current source which maintains J as the controlled variable. This equation, like (10.3.5), holds for processes which may be irreversible provided the interaction between the system and the magnetic field is reversible.

First law relations such as this also hold for the other potentials, U_Φ, H_Φ and H_J. For U_Φ there is no implied work interaction with the magnetic field, when external work, W, is performed on the system. In this case the first law,

$$\Delta U_\Phi = Q + W, \tag{10.3.17}$$

holds for both reversible and irreversible processes. Because U_Φ satisfies a simple first law relation it is the appropriate thermodynamic potential for an isolated system. However, it is seldom used because it is usual to take $\mathbf{B}_0$ to mean the applied field defined by J.

When the medium is isotropic, $\mathbf{B}_0$ and $\mathbf{M}$ are collinear and we may use scalar values. In terms of the energy, entropy, and mole numbers per unit volume,

$$u_J = u_J(s, M, n_1, n_2, \ldots, n_r), \tag{10.3.18}$$

and

$$\delta u_J = T \delta s + B_0 \delta M + \sum_{i=1}^{r} \mu_i \delta n_i, \tag{10.3.19}$$

where

$$T = \left(\frac{\partial u_J}{\partial s}\right)_{M, n_i}, \qquad B_0 = \left(\frac{\partial u_J}{\partial M}\right)_{s, n_i}, \qquad \mu_i = \left(\frac{\partial u_J}{\partial n_i}\right)_{s, M, n_j, j \neq i}. \tag{10.3.20}$$

If B_0 is identified with the flux of the source circuit, the corresponding equations for u_Φ are simply obtained. The magnetic work term then becomes $-M \delta B_0$, and the subsequent relations must be modified accordingly.

The magnetic Helmholtz, Gibbs, and grand potentials may also be defined in a straightforward way. Consider the Helmholtz and Gibbs potentials corresponding to u_J. First the Helmholtz potential is

$$f_J = u_J - Ts, \tag{10.3.21}$$

and we find

$$f_J = f_J(T, M, n_1, n_2, \ldots, n_r),\qquad(10.3.22)$$

for which

$$\delta f_J = -s\,\delta T + B_0\,\delta M + \sum_{i=1}^{r} \mu_i\,\delta n_i,\qquad(10.3.23)$$

where

$$s = -\left(\frac{\partial f_J}{\partial T}\right)_{M,n_i},\qquad B_0 = \left(\frac{\partial f_J}{\partial M}\right)_{T,n_i},\qquad \mu_i = \left(\frac{\partial f_J}{\partial n_i}\right)_{T,M,n_j,j\neq i}.$$

$$(10.3.24)$$

The corresponding Gibbs free energy per unit volume is

$$g_J = u_J - Ts - MB_0 = g_J(T, B_0, n_1, n_2, \ldots, n_r),\qquad(10.3.25)$$

and hence,

$$\delta g_J = -s\,\delta T - M\,\delta B_0 + \sum_{i=1}^{r} \mu_i\,\delta n_i,\qquad(10.3.26)$$

where,

$$s = -\left(\frac{\partial g_J}{\partial T}\right)_{B_0,n_i},\qquad M = -\left(\frac{\partial g_J}{\partial B_0}\right)_{T,n_i},\qquad \mu_i = \left(\frac{\partial g_J}{\partial n_i}\right)_{T,B_0,n_j,j\neq i}.$$

$$(10.3.27)$$

Remember that when we use the extensive functions, U_Φ, U_J, and their related potentials, the volume should normally be included as an additional deformation variable.

General forms

To generalize we shall denote the extensive variables of a system, Σ, by $(X_1, X_2, \ldots, X_t)$. Suppose Σ is in contact with a reservoir such that a quantity, i, having the extensive parameter X_i, is exchanged with a reservoir. We shall denote the corresponding intensive parameter in the energy representation by

$$P_i = \left(\frac{\partial U}{\partial X_i}\right)_{X_j,j\neq i}.\qquad(10.3.28)$$

Notice that in this notation, the variable P_i conjugate to the volume is $-P$, the negative of the usual pressure. The corresponding Legendre transform for the energy is

$$\Psi(X_1, X_2, \ldots, P_i, \ldots, X_t) = U(X_1, X_2, \ldots, X_i, \ldots, X_t) - X_i P_i.$$

$$(10.3.29)$$

Subject to the condition $P_i = P_{ri}$, Ψ represents the work potential for the system. This potential can be realized only when the work process is reversible within both the system and the reservoir. Further, the equilibrium value of any unconstrained parameter is such that $\Psi(X_1, X_2, \ldots, P_i, \ldots, X_t)$ is a minimum, given $P_i = P_{ri}$.

The notation used in (10.3.29) can be extended to multiple Legendre transforms. The relation between Ψ and the energy, U, may be indicated by explicitly showing the independent variables whilst observing the notation that X_i denotes an extensive variable and P_i indicates the corresponding intensive variable.

Legendre transforms of the entropy

The Legendre transforms of the entropy fundamental relation are analogous to the transforms of the energy, but their physical applications are less obvious. Here we shall briefly introduce two entropy transforms, the Massieu and the Planck functions (we use the names adopted by Guggenheim (1967)), which actually predate the energy-based potentials (Massieu 1869). The significance of these functions is that U and H can be obtained as derivatives, and they are directly related to F and G. Another entropic potential is the subject of question 10.3.10.

Recollect the Gibbs equation for S, eqn (9.5.21),

$$dS = \beta \, dU + \gamma \, dV + \sum_{i=1}^{r} \varphi_i \, dN_i, \qquad (10.3.30)$$

where $\beta = 1/T$, $\gamma = P/T$, and $\varphi_i = -\mu_i/T$ are the entropy intensive parameters. The Massieu function, J, is the Legendre transform of S in which β replaces U as an independent variable. Thus,

$$J = S - \beta U \qquad (10.3.31)$$

and the corresponding Gibbs equation is

$$dJ = -U \, d\beta + \gamma \, dV + \sum_{i=1}^{r} \varphi_i dN_i. \qquad (10.3.32)$$

We see that $J = J(\beta, V, N_1, \ldots, N_r)$. The Massieu function is the entropic potential which corresponds to the Helmholtz function.

The Planck function, Y, which is the entropic potential corresponding to the Gibbs function, is defined by

$$Y = S - \beta U - \gamma V. \qquad (10.3.33)$$

In this case we have,

$$dY = -U \, d\beta - V \, d\gamma + \sum_{i=1}^{r} \varphi_i \, dN_i, \qquad (10.3.34)$$

showing that Y can be expressed in terms of its natural variables as $Y = Y(\beta, \gamma, N_1, \ldots, N_r)$. Further properties of the entropic potentials are developed in the following questions.

Questions

10.3.1.   Obtain the Euler forms for the thermodymamic potentials H, G, and Ψ_i of a simple fluid. For a single-component fluid show that

$$G = N\mu \tag{10.3.35}$$

and

$$\Psi = -PV. \tag{10.3.36}$$

10.3.2.   The enthalpy fundamental relation for a particular simple system is

$$H = 2NP^{0.5}\sigma^{0.5}\exp\left(\frac{S}{2NR} - 1\right). \tag{10.3.37}$$

Here σ is a physical constant determined by the particles of the system:

$$\sigma = \frac{\tilde{N}_A^2 h^2}{2\pi mg}, \tag{10.3.38}$$

where $\tilde{N}_A$ is Avogadro's number, h is Planck's constant, m is the particle mass, and $g = 2s + 1$, where s is the particle spin. Show that

$$U = \frac{\sigma N^2}{V}\exp\left(\frac{S}{NR} - 2\right), \tag{10.3.39}$$

$$F = -NRT\left[1 + \ln\left(\frac{RTV}{N\sigma}\right)\right], \tag{10.3.40}$$

$$G = -NRT\ln\left(\frac{R^2 T^2}{P\sigma}\right), \tag{10.3.41}$$

$$\Psi = -\frac{R^2 VT^2}{\sigma}\exp\left(\frac{\mu}{RT}\right). \tag{10.3.42}$$

Identify the system from the examples given at the beginning of Section 9.3.

10.3.3.   Obtain the following potentials for a mixture of ideal gases. Start either from the entropy representation, eqn (9.3.2), or the Helmholtz free energy, eqns (10.2.40) and (10.2.41).

(a)
$$H = \frac{5}{3} \left(\frac{3P}{2} \right)^{0.4} N^{0.6} \exp \left(\frac{2S}{5NR} - 1 \right) \prod_{i=1}^{r} (N_i \sigma_i)^{\frac{2N_i}{5N}}.$$

(10.3.43)

(b)
$$G = -RT \sum_{i=1}^{r} N_i \ln \left(\frac{N(1.5)^{1.5}(RT)^{2.5}}{N_i \sigma_i P} \right).$$

(10.3.44)

Evaluate μ_i by differentiating G and confirm that $G = \sum_{i=1}^{r} N_i \mu_i$.

(c)
$$\Psi_j = -\frac{(1.5RT)^{2.5} V}{1.5\sigma_j} \exp \left(\frac{\mu_j}{RT} \right)$$
$$-RT \sum_{i \neq j} N_i \left\{ 1 + \ln \left[\frac{(1.5RT)^{1.5} V}{N_i \sigma_i} \right] \right\}.$$

(10.3.45)

10.3.4. The energy and enthalpy of a particular body subject to a magnetic field are given by (9.5.43) and (9.5.47). The field in the absence of the body is identified by the source coil current, J, being constant. Obtain the following expressions for the magnetic Helmholtz and Gibbs functions per unit volume:

$$f_J = n(u_0 - Ts_0) + \frac{\mu_0 T M^2}{2nc_c} + \frac{nc_0}{T_0} \left(1 - \frac{T}{2T_0} - \frac{T_0}{2T} \right)$$

(10.3.46)

$$g_J = n(u_0 - Ts_0) - \frac{nc_c B^2}{2\mu_0 T} + \frac{nc_0}{T_0} \left(1 - \frac{T}{2T_0} - \frac{T_0}{2T} \right).$$

(10.3.47)

10.3.5. Demonstrate the following Gibbs–Helmholtz equations for a simple fluid mixture:

$$U = H - P \left(\frac{\partial H}{\partial P} \right)_{S, N_j}$$

(10.3.48)

$$F = G - P \left(\frac{\partial G}{\partial P} \right)_{T, N_j}$$

(10.3.49)

$$G = F - V \left(\frac{\partial F}{\partial V} \right)_{T, N_j}.$$

(10.3.50)

10.3.6. Show that the Massieu function can be written

$$J = -F/T.$$

(10.3.51)

Demonstrate that

$$U = -\left(\frac{\partial J}{\partial \beta}\right)_{V,N} = \left(\frac{\partial (\beta F)}{\partial \beta}\right)_{V,N} = -T^2\left(\frac{\partial (F/T)}{\partial T}\right)_{V,N},$$

$$(10.3.52)$$

and hence establish the Gibbs–Helmholtz equation given in (10.2.39).

10.3.7. (a) Show that the Gibbs equation for S can be expressed in the form

$$dS = \beta\, dH + \lambda\, dP + \varphi\, dN, \qquad (10.3.53)$$

where $\beta = 1/T$, $\lambda = -V/T$, and $\varphi = -\mu/T$.

 (b) Show that the Planck function is given by

$$Y = S - \beta H, \qquad (10.3.54)$$

and that

$$Y = -G/T. \qquad (10.3.55)$$

 (c) Hence establish that the Gibbs equation for Y can be written either as in (10.3.34) or in the form

$$dY = -H\, d\beta + \lambda\, dP + \varphi\, dN. \qquad (10.3.56)$$

 (d) Use the above results to establish the following Gibbs–Helmholtz equations for a single-component system. Recollect $\gamma = P/T$.

$$H = \left(\frac{\partial \beta G}{\partial \beta}\right)_{P,N} = -T^2\left(\frac{\partial (G/T)}{\partial T}\right)_{P,N}$$

$$= G - T\left(\frac{\partial G}{\partial T}\right)_{P,N}, \qquad (10.3.57)$$

$$U = \left(\frac{\partial \beta G}{\partial \beta}\right)_{\gamma,N} = -T^2\left(\frac{\partial (G/T)}{\partial T}\right)_{P/T,N}$$

$$= G - T\left(\frac{\partial G}{\partial T}\right)_{P/T,N}. \qquad (10.3.58)$$

10.3.8. The Gibbs free energy for a certain simple system is

$$G = -\frac{NT^4}{64AP^2}, \qquad (10.3.59)$$

where A is a positive constant. Use equations (10.3.57) and (10.3.58) to obtain U and H. Calculate S and by eliminating T and P show that

$$U = \frac{AS^4}{NV^2},$$ (10.3.60)

and

$$H = 3\left(\frac{AS^4P^2}{4N}\right)^{1/3}.$$ (10.3.61)

10.3.9.   A system consists of 8 moles of the fluid whose fundamental relations are given in question 10.3.8. The system undergoes a reversible process whilst in equilibrium with a heat reservoir at temperature T and interacting with two constant pressure reservoirs in succession, together with an external work source. The pressures of the initial and final reservoirs are $P/3$ and P, respectively.

(a)   Show that the work done on the system by the external work source is,

$$W = \frac{T^4}{AP^2}.$$ (10.3.62)

(b)   Obtain the work done on the system by the initial and final constant pressure reservoirs, W_i and W_f:

$$W_i = \frac{9T^4}{4AP^2},$$ (10.3.63)

$$W_f = -\frac{T^4}{4AP^2}.$$ (10.3.64)

(c)   Show that the heat input to the system by the heat reservoir is

$$Q = -\frac{4T^4}{AP^2},$$ (10.3.65)

and confirm that $\Delta U = W + W_i + W_f + Q$.

(d)   Suppose the system is initially in equilibrium with a heat reservoir at a different temperature from the final one. In the process the system exchanges heat only when it is in equilibrium with one reservoir or the other. Can the work done on the system be expressed as a function of the initial and final states alone? Explain

10.3.10. (a)   The entropic potential which corresponds to the grand potential is the Kramers function (Kubo 1976),

$$X = S - \beta U - \varphi N,$$ (10.3.66)

where $\beta = 1/T$ and $\varphi = -\beta\mu$. Take the approach used in question 10.3.6 to establish the Gibbs–Helmholtz equation:

$$U = -\left(\frac{\partial X}{\partial \beta}\right)_{V,\varphi} = \left(\frac{\partial \beta\Psi}{\partial \beta}\right)_{V,\varphi}$$

$$= -T^2\left(\frac{\partial(\Psi/T)}{\partial T}\right)_{V,\mu/T} = \Psi - T\left(\frac{\partial\Psi}{\partial T}\right)_{V,\mu/T}.$$

$$(10.3.67)$$

(b) The grand potential of a certain single-component simple system is

$$\Psi = -\frac{VR^4T^4}{\sigma}\exp\left(\frac{\mu}{RT}\right), \qquad (10.3.68)$$

where σ is a physical constant. Obtain U, S, and N in terms of T, V, and μ by differentiation. By eliminating T and μ show that the fundamental relation in the energy representation is given by (9.3.11).

(c) For the system considered in (b) show that

$$H = 4NP^{0.25}\sigma^{0.25}\exp\left(\frac{S}{4NR} - 1\right), \qquad (10.3.69)$$

$$F = -NRT\left(1 + \ln\left(\frac{R^3T^3V}{N\sigma}\right)\right), \qquad (10.3.70)$$

$$G = -NRT\ln\left(\frac{R^4T^4}{P\sigma}\right). \qquad (10.3.71)$$

10.3.11. The grand potential for a single-component ideal gas in r dimensions contained in a hypercube of volume $V = a^r$ is

$$\Psi = -(RT)^{(r+2)/2}\left(\frac{r}{2}\right)^{r/2}\frac{V}{\sigma}\exp\left(\frac{\mu}{RT}\right), \qquad (10.3.72)$$

where σ is given by (9.3.10a). Show that the fundamental relation in the energy representation is given by (9.3.10)

10.3.12. A system Σ interacts with a number of reservoirs, including a heat reservoir, although it is not necessarily in equilibrium with them. Denote the extensive variables of the system, $X_1, X_2, \ldots, X_t$, and the conjugate intensive variables $P_1, P_2, \ldots, P_t$. Corresponding to certain quantities, $X_j, X_k, \ldots$, there are reservoirs, $\Sigma_{rj}, \Sigma_{rk}, \ldots$, whose intensive parameters are $P_{rj}, P_{rk}, \ldots$, respectively. Define the availability function for Σ by

$$A(X_1, X_2, \ldots, X_t, P_{rj}, P_{rk}, \ldots) = U(X_1, X_2, \ldots, X_t)$$
$$- X_j P_{rj} - X_k P_{rk} - \cdots$$
$$(10.3.73)$$

$j, k, \ldots$, being the reservoir labels. This represents a generalization of the availability function defined in Section 7.2.

Show that the work done on the system in any process satisfies $W \geqslant \Delta A$, where the equality holds when the process is reversible. Here W includes the work done on any auxiliary devices which may be necessary to couple the system to the reservoirs to ensure the process is executed reversibly, as in Fig. 7.4, for example. Notice that if Σ is in equilibrium with the reservoirs the availability function, A, reduces to the Legendre transform of U, as in equation (10.3.29).

10.4 Derivative relations

The Maxwell relations

The energy fundamental relation for a simple system, $U(S, V, N_1, \ldots, N_r)$, and its derivatives are continuous functions of the independent variables. In this situation the mixed second-order derivatives of U are of particular interest, because the order in which they are determined is irrelevant (see Appendix D). Thus, for instance,

$$\frac{\partial}{\partial V}\left(\frac{\partial U}{\partial S}\right) = \frac{\partial}{\partial S}\left(\frac{\partial U}{\partial V}\right). \tag{10.4.1}$$

On substituting for the first derivatives in terms of T and P, we have

$$\frac{\partial}{\partial V}\left(\frac{\partial U}{\partial S}\right) = \left(\frac{\partial T}{\partial V}\right)_{S,N_j} \quad \text{and} \quad \frac{\partial}{\partial S}\left(\frac{\partial U}{\partial V}\right) = -\left(\frac{\partial P}{\partial S}\right)_{V,N_j},$$

and hence

$$\left(\frac{\partial T}{\partial V}\right)_{S,N_j} = -\left(\frac{\partial P}{\partial S}\right)_{V,N_j}. \tag{10.4.2}$$

This is an example of a number of such thermodynamic equalities, known as the *Maxwell relations* (Maxwell 1872). All derive from the equality of the mixed second-order derivatives of thermodynamic potentials. Figure 10.4 illustrates a method for writing the Maxwell relations by inspection starting from the relevant Gibbs equation.

The Maxwell relations provide a useful link between the derivatives of thermodynamic variables because they allow quantities which may be inaccessible to direct measurement to be expressed in terms of parameters

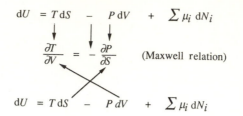

Fig. 10.4 Illustrating the connection between the Gibbs equation for U and one of the Maxwell relations which derive from it.

which are measurable. The Maxwell relations are especially useful for eliminating the entropy from derivative equations. Other Maxwell relations may be obtained from other forms of the fundamental relation and the corresponding Gibbs equation. For example, starting with the Gibbs equation for F,

$$dF = -S\,dT - P\,dV + \sum_{i=1}^{r} \mu_i\,dN_i, \qquad (10.4.3)$$

and applying the procedure illustrated in Fig. 10.4, we have:

$$\left(\frac{\partial S}{\partial V}\right)_{T,N_j} = \left(\frac{\partial P}{\partial T}\right)_{V,N_j}, \qquad \text{where } j = 1, \ldots, r, \qquad (10.4.4)$$

$$\left(\frac{\partial S}{\partial N_i}\right)_{T,V,N_j} = -\left(\frac{\partial \mu_i}{\partial T}\right)_{V,N_i,N_j}, \qquad \text{where } j \neq i, \qquad (10.4.5)$$

$$\left(\frac{\partial P}{\partial N_i}\right)_{T,V,N_j} = -\left(\frac{\partial \mu_i}{\partial V}\right)_{T,N_i,N_j}, \qquad \text{where } j \neq i, \qquad (10.4.6)$$

$$\left(\frac{\partial \mu_k}{\partial N_i}\right)_{T,V,N_k,N_j} = \left(\frac{\partial \mu_i}{\partial N_k}\right)_{T,V,N_i,N_j}, \qquad \text{where } j \neq i, j \neq k, i \neq k.$$
$$(10.4.7)$$

Other Maxwell relations are obtained from the potentials H, G, and Ψ in a similar way. The following derive from H and G, respectively,

$$\left(\frac{\partial T}{\partial P}\right)_{S,N_j} = \left(\frac{\partial V}{\partial S}\right)_{P,N_j}, \qquad (10.4.8)$$

$$\left(\frac{\partial S}{\partial P}\right)_{T,N_j} = -\left(\frac{\partial V}{\partial T}\right)_{P,N_j}. \qquad (10.4.9)$$

In addition, from (9.5.17) and the Gibbs equations for the Legendre transforms of U_J, we obtain Maxwell relations for a body subject to a magnetic field. Here we take $\mathbf{B}_0$ and $\mathbf{I}$ to be collinear.

$$\left(\frac{\partial T}{\partial I}\right)_{S,V,N_j} = \left(\frac{\partial B_0}{\partial S}\right)_{V,I,N_j}, \qquad (10.4.10)$$

$$\left(\frac{\partial T}{\partial B_0}\right)_{S,V,N_j} = -\left(\frac{\partial I}{\partial S}\right)_{V,B_0,N_j}. \qquad (10.4.11)$$

These equations may be used when B_0 is identified by either Φ or J.

Manipulating derivatives

In order to evaluate a derivative in terms of measured quantities it may be necessary to express it in a new form, perhaps by using a Maxwell relation. For example, it is more convenient to measure the molar heat capacity of a solid at constant pressure rather than constant volume. On the other hand, it is usually simpler to calculate c_v instead of c_p. A relation between c_p and c_v is therefore required. Unfortunately it is rather easy to go round in circles when you try to establish such a thermodynamic relation in terms of a given set of measurable quantities, such as α, κ_T and c_p (see 7.3). There are a number of well-developed procedures (Margenau and Murphy 1956) which deal with this problem, but for many systems the following guidelines (Tisza 1966; Callen 1985) are sufficient.

1. If the derivative contains a thermodynamic potential, bring it to the numerator and eliminate the potential using a Gibbs equation. The chemical potential may be treated as the Gibbs potential of a system with $N = 1$.

2. If the derivative contains the entropy, bring it to the numerator and eliminate it if possible using a Maxwell relation. Otherwise write the derivative of S in the form $\partial S/\partial T$ which can be re-expressed as a heat capacity.

3. Bring V to the numerator and eliminate in terms of α or κ_T.

In the following we will use some standard properties of partial derivatives which have been summarized in Appendix D. We will also find it convenient to use the Jacobian method, a useful extension of the procedures for partial derivatives, which is also described in Appendix D.

Example

When the volume of a simple fluid system changes in a reversible adiabatic process the associated temperature change is detemined by the derivative $(\partial T/\partial V)_S$, taking N to be constant. Obtain an expression for this function in terms of c_v, α, and κ_T. (See (7.3.9), (7.3.11), and (7.3.12).)

Solution

The derivative contains no potentials, so we proceed to the second rule. First bring S to the numerator using the quotient rule, (D. 16),

$$\left(\frac{\partial T}{\partial V}\right)_S = -\frac{\left(\frac{\partial S}{\partial V}\right)_T}{\left(\frac{\partial S}{\partial T}\right)_V}.$$

The denominator is not of the form of a Maxwell relation. Instead we eliminate the entropy in terms of c_v using

$$\left(\frac{\partial S}{\partial T}\right)_V = \frac{Nc_v}{T}. \qquad (10.4.12)$$

We can eliminate the entropy from the numerator by applying the Maxwell relation (10.4.4)

$$\left(\frac{\partial S}{\partial V}\right)_T = \left(\frac{\partial P}{\partial T}\right)_V,$$

which can be further reduced by bringing V to the numerator under the third rule,

$$\left(\frac{\partial P}{\partial T}\right)_V = -\frac{\left(\frac{\partial V}{\partial T}\right)_P}{\left(\frac{\partial V}{\partial P}\right)_T} = \frac{\alpha}{\kappa_T}. \qquad (10.4.13)$$

Hence we find

$$\left(\frac{\partial T}{\partial V}\right)_S = -\frac{T\alpha}{Nc_v\kappa_T}. \qquad (10.4.14)$$

Example

When a fluid undergoes an adiabatic throttling process (see Section 5.6) the enthalpy is constant. The temperature change associated with this process is represented by the Joule–Thomson coefficient which is defined as

$$\eta = \left(\frac{\partial T}{\partial P}\right)_H. \qquad (10.4.15)$$

Here, and in the following, assume N is fixed. Obtain an expression for η.

Solution

Apply the first rule by bringing H to the numerator, using (D. 16),

$$\eta = \left(\frac{\partial T}{\partial P}\right)_H = -\frac{\left(\dfrac{\partial H}{\partial P}\right)_T}{\left(\dfrac{\partial H}{\partial T}\right)_P}.$$

Now,

$$\left(\frac{\partial H}{\partial T}\right)_P = Nc_p$$

and from the Gibbs equation for H,

$$dH = T\,dS + V\,dP$$

we obtain

$$\left(\frac{\partial H}{\partial P}\right)_T = T\left(\frac{\partial S}{\partial P}\right)_T + V.$$

The entropy may now be eliminated using a Maxwell relation based on the Gibbs potential, eqn (10.4.9),

$$\left(\frac{\partial H}{\partial P}\right)_T = -T\left(\frac{\partial V}{\partial T}\right)_P + V = -TV\alpha + V,$$

and hence we obtain the required expression:

$$\eta = \frac{V}{Nc_p}\,(T\alpha - 1). \qquad (10.4.16)$$

Example

Establish the following equation for a simple fluid system:

$$c_p = c_v + \frac{TV\alpha^2}{N\kappa_T}. \qquad (10.4.17)$$

Solution

On the left-hand side, c_p involves S expressed as a function of T and P, whereas the independent variables are T and V on the right-hand side. The equation is therefore an expression of a change of variable, rather like (D.29). However, it is convenient to start from first principles and we consider first S as a function of T and V. This is not a fundamental relation, but T and V are sufficient to identify the thermodynamic state and S is therefore an exact function of these variables. We have

$$dS = \left(\frac{\partial S}{\partial T}\right)_V dT + \left(\frac{\partial S}{\partial V}\right)_T dV$$

and hence

$$\left(\frac{\partial S}{\partial T}\right)_P = \left(\frac{\partial S}{\partial T}\right)_V + \left(\frac{\partial S}{\partial V}\right)_T \left(\frac{\partial V}{\partial T}\right)_P.$$

Multiplying by T and substituting for c_p and c_v,

$$Nc_p = Nc_v + T\left(\frac{\partial S}{\partial V}\right)_T \left(\frac{\partial V}{\partial T}\right)_P. \tag{10.4.18}$$

Now,

$$\left(\frac{\partial V}{\partial T}\right)_P = V\alpha,$$

and, using the Maxwell relation, (10.4.4), together with (10.4.13),

$$\left(\frac{\partial S}{\partial V}\right)_T = \left(\frac{\partial P}{\partial T}\right)_V = \frac{\alpha}{\kappa_T}. \tag{10.4.19}$$

Equation (10.4.17) follows on substituting in (10.4.18) and rearranging.

Example

Consider a single-component simple system. Suppose we express U as a function of S, V, and N, taking N to be constant. Products such as

$$U_{SS}U_{VV} - (U_{SV})^2 = \left(\frac{\partial^2 U}{\partial S^2}\right)\left(\frac{\partial^2 U}{\partial V^2}\right) - \left(\frac{\partial^2 U}{\partial S \partial V}\right)^2$$

arise in the theory of stability of thermodynamic systems (12.2). Express $U_{SS}U_{VV} - (U_{SV})^2$ in terms of c_v and κ_T.

Solution

The expression may be reduced using the Jacobian method (Appendix D). This is suggested by the form of the equation, especially when we express the first derivatives of U in terms of T and P:

$$U_{SS}U_{VV} - (U_{SV})^2 = -\left(\frac{\partial T}{\partial S}\right)_V\left(\frac{\partial P}{\partial V}\right)_S + \left(\frac{\partial P}{\partial S}\right)_S\left(\frac{\partial T}{\partial V}\right)_S = -\frac{\partial(T, P)}{\partial(S, V)}.$$

The reduction of the Jacobian follows the same strategy set out for partial derivatives, since we aim to get T, P, and N as the independent variables. In this respect (D.33) and (D.34) are especially useful. First we convert the Jacobian to a product of partial derivatives, by introducing $\partial(T, V)$,

$$\frac{\partial(T, P)}{\partial(S, V)} = \frac{\partial(T, P)}{\partial(T, V)}\frac{\partial(T, V)}{\partial(S, V)} = \left(\frac{\partial P}{\partial V}\right)_T\left(\frac{\partial T}{\partial S}\right)_V = \frac{-T}{Nc_v V\kappa_T}.$$

Thus we find

$$U_{SS}U_{VV} - (U_{SV})^2 = \frac{T}{Nc_v V\kappa_T}. \qquad (10.4.20)$$

This relation may also be obtained by applying the rules for the manipulation of the partial derivatives, but it takes longer.

Questions

10.4.1.   For a simple fluid, demonstrate that

$$\left(\frac{\partial T}{\partial V}\right)_S = (1 - \gamma)\left(\frac{\partial T}{\partial V}\right)_P = \frac{(1 - \gamma)}{V\alpha}, \qquad (10.4.21)$$

where $\gamma = c_p/c_v$ and α is the volume expansivity. Hence, for a gas which satisfies $PV = NRT$, show that

$$\left(\frac{\partial T}{\partial V}\right)_S = (1 - \gamma)\frac{T}{V}, \qquad (10.4.22)$$

and confirm that this yields (3.5.7).

10.4.2.   The magnetization of a dilute paramagnetic salt is given by the Curie law:

$$M = \frac{nc_c B}{\mu_0 T},$$

where n is the molar concentration (N/V), $nc_c = c$ is the Curie constant, and B is the field in the absence of the medium, given the current of the source coil is fixed. Consider an adiabatic reversible process during which B changes. Show that the temperature and magnetic field are related by

$$\left(\frac{\partial T}{\partial B}\right)_S = \frac{c_c B}{\mu_0 T c_B}, \qquad (10.4.23)$$

where c_B is the molar heat capacity at constant B.

10.4.3.   Establish the following relations for a simple fluid system, taking N to be constant.

(a) $$\left(\frac{\partial T}{\partial P}\right)_S = \frac{TV\alpha}{Nc_p}. \qquad (10.4.24)$$

(b) $$\left(\frac{\partial T}{\partial P}\right)_V = \frac{\kappa_T}{\alpha}. \qquad (10.4.25)$$

(c) $$\left(\frac{\partial T}{\partial P}\right)_S = \frac{\gamma - 1}{\gamma}\left(\frac{\partial T}{\partial P}\right)_V. \qquad (10.4.26)$$

Here $\gamma = c_p/c_v$. Show that (10.4.26) reduces to (3.5.15) for an ideal gas.

10.4.4. Establish the following derivative relations for a simple fluid. Take N to be constant.

$$\left(\frac{\partial c_v}{\partial V}\right)_T = \frac{T}{N}\left(\frac{\partial^2 P}{\partial T^2}\right)_V \qquad (10.4.27)$$

$$\left(\frac{\partial c_p}{\partial P}\right)_T = \frac{T}{N}\left(\frac{\partial^2 V}{\partial T^2}\right)_P. \qquad (10.4.28)$$

10.4.5. By expressing S as a function of the coordinates (T, V), (P, T), and (P, V) demonstrate the following relations, known as the first, second, and third T-dS equations, respectively.

$$T\,dS = Nc_v\,dT + \frac{T\alpha}{\kappa_T}\,dV \qquad (10.4.29)$$

$$T\,dS = Nc_p\,dT - TV\alpha\,dP \qquad (10.4.30)$$

$$T\,dS = \frac{Nc_p}{V\alpha}\,dV + \frac{Nc_v\kappa_T}{\alpha}\,dP. \qquad (10.4.31)$$

In particular, show that

$$c_p - c_v = vT\kappa_T\left\{\left(\frac{\partial P}{\partial T}\right)_v\right\}^2. \qquad (10.4.32)$$

10.4.6. The Joule coefficient expresses the rate of temperature change with volume when a fluid undergoes a Joule free-expansion process. The definition is analogous to equation (10.4.15) for the Joule–Thomson coefficient:

$$\mu = \left(\frac{\partial T}{\partial V}\right)_U.$$

Show that

$$\mu = \frac{-1}{Nc_v}\left(\frac{T\alpha}{\kappa_T} - P\right). \qquad (10.4.33)$$

10.4.7. Show that

$$\left(\frac{\partial P}{\partial V}\right)_S = \frac{c_p}{V\kappa_T c_v},$$

and hence that

$$\kappa_T c_v = \kappa_s c_p. \qquad (10.4.34)$$

10.4.8. By expressing V as a function of T and P (cf. the T–dS equations, question 10.4.5), show that

$$\kappa_s = \kappa_T - \frac{TV\alpha^2}{Nc_p}.$$ (10.4.35)

10.4.9. (a) Establish the Jacobian relation,

$$\frac{\partial(T, S)}{\partial(P, V)} = 1.$$ (10.4.36)

(b) Let $F(T, V, N)$ be the Helmholtz free energy of a simple system. Take N to be constant and show that

$$\left(\frac{\partial^2 F}{\partial T^2}\right)\left(\frac{\partial^2 F}{\partial V^2}\right) - \left(\frac{\partial^2 F}{\partial T \partial V}\right)^2 = -\frac{c_p}{vT\kappa_T}.$$ (10.4.37)

(c) For a fixed quantity of fluid show that

$$\left(\frac{\partial^2 U}{\partial S^2}\right)\left(\frac{\partial^2 U}{\partial V^2}\right) - \left(\frac{\partial^2 U}{\partial S \partial V}\right)^2 = \frac{T}{Nc_p V\kappa_s}.$$ (10.4.38)

10.4.10. Equation (10.4.16) for the Joule–Thomson coefficient may be used to establish a relation between the empirical temperature, θ, determined by an uncorrected constant volume gas thermometer, and the thermodynamic temperature, T. (The difference between θ and T is due to non-ideal behaviour of the gas in the thermometer.)

(a) When the Joule–Thomson coefficient and molar heat capacity at constant pressure are measured using the thermometer we obtain not η and c_p, but

$$\eta' = \left(\frac{\partial\theta}{\partial P}\right)_H \quad \text{and} \quad c_p' = \frac{1}{N}\left(\frac{\partial H}{\partial\theta}\right)_P,$$

instead. Show that

$$\eta' c_p' = -\frac{1}{N}\left(\frac{\partial H}{\partial P}\right)_\theta,$$ (10.4.39)

and hence, provided θ and T are in a one–one relationship, $\eta' c_p' = \eta c_p$.

(b) In addition, instead of measuring α we obtain α' given by

$$\alpha' = \frac{1}{V}\left(\frac{\partial V}{\partial\theta}\right)_P.$$

Show that

$$\alpha = \alpha' \left(\frac{\partial \theta}{\partial T} \right)_P. \tag{10.4.40}$$

(c) Since the empirical temperature is given by $Pv = \text{const} \cdot \theta$, it follows that

$$\left(\frac{\partial P}{\partial \theta} \right)_v = \frac{P}{\theta}.$$

Hence show that

$$\alpha' = \frac{P}{\theta} \kappa_T. \tag{10.4.41}$$

(d) Use (10.4.16) together with (10.4.39)–(10.4.41) to establish the following relation between T and θ, employing only measurable quantities:

$$\eta' c_p' = \frac{V}{N} \left\{ \frac{TP\kappa_T}{\theta} \left(\frac{\partial \theta}{\partial T} \right)_P - 1 \right\}. \tag{10.4.42}$$

Part IV
Applications

11

Equations of state

11.1 Finding the fundamental relation

An equation which links the thermodynamic parameters of a system most accessible to measurement, such as P, v, and T, is normally called an *equation of state*. For a simple fluid system the equation of state typically has the form,

$$\phi(P, v, T) = 0. \tag{11.1.1}$$

We shall distinguish two uses for the term *equation of state*. First, it may represent a real equation, like the ideal gas law, $Pv = RT$. Such an equation is effectively a thermodynamic model for the system. It would not be surprising to find that a particular model equation is reasonable under some conditions but is unacceptable under others. We shall normally assume that model equations of state are well-behaved continuous functions of the independent variables with continuous derivatives, even though the equation may not be physically valid for all states.

Second, the term *equation of state* may be used to denote the actual equilibrium property relation for a system. We assume that a real equation can always be established for states of equilibrium, but we should not expect such equations always to be continuous with continuous derivatives. For example, phase changes are directly associated with discontinuities in the equations of state of a system.

Where there is a need to distinguish these terms, we shall refer to the first as a *model* equation of state and the second as an *equilibrium* equation of state.

An equation of state is not normally a fundamental relation in the usual sense. But the fundamental relation of a system is of primary interest since it represents the thermodynamic properties in a systematic way. It is therefore important to establish connections between the equation of state of a system and its fundamental relations.

Consider a simple fluid system in which the equation of state has T and v as the independent variables. Given these variables the natural form of the fundamental relation will be the Helmholtz potential, F, or $f = F/N$. Here we shall deal in the molar quantities for simplicity. Let s_0 and u_0, respectively, be the molar entropy and molar energy at temperature T_0 and molar volume v_0. It is convenient to establish u and s as functions of T

and v first, and then to find f. We calculate $u - u_0$ in two steps using

$$u(T, v) - u(T_0, v_0) = u(T, v) - u(T, v_0) + u(T, v_0) - u(T_0, v_0).$$
(11.1.2)

Here,

$$u(T, v) - u(T, v_0) = \int_{v_0}^{v} \left(\frac{\partial u(T, v')}{\partial v'} \right)_T dv'.$$
(11.1.3)

Now since $du = T\, ds - P\, dv$ we have

$$\left(\frac{\partial u}{\partial v} \right)_T = T \left(\frac{\partial s}{\partial v} \right)_T - P = T \left(\frac{\partial P}{\partial T} \right)_v - P,$$
(11.1.4)

where we have used (10.4.4), a Maxwell relation. Hence,

$$u(T, v) - u(T, v_0) = \int_{v_0}^{v} \left\{ T \left(\frac{\partial P(T, v')}{\partial T} \right)_{v'} - P(T, v') \right\} dv'.$$
(11.1.5)

In addition,

$$u(T, v_0) - u(T_0, v_0) = \int_{T_0}^{T} \left(\frac{\partial u(T', v_0)}{\partial T'} \right)_{v_0} dT'$$
(11.1.6)

$$= \int_{T_0}^{T} c_v(T', v_0)\, dT',$$
(11.1.7)

where c_v is the molar heat capacity at constant V. Thus we obtain an expression for the internal energy using the equation of state, $P(T, v)$, and the molar heat capacity, c_v, expressed as a function of T at volume, v_0:

$$u(T, v) = u(T_0, v_0) + \int_{T_0}^{T} c_v(T', v_0)\, dT'$$

$$+ \int_{v_0}^{v} \left\{ T \left(\frac{\partial P(T, v')}{\partial T} \right)_{v'} - P(T, v') \right\} dv'.$$
(11.1.8)

This shows that when $P \propto T$, $u(T, v)$ will be independent of v, as happens in an ideal gas.

Similarly, the entropy can be expressed as

$$s(T, v) - s(T_0, v_0) = s(T, v) - s(T, v_0) + s(T, v_0) - s(T_0, v_0)$$
(11.1.9)

$$= \int_{v_0}^{v} \left(\frac{\partial s(T, v')}{\partial v'} \right)_T dv' + \int_{T_0}^{T} \left(\frac{\partial s(T', v_0)}{\partial T'} \right)_{v_0} dT'.$$
(11.1.10)

Now the first term is transformed using the Maxwell relation (10.4.4) and the second is simply given by

$$\left(\frac{\partial s}{\partial T}\right)_v = \frac{c_v}{T}.\qquad(11.1.11)$$

Hence,

$$s(T, v) = s(T_0, v_0) + \int_{v_0}^{v} \left(\frac{\partial P(T, v')}{\partial T}\right)_{v'} dv' + \int_{T_0}^{T} \frac{c_v(T', v_0)}{T'} dT'.\quad(11.1.12)$$

The Helmholtz free energy is now obtained from eqns (11.1.8) and (11.1.12)

$$f(T, v) = u(T_0, v_0) - Ts(T_0, v_0) + \int_{T_0}^{T} c_v(T', v_0)\, dT'$$

$$- T \int_{T_0}^{T} \frac{c_v(T', v_0)}{T'} dT' - \int_{v_0}^{v} P(T, v')\, dv'.\qquad(11.1.13)$$

Thus the equation of state and the molar heat capacity measured at one volume are sufficient to establish the free energy. This emphasizes the importance of the equation of state.

Questions

11.1.1. (a) Show that

$$c_v(T, v) = c_v(T, v_0) + T \int_{v_0}^{v} \left(\frac{\partial^2 P(T, v')}{\partial T^2}\right)_{v'} dv'.\quad(11.1.14)$$

(b) Obtain an expression for $u(T, v) - u(T_0, v_0)$ using

$$u(T, v) - u(T_0, v_0) = u(T, v) - u(T_0, v) + u(T_0, v)$$
$$- u(T_0, v_0),$$

and show that the final expression agrees with (11.1.8). Demonstrate that the analogous equation for the entropy yields eqn (11.1.12).

11.1.2. (a) Starting with the Gibbs equation, $df = -s\,dT - P\,dv$, show that

$$f(T, v) = f(T, v_0) - \int_{v_0}^{v} P(T, v')\, dv'\qquad(11.1.15)$$

$$= f(T_0, v_0) - \int_{T_0}^{T} s(T', v_0)\, dT' - \int_{v_0}^{v} P(T, v')\, dv'$$

$$(11.1.16)$$

where $f(T_0, v_0) = u_0 - T_0 s_0$.

(b) Show that (11.1.16) reduces to (11.1.13). (Hint: integrate the entropy term by parts and use (11.1.12).)

11.1.3. Demonstrate that the molar Gibbs function of a fluid is given in terms of its natural variables, T and P, by

$$g(T, P) = h(T_0, P_0) - Ts(T_0, P_0) + \int_{T_0}^{T} c_p(T', P_0) \, dT'$$

$$- T \int_{T_0}^{T} \frac{c_p(T', P_0)}{T'} \, dT' + \int_{P_0}^{P} v(T, P') \, dP'.$$

$$(11.1.17)$$

Note that $g(T, P) = \mu(T, P)$ for a system consisting of a single chemical component.

11.2 Equations of state for fluids

The first equation of state to provide a reasonable model for both liquid and gas properties was proposed by van der Waals:

$$\left(P + \frac{a}{v^2}\right)(v - b) = RT,$$

$$(11.2.1)$$

where $v = V/N$ is the molar volume and a and b are constants. Van der Waals established the equation in 1873 on the basis of the kinetic theory of gases and the theory of capillary attraction (Rowlinson 1988). Qualitatively the term a/v^2 represents the effect of long-range attractive molecular forces, which tend to increase the density of the gas for a given external pressure. The term b represents the influence of short-range repulsive forces which effectively reduce the available volume at high density. The significance of the equation is that it provides an illustration of real fluid phenomena, such as phase change and critical point behaviour, within a relatively simple analytic form. Its continuing value is that it represents a prototype for a class of theories of the fluid state, called mean field theories. Thus the van der Waals equation provides a point of reference for theories of fluid phenomena. However, the equation is generally not satisfactory for quantitative calculations of real gas properties.

The gross features of the van der Waals equation are illustrated in Fig. 11.1, which shows isotherms for argon in P–v coordinates. Very small values of v (that is, $v < b$) have been excluded and the constants a and b have been selected to reproduce the critical point; see question 11.2.3. The isotherms have two forms, being monotonic at high temperature, but at low temperature there are two turning points, d and b. These are associated with a region of instability in the equation of state, which is related to the liquid–vapour phase change. The fact that the region between d and b is unstable is intuitively obvious from the fact that P is an increasing function

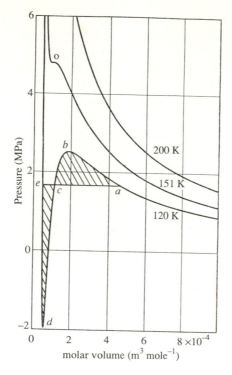

Fig. 11.1 Isotherms for a van der Waals model of argon. The critical point, o, is the point of inflection in the critical isotherm (151 K). The states lying on the isotherm between the points *a* and *e* are unstable or metastable. Instead, the equilibrium states lie on the horizontal line *a–c–e*. The position of the line *a–c–e* is determined by Maxwell's equal area rule, which requires that the shaded areas a–b–c–a and c–d–e–c should be equal.

of v. Thus the van der Waals equation is an example of a model which is analytically well behaved whilst having regions in which it cannot represent equilibrium states.

Van der Waals showed that the liquid–vapour phase change can be represented by replacing the isotherm, *a–b–c–d–e* by the horizontal line as illustrated. In this region the stable states are inhomogeneous mixtures of the liquid and vapour phases. State *a* is said to be the saturated vapour state; state *e* is the saturated liquid state. However, the equation of van der Waals itself does not directly define the pressure of the phase change for a given isotherm.

This difficulty was resolved by Maxwell who showed that the pressure should be selected so that the areas of the loops cut off the isotherms by the line *a–c–e* should be equal. This is known as *Maxwell's equal area rule*. We can understand this rule on the basis of the second law by considering a hypothetical thermodynamic cycle passing through the states a–b–c–d–e–c–a, shown in Fig. 11.1. Since the cycle is isothermal, the net work done, $\int P \, dv$, must be zero, in which case the areas of the loops will be equal. Of course, this argument may be criticized because it treats the unstable states along the path b–c–d as if they were physically real. On the other hand, the equal area rule applies to the model, not the equilibrium equation of state. The question of the intrinsic stability of phases is discussed in Section 12.2. In Section 12.4 we shall show that the equal area rule applies to any model equation, provided it is a well-behaved function of its variables.

The boundary between the single-phase isotherms and those exhibiting the phase change instability is defined by the critical isotherm. The critical state (labelled o in Fig. 11.1) is the point on this isotherm which is also on the boundary of the mixed-phase region. This state is identified by being the point of inflection for which the gradient of the isotherm is zero. Thus in a van der Waals gas the critical state is defined by the conditions

$$\left(\frac{\partial P}{\partial v}\right)_T = 0, \tag{11.2.2}$$

$$\left(\frac{\partial^2 P}{\partial v^2}\right)_T = 0. \tag{11.2.3}$$

These criteria are not sufficient to identify the critical point in general, however. To illustrate, we note that the critical isotherm is actually much flatter at the critical point than the van der Waals equation indicates. Now the van der Waals equation is a cubic in v, as we can see by writing (11.2.1) as

$$Pv^3 - (bP + RT)v^2 + av - ab = 0. \tag{11.2.4}$$

It turns out that more realistic models of the isotherms can be achieved by using higher-order equations of state. The intrinsically unstable region then contains more turning points and the critical point is then characterized by requiring higher derivatives to be zero.

As examples of higher-order equations of state we briefly note the *virial equations* of which there are two forms:

$$\frac{Pv}{RT} = 1 + \frac{B}{v} + \frac{C}{v^2} + \frac{D}{v^3} + \cdots \tag{11.2.5}$$

and

$$\frac{Pv}{RT} = 1 + B'P + C'P^2 + D'P^3 + \cdots. \tag{11.2.6}$$

Here the coefficients, B, C, ..., and B', C', ..., called the *virial coefficients*, are functions of T. The virial equations are of particular interest because the coefficients can be calculated from the interatomic potential function (for example, see Reif (1965)). They are useful for estimating small deviations from ideal gas behaviour, in which case the second virial coefficient, B or B', may suffice.

No model equation of state is entirely satisfactory over a wide range of temperature and pressure, especially near the critical point where the thermodynamic representation fails due to large-scale fluctuations. On the other hand, for fluids which have important engineering uses, water for example, comprehensive empirical equations of state have been developed for computational purposes (Appendix B). When only limited data exists for a fluid, relatively simple cubic equations are often useful for estimating its thermodynamic properties in the gas and liquid phase (Reid *et al.* 1986).

Questions

11.2.1. (a) Show that the virial coefficients for the volume and pressure virial equations are related by

$$B' = B/RT, \qquad (11.2.7)$$

$$C' = \frac{C - B^2}{R^2 T^2} . \qquad (11.2.8)$$

(b) The Boyle line for a gas is the locus of states such that

$$\left(\frac{\partial (Pv)}{\partial P} \right)_T = 0. \qquad (11.2.9)$$

Thus Pv is independent of P, being dependent on T only, so the gas then satisfies Boyle's law. Show that the second virial coefficient, $B = 0$, on the Boyle line.

(c) At the inversion temperature the Joule–Thomson coefficient of a gas, given by (10.4.16), is zero. Show that the inversion temperature can be expressed in terms of the second virial coefficient by

$$\frac{dB}{dT} = \frac{B}{T} . \qquad (11.2.10)$$

11.2.2. (a) Show that the molar heat capacity at constant volume is independent of V for a van der Waals gas.

(b) Assume that c_v is independent of T for a certain van der Waals gas. Obtain u and s as functions of T and v. (Lower case symbols denote molar quantities.) Hence show that in an adiabatic free-expansion process,

$$\left(\frac{\partial T}{\partial v} \right)_u = -\frac{a}{c_v v^2} , \qquad (11.2.11)$$

and in an adiabatic reversible process,

$$T(v - b)^{R/c_v} = \text{const.} \tag{11.2.12}$$

(c) Show that the Helmholtz free energy of a van der Waals gas is given by

$$F(T, V, N) = Nu_0 - NTs_0 + Nc_v \left\{ T - T_0 - T \ln\left(\frac{T}{T_0}\right) \right\}$$

$$- Na\left(\frac{1}{v} - \frac{1}{v_0}\right) - NRT\ln\left(\frac{v - b}{v_0 - b}\right). \tag{11.2.13}$$

Here u_0 and s_0 denote the molar energy and entropy at the state identified by (T_0, v_0).

11.2.3. (a) Show that the pressure, molar volume, and temperature of a van der Waals gas at the critical point are given, respectively, by

$$P_c = \frac{a}{27b^2}, \qquad v_c = 3b, \qquad T_c = \frac{8a}{27Rb}. \tag{11.2.14}$$

(b) The molar volume of a gas of non-interacting atoms is v_c and the temperature is T_c. Demonstrate that the pressure of a gas is given by

$$P = \frac{8P_c}{3}. \tag{11.2.15}$$

(c) Define the reduced pressure, reduced molar volume, and reduced temperature by

$$P_r = \frac{P}{P_c}, \qquad v_r = \frac{v}{v_c}, \qquad T_r = \frac{T}{T_c}. \tag{11.2.16}$$

Show that the equation of state for a van der Waals gas can be expressed in terms of the reduced variables as

$$P_r = \frac{8T_r}{3v_r - 1} - \frac{3}{v_r^2}. \tag{11.2.17}$$

Thus any gas which conforms to the van der Waals equation may be represented by equation (11.2.17). The equation is an example of the *principle of corresponding states* (Guggenheim 1967), according to which relations involving P, v and T should have the same form for all gases when expressed in reduced variables. It holds particularly well in the gas and saturated regions for a number of gases, such as the heavier noble gases, as well as O_2 and N_2. The critical temperature of H_2 was correctly estimated on the basis of the principle of corresponding states ten years before the gas was first liquefied (Rowlinson 1988).

11.2.4.   (a)   Denote the reduced variables (eqn 11.2.16) for the Boyle line (eqn (11.2.9)) by P_{br}, v_{br}, and T_{br}. Show that the equations for the Boyle line of a van der Waals gas are

$$P_{br} = \frac{3(3v_{br} - 2)}{v_{br}^2}, \qquad T_{br} = \frac{3(3v_{br} - 1)^2}{8v_{br}^2}. \qquad (11.2.18)$$

(b)   Show that when v is fixed,

$$T_r \gtrless T_{br} \Rightarrow \left(\frac{\partial(Pv)}{\partial P}\right)_T \lessgtr 0. \qquad (11.2.19)$$

11.2.5.   (a)   Define the reduced isothermal compressibility of a gas by

$$\kappa_{Tr} = -\frac{1}{v_r}\left(\frac{\partial v_r}{\partial P_r}\right)_{T_r}, \qquad (11.2.20)$$

where P_r and v_r are the reduced pressure and reduced molar volume, respectively. For a van der Waals gas show that

$$\kappa_T = \frac{27b^2}{a}\kappa_{Tr}, \qquad (11.2.21)$$

and

$$\kappa_{Tr} = \frac{v_r^2(3v_r - 1)^2}{6(4v_r^3 T_r - (3v_r - 1)^2)}. \qquad (11.2.22)$$

(b)   Define the reduced volume expansivity of a gas by

$$\alpha_r = \frac{1}{v_r}\left(\frac{\partial v_r}{\partial T_r}\right)_{P_r}. \qquad (11.2.23)$$

Show that

$$\alpha = \frac{27Rb}{8a}\alpha_r, \qquad (11.2.24)$$

for a van der Waals gas, and that

$$\alpha_r = \frac{8\kappa_{Tr}}{3v_r - 1}. \qquad (11.2.25)$$

Note that both α and κ_T diverge at the critical point; $\kappa_T \rightarrow \infty$ as a trivial consequence of (11.2.2) and the divergence of α is in accord with observation. However (see section 12.5) the van der Waals expressions for α and κ_T in the neighbourhood of the critical point are inconsistent with the behaviour of real gases.

11.2.6. (a) Show that the inversion line (question 11.2.1) for a van der Waals gas is given by

$$T_r = \frac{3(3v_r - 1)^2}{4v_r^2}, \qquad P_r = \frac{18}{v_r} - \frac{9}{v_r^2}. \qquad (11.2.26)$$

The reduced variables T_r, v_r, and P_r are defined by (11.2.16).

(b) For a given temperature, show that the Joule–Thomson coefficient satisfies $\eta > 0$ for pressures higher than the inversion pressure, and $\eta < 0$ below the inversion pressure.

(c) Show that the maximum inversion pressure is $9P_c$ and is obtained at $T = 3T_c$.

11.2.7. (a) For a van der Waals gas show that the molar heat capacities at constant volume and pressure, c_v and c_p, are related by

$$c_p - c_v = \frac{R}{\left\{ 1 - \dfrac{(3v_r - 1)^2}{4v_r^3 T_r} \right\}}, \qquad (11.2.27)$$

where v_r and T_r, the reduced volume and temperature, are given by (11.2.16). (Hint: see (10.4.17).)

(b) Show that $c_p \to \infty$ at the critical point of a van der Waals gas.

11.3   The Grüneisen constant

The Grüneisen constant originally derived from the observation that the ratio, α/c_p, is almost independent of temperature for many metals (Roberts 1940). Subsequently a related result was established on the basis of the Debye theory of solids (question 10.2.4). This is now used to identify the Grüneisen constant, rather than the original ratio, α/c_p. According to the Debye–Grüneisen theory (Rosenberg 1963), the molar entropy of a solid can be written as

$$s = s(T/T_D), \qquad (11.3.1)$$

where T_D, the *Debye temperature,* is a function of v, the molar volume. This function may be characterized by the Grüneisen constant, γ_g, given by

$$\gamma_g = -\frac{v}{T_D} \frac{dT_D}{dv}. \qquad (11.3.2)$$

We shall now show that γ_g is closely related to the original ratio of Grüneisen. First, consider the derivative, $(\partial T/\partial v)_s$. By the quotient rule, eqn (D.16), we have

$$\left(\frac{\partial T}{\partial v}\right)_s = -\frac{\left(\dfrac{\partial s}{\partial v}\right)_T}{\left(\dfrac{\partial s}{\partial T}\right)_v}, \tag{11.3.3}$$

and writing $s = s(x)$, where $x = T/T_D$, we obtain

$$\left(\frac{\partial s}{\partial v}\right)_T = \frac{ds}{dx}\left(\frac{\partial x}{\partial v}\right)_T = \frac{ds}{dx}\frac{dx}{dT_D}\frac{dT_D}{dv} = \frac{ds}{dx}\frac{\gamma_g T}{v T_D}, \tag{11.3.4}$$

$$\left(\frac{\partial s}{\partial T}\right)_v = \frac{ds}{dx}\left(\frac{\partial x}{\partial T}\right)_v = \frac{ds}{dx}\frac{1}{T_D}. \tag{11.3.5}$$

Hence,

$$\left(\frac{\partial T}{\partial v}\right)_s = -\frac{\gamma_g T}{v}. \tag{11.3.6}$$

Now $(\partial T/\partial v)_s$ is given by (10.4.14), re-expressed using the molar volume, as

$$\left(\frac{\partial T}{\partial v}\right)_s = -\frac{T\alpha}{c_v \kappa_T}. \tag{11.3.7}$$

While this relation has been obtained for a fluid, it is also applicable to an isotropic solid subject to uniform fluid pressure. Thus, from (11.3.6) and (11.3.7) we have

$$\gamma_g = \frac{v\alpha}{c_v \kappa_T}. \tag{11.3.8}$$

In the following we shall take this relation rather than (11.3.2) to be the definition of γ_g. Hence we may obtain the Grüneisen constant for systems in which the Debye–Grüneisen model of a solid is inapplicable. For many solids, metals especially, γ_g is constant over a wide temperature range, the value being typically between 1 and 3. For instance, at room temperature γ_g is 2.7, 2.0, and 2.4 for lead, copper, and silver, respectively (Rosenberg 1963). The Grüneisen constant is a useful parameter for characterizing the properties of fluids and solids at high temperatures and pressures.

Questions

11.3.1.   Assume that γ_g as defined in (11.3.8) is a constant. Show that the equation for the molar entropy must have the form

$$s = s(y), \tag{11.3.9}$$

where $y = Tv^{\gamma_g}$. (Hint: (11.3.6) must hold. If s were also a function of another variable then (11.3.6) would fail.)

11.3.2. Given the definition (11.3.8), show that the Grüneisen constant is given by

$$\gamma_g = v \left(\frac{\partial P}{\partial u} \right)_v, \tag{11.3.10}$$

where u is the molar energy. Consider a simple system in which γ_g is constant. Show that the system has the following equation of state:

$$Pv = \gamma_g u + \phi(v), \tag{11.3.11}$$

where ϕ is a function of v. This is known as the *Mie–Grüneisen equation of state*.

11.3.3. An *ideal quantum gas* is a simple system for which the equation of state has the form (Landsberg 1961),

$$Pv = \gamma_g u, \tag{11.3.12}$$

where γ_g is a constant.

(a) Show that γ_g satisfies the definition for the Grüneisen constant, eqn (11.3.8). This means that (11.3.9) must hold for an ideal quantum gas and therefore in an isentropic process,

$$y = Tv^{\gamma_g} = \text{const}. \tag{11.3.13}$$

(b) Show that $\gamma_g = 2/r$ for a classical ideal gas of structureless particles in r dimensions; see Section 9.3, example (10).

(c) Show that $\gamma_g = 1/3$ for a photon gas and for a relativistic ideal gas; see Section 9.3, examples (3), (11), and (12).

(d) For a van der Waals gas show that

$$\gamma_g = \frac{Rv}{c_v(v - b)}. \tag{11.3.14}$$

11.3.4. For a simple fluid system show that $\gamma = c_p/c_v$ is given by

$$\gamma = 1 + T\alpha\gamma_g, \tag{11.3.15}$$

where γ_g is the Grüneisen constant and α is the volume expansivity. Show that $\gamma = 1 + \gamma_g$ given that the ideal gas equation, $Pv = RT$, also holds. (Note that $\gamma = \kappa_T/\kappa_s$; see eqn (7.3.14).)

11.3.5. Equation (11.3.9) shows that Tv^{γ_g} is constant for an isentropic process in a simple system for which γ_g is constant. Show that if the system is also an ideal quantum gas (see question 11.3.3) then

$$Pv^{1+\gamma_g} = \text{const} \tag{11.3.16}$$

and

$$P^{-\gamma_g}T^{1+\gamma_g} = \text{const.} \tag{11.3.17}$$

(Hint: begin with the Gibbs equation for u.)

11.4   High-pressure processes

Many processes of interest in geophysics take place at very high pressure. At the centre of the earth for instance the pressure is in the region of 3.5×10^{11} Pa. These pressures are at the limit of laboratory-based studies employing diamond-anvil high-pressure cells (Jayaraman 1986). Although the detailed description of solid deformation processes (McLellan 1980) is outside the scope of this book, many systems subject to high hydrostatic pressures may be treated as if they were simple fluids. A useful equation of state at such pressures is the Murnaghan equation which is based on the assumption that the isothermal bulk modulus, B_T, is a linear function of pressure at constant temperature. Recollect that

$$B_T = -V\left(\frac{\partial P}{\partial V}\right)_T = \frac{1}{\kappa_T}, \tag{11.4.1}$$

where κ_T is the isothermal compressibility. We shall assume B_T is a function of P and T of the form

$$B_T(P, T) = B_T(0, T) + \beta_T P, \tag{11.4.2}$$

where β_T is the isothermal pressure derivative of B_T evaluated at $P = 0$:

$$\beta_T = \left(\frac{\partial B_T(0, T)}{\partial P}\right)_T. \tag{11.4.3}$$

The *Murnaghan equation of state* follows from (11.4.2) and (11.4.3):

$$P(V, T) = \frac{B_T(0, T)}{\beta_T}\left\{\left(\frac{V_0}{V}\right)^{\beta_T} - 1\right\}. \tag{11.4.4}$$

Here V_0 is the volume at very low pressure, effectively at $P = 0$. Typically the equation holds for solids subject to volume changes of up 50%, provided the system does not undergo a change of phase. The Murnaghan equation highlights the convenience of the bulk modulus, rather than the compressibility, as a thermodynamic parameter in high-pressure processes. However, the equation is restricted to isothermal processes only.

To characterize the temperature dependence of the bulk modulus, the *Anderson–Grüneisen constant*, δ_a, a dimensionless ratio, is useful:

$$\delta_a = -\frac{1}{\alpha B_T}\left(\frac{\partial B_T}{\partial T}\right)_P, \tag{11.4.5}$$

where α is the volume expansivity. The value of δ_a is typically in the range 3–7 for materials of geophysical interest. Guidelines such as this may be used to decide whether a particular approximation is acceptable, or if uncertain numerical data is reasonable. In this context the *Dugdale–MacDonald equation* (Dugdale and MacDonald 1953), an approximate relation between the Grüneisen constant and β_T, is also valuable:

$$\beta_T \approx 2\gamma_g + 1. \tag{11.4.6}$$

Example

Sodium is a particularly light metal which has comparatively large values for the volume expansivity and isothermal compressibility. The following parameters, taken from Anderson (1966) and Kaye and Laby (1959), have been determined at $T = 295$ K, $P = 10^5$ Pa.

$$\alpha = 2.01 \times 10^{-4} \, \text{K}^{-1}, \qquad\qquad \left(\frac{\partial \alpha}{\partial T}\right)_P = 9.0 \times 10^{-8} \, \text{K}^{-2},$$

$$B_T = 6.18 \times 10^9 \, \text{Pa}, \qquad\qquad \beta_T = 3.59,$$

$$\left(\frac{\partial B_T}{\partial T}\right)_P = -3.93 \times 10^6 \, \text{Pa K}^{-1}, \qquad c_p = 28.41 \, \text{J mol}^{-1}\text{K}^{-1},$$

$$v = 2.37 \times 10^{-5} \, \text{m}^3 \, \text{mol}^{-1}, \qquad\qquad \text{molar mass} = 2.299 \times 10^{-2} \, \text{kg mol}^{-1}.$$

(a) Determine c_v, the Grüneisen constant, and the Anderson–Grüneisen constant.

(b) Calculate the change in the volume expansivity, and in the bulk modulus, when the pressure increases from 10^5 Pa to 10^8 Pa isothermally.

(c) The pressure on a body consisting of 100 g of sodium increases from 10^5 Pa to 10^8 Pa quasistatically. Determine the volume change, given that the process is carried out (i) isothermally at 295 K, and (ii) adiabatically starting at 295 K. In the latter case calculate the final temperature.

(d) Obtain the heat and work input in these processes.

Solution (a)

We use (10.4.17) to determine c_v:

$$c_v = c_p - Tv\alpha^2 B_T = 28.41 - 1.75 = 26.66 \, \text{J mol}^{-1}\text{K}^{-1}$$

and from (11.3.8),

$$\gamma_g = \frac{v\alpha B_T}{c_v} = 1.10.$$

Note that $2\gamma_g + 1 = 3.2$, which is reasonably close to the value given for β_T, as we would expert from (11.4.6).

The Anderson–Grüneisen constant is obtained directly from the definition (11.4.5)

$$\delta_a = \frac{3.93 \times 10^6}{2.01 \times 10^{-4} \times 6.18 \times 10^9} = 3.16.$$

Solution (b)

In this process the maximum pressure is relatively small. Consequently we may use the following linear approximations to determine the changes, $\delta\alpha$ and δB_T:

$$\delta\alpha = \left(\frac{\partial\alpha}{\partial P}\right)_T \delta P, \quad \delta B_T = \left(\frac{\partial B_T}{\partial P}\right)_T \delta P.$$

Now $(\partial B_T/\partial P)_T = \beta_T$ is given. On the other hand, $(\partial\alpha/\partial P)_T$ must be obtained from the data supplied indirectly:

$$\left(\frac{\partial\alpha}{\partial P}\right)_T = \left(\frac{\partial}{\partial P}\frac{1}{V}\left(\frac{\partial V}{\partial T}\right)_P\right)_T = \left(\frac{\partial}{\partial P}\left(\frac{\partial(\ln V)}{\partial T}\right)_P\right)_T.$$

From the equality of the mixed second derivatives (Appendix D) we obtain

$$\left(\frac{\partial\alpha}{\partial P}\right)_T = \left(\frac{\partial}{\partial T}\left(\frac{\partial(\ln V)}{\partial P}\right)_T\right)_P = -\left(\frac{\partial\kappa_T}{\partial T}\right)_P, \tag{11.4.7}$$

or, in terms of $B_T = 1/\kappa_T$,

$$\left(\frac{\partial\alpha}{\partial P}\right)_T = \frac{1}{B_T^2}\left(\frac{\partial B_T}{\partial T}\right)_P. \tag{11.4.8}$$

Thus on substituting the values given,

$$\delta\alpha = -1.03 \times 10^{-5}\,\mathrm{K}^{-1}, \qquad \delta B_T = 3.59 \times 10^8\,\mathrm{Pa},$$

representing changes of 5.1% and 5.8% in α and B_T, respectively. Hence, to a reasonable approximation we may simply treat α and B_T as constant during the isothermal process, using their mean values, $\bar\alpha$ and $\bar B_T$, for numerical calculations:

$$\bar\alpha = 1.96 \times 10^{-4}\,\mathrm{K}^{-1}, \qquad \bar B_T = 6.35 \times 10^9\,\mathrm{Pa}.$$

Solution (c)

Initially, $V = 0.1v/(\text{molar mass}) = 1.031 \times 10^{-4}\,\mathrm{m}^3$. To find the final volume we first show that the linear approximation is reasonable. For the isothermal process the change in volume is, to first order,

$$\delta V_T = \left(\frac{\partial V}{\partial P}\right)_T \delta P = -\frac{V\delta P}{B_T} = -1.62 \times 10^{-6}\,\mathrm{m}^3.$$

Note that $\delta V_T/V = 0.0157$, so the linear approximation is sufficient. For the isentropic process we proceed in a similar manner, using the mean isentropic bulk modulus, $\bar{B}_S$ instead of $\bar{B}_T$. Now,

$$\frac{B_S}{B_T} = \frac{\kappa_T}{\kappa_s} = \frac{c_p}{c_v} = \frac{28.41}{26.66} = 1.066$$

and assuming that the same ratio applies to the mean values, we have

$$\bar{B}_S = 1.066 \times 6.35 \times 10^9 = 6.77 \times 10^9 \, \text{Pa}.$$

Hence the volume change in the isentropic process is

$$\delta V_S = -\frac{V \delta P}{B_S} = -1.52 \times 10^{-6} \, \text{m}^3.$$

In this case $\delta V_S/V = 0.0148$.

To calculate δT we could use the first T-dS equation (10.4.29). Equivalently, we shall use the fact that Tv^{γ_g} is constant in an isentropic process, by (11.3.9), and hence

$$\frac{\delta T}{T} = -\gamma_g \frac{\delta V_S}{V},$$

yielding

$$\delta T = 295 \times 1.10 \times 0.0148 = 4.80 \, \text{K}.$$

Solution (d)

The work done on the system is $W = -\int P \, dV$ and in the isothermal process we have

$$dV = \left(\frac{\partial V}{\partial P}\right)_T dP = -\frac{V}{B_T} dP.$$

Here V and B_T can be treated as constants, $\bar{V}_T$ and $\bar{B}_T$, where $\bar{V}_T = 1.023 \times 10^{-4} \, \text{m}^3$. Thus the work done in the isothermal process, W_T, is

$$W_T = \frac{\bar{V}_T}{\bar{B}_T} \int_{P_0}^P P' \, dP' = \frac{\bar{V}_T(P^2 - P_0^2)}{2\bar{B}_T}. \tag{11.4.9}$$

Hence we find

$$W_T = 81 \, \text{J}.$$

Similarly, for the isentropic process we use (11.4.9), replacing the isothermal mean values by isentropic mean values. In particular, $\bar{V}_S = 1.023 \times 10^{-4} \, \text{m}^3$, and we obtain W_S, the isentropic work,

$$W_S = 76 \, \text{J}.$$

The heat input to the system in the isothermal process is obtained from

$$\dce Q_T = T\,dS = T\left(\frac{\partial S}{\partial P}\right)_T dP = - T\left(\frac{\partial V}{\partial T}\right)_P dP = - TV\alpha\,dP.$$

Now replace V and α by their isothermal mean values and integrate over P:

$$Q_T = - T\bar{V}_T\bar{\alpha}(P - P_0). \tag{11.4.10}$$

Hence we find

$$Q_T = - 592 \text{ J.}$$

Notice that $\Delta U = Q_T + W_T = -511$ J in the isothermal process. However, for larger compressions ΔU becomes positive because W_T is quadratic in P whereas Q_T is linear. (See question 11.4.6.)

Questions

11.4.1. Use equation (11.4.8) to demonstrate that

$$\left(\frac{\partial \ln(\alpha)}{\partial P}\right)_T = -\frac{\delta_a}{B_T}. \tag{11.4.11}$$

Consider two states at pressures P_0 and P, but having the same temperature. Assume that δ_a is constant and show that

$$\frac{\alpha(P)}{\alpha(P_0)} = \left(\frac{V(P)}{V(P_0)}\right)^{\delta_a}. \tag{11.4.12}$$

11.4.2. Demonstrate that

$$\left(\frac{\partial \ln(B_T)}{\partial T}\right)_P = - \alpha\,\delta_a. \tag{11.4.13}$$

Consider two states at temperatures T and T_0, but having the same pressure. Assume that α and δ_a are independent of T and show that

$$B_T(T) = B_T(T_0)\exp(\alpha\,\delta_a(T_0 - T)). \tag{11.4.14}$$

11.4.3. Show that the Murnaghan equation of state (11.4.4) is a solution of

$$\left(\frac{\partial B_T(0, T)}{\partial P}\right)_T = \beta_T, \tag{11.4.15}$$

where β_T is independent of P.

11.4.4. In the limit $P \to 0$ Dugdale and MacDonald (1953) have shown that the Grüneisen constant is given by the approximate relation

$$\gamma_g \approx -1 - \frac{\dfrac{v}{2}\left(\dfrac{\partial^2 P}{\partial v^2}\right)_T}{\left(\dfrac{\partial P}{\partial v}\right)_T}.$$

Use this equation to establish (11.4.6).

11.4.5. Equation (11.4.2) expresses B_T as a function of P. Show that B_T can be expressed as a function of V as

$$B_T(V, T) = B_T(V_0, T)\left(\frac{V_0}{V}\right)^{\beta_T} \qquad (11.4.16)$$

11.4.6. (a) A solid undergoes an isothermal volume change, $V_0 \to V$, due to external fluid pressure. Given that the Murnaghan equation (11.4.4) applies, show that the work, W_T, and heat, Q_T, input to the system are given by

$$W_T = \frac{B_{T0} V_0}{\beta_T}\left\{\frac{1}{(\beta_T - 1)}\left[\left(\frac{V_0}{V}\right)^{\beta_T - 1} - 1\right] + \frac{V}{V_0} - 1\right\},$$

$$\qquad (11.4.17)$$

$$Q_T = \frac{T\alpha_0 B_{T0} V_0}{1 + \delta_a - \beta_T}\left\{\left(\frac{V}{V_0}\right)^{1 + \delta_a - \beta_T} - 1\right\}, \qquad (11.4.18)$$

where α_0 is the volume expansivity measured in the initial state and B_{T0} is the corresponding bulk modulus.

(b) Let $V = (1 - \varepsilon)V_0$ where $\varepsilon \ll 1$. Evaluate W_T and Q_T to second order in ε and show that the corresponding change in the internal energy, ΔU, is given by

$$\Delta U = \frac{\varepsilon B_{T0} V_0}{2}\left[\varepsilon\{1 + T\alpha_0(\delta_a - \beta_T)\} - 2T\alpha_0\right]. \quad (11.4.19)$$

11.4.7. (a) A body consists of $10^{-3}\,\text{m}^3$ of MgO at $T = 295\,\text{K}$ and $P = 10^5\,\text{Pa}$. It is subjected to isothermal isotropic fluid compression, the final volume being $7 \times 10^{-4}\,\text{m}^3$. Use the Murnaghan equation and equations (11.4.17) and (11.4.18) to determine the final pressure, the work done, the heat input, and the change in internal energy. The following property values (Anderson 1966) for MgO may be used:

$$\alpha = 3.15 \times 10^{-5}\,\text{K}^{-1}, \qquad \left(\frac{\partial \alpha}{\partial T}\right)_P = 8.7 \times 10^{-8}\,\text{K}^{-2},$$

$$B_T = 1.62 \times 10^{11}\,\text{Pa}, \qquad \beta_T = 4.54,$$

$$\gamma_g = 1.60, \qquad\qquad \left(\frac{\partial B_T}{\partial T}\right)_P = -2.80 \times 10^7 \, \text{Pa K}^{-1}.$$

(b) Determine the final temperature if the system is compressed to the same final volume quasistatically and adiabatically. Assume γ_g is constant.

(c) Use (11.4.19) to determine the degree of compression for which $\Delta U = 0$ in MgO in an isothermal process starting at the initial condition given in (a).

11.5 Simple processes in rubber

Vulcanized natural rubber consists of long polymer molecules having typically thousands of linked structural units, called monomers. Because the adjacent monomers are able to rotate freely the natural configuration of a molecule is a crumpled coil-like structure. The internal energy of the system is the kinetic energy of the monomers due to vibration and rotation. From a microscopic point of view, the molecular configuration is more highly ordered in the stretched state than in the natural state which means that the stretched state corresponds to a state of lower entropy. Callen (1985) describes a simple molecular model for a polymer. In addition, at a given temperature the energy of the system is not strongly dependent on the degree of stretching. Consequently, as we shall see, the elasticity of rubber, and polymers generally, may be attributed to the entropy of the system rather than the energy.

Example

A cylindrical sample of rubber is subject to an axial force, τ. Let L_0 be the natural length of the cylinder, and let L be the stretched length (Fig. 11.2). For this system the equation of state may be expressed by the model (Treloar 1973)

$$\tau = AT\left\{\frac{L}{L_0} - \left(\frac{L_0}{L}\right)^2\right\}, \qquad\qquad (11.5.1)$$

where A is a constant for the system and T is the temperature. Assume the heat capacity of the system at $L = L_0$ is constant, C_L. We use the upper case letter since this is an extensive quantity

$$C_L = T\left(\frac{\partial S}{\partial T}\right)_L. \qquad\qquad (11.5.2)$$

Obtain a fundamental relation for the system.

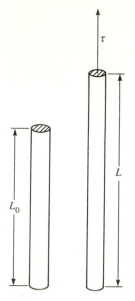

Fig. 11.2 The unstretched and stretched states of a cylinder of rubber.

Solution

Equation (11.5.1) is analogous to the P-V-T equation of state for a fluid given the correspondence $V \to L$, $P \to -\tau$. We can therefore write down the Helmholtz free energy directly using (11.1.13):

$$F = U_0 - TS_0 + \int_{T_0}^{T} C_L \, dT' - T \int_{T_0}^{T} \frac{C_L}{T'} \, dT' + \int_{L_0}^{L} \tau(T, L') \, dL'$$

$$= F_0(T) - TC_L \ln\left(\frac{T}{T_0}\right) + AT\left\{\frac{L^2}{2L_0} + \frac{L_0^2}{L}\right\}, \tag{11.5.3}$$

where

$$F_0(T) = U_0 - T_0 C_L - T(S_0 - C_L + 1.5AL_0). \tag{11.5.4}$$

From F we obtain U and S by the standard relations

$$S = -\left(\frac{\partial F}{\partial T}\right)_L = S_0 + C_L \ln\left(\frac{T}{T_0}\right) - A\left\{\frac{L^2}{2L_0} + \frac{L_0^2}{L} - \frac{3L_0}{2}\right\} \tag{11.5.5}$$

and

$$U = F + TS = U_0 + (T - T_0)C_L. \tag{11.5.6}$$

Thus given $T = T_0$ and $L = L_0$, then $S = S_0$, $U = U_0$, in which case $F = U_0 - T_0 S_0$.

We see that U is a function of T alone in this approximation and in this respect rubber is like an ideal gas. From a thermodynamic viewpoint this happens because $\tau \propto T$, as we noted in Section 11.1. However, this analogy goes further. It is often said that the elasticity of rubber is due to 'entropic' forces. We can illustrate what this means using the relation $\tau = (\partial F/\partial L)_T$ as follows. Since $F = U - TS$, we have (treating U and S as functions of T and L),

$$\tau = \left(\frac{\partial U}{\partial L}\right)_T - T\left(\frac{\partial S}{\partial L}\right)_T = -T\left(\frac{\partial S}{\partial L}\right)_T, \qquad (11.5.7)$$

since U is independent of L given T is fixed. Thus the force depends on how S, rather than U, varies with L at constant T. An ideal gas is like that too. The pressure of an ideal gas is due to the kinetic rather than the potential energy of its molecules. Similarly the elasticity of a polymer derives from the molecular kinetic energy of rotation rather than the inter-molecular potential energy. This situation is in contrast to a crystalline solid, such as a metallic wire (see question 11.5.6).

Questions

11.5.1. Show that the heat capacity at constant extension, C_L, and the heat capacity at constant tension, C_τ, are related by

$$C_L = C_\tau - T\frac{\left(\dfrac{\partial \tau}{\partial T}\right)_L^2}{\left(\dfrac{\partial \tau}{\partial L}\right)_T} = C_\tau - \frac{A}{L_0 L}\frac{(L^3 - L_0^3)^2}{L^3 + 2L_0^3}. \qquad (11.5.8)$$

11.5.2. A rubber cylinder is subject to an axial change in length (which may be stretching or compression). The process is adiabatic and quasistatic. Show that

$$\left(\frac{\partial T}{\partial L}\right)_S = -\frac{T}{C_L}\left(\frac{\partial S}{\partial L}\right)_T = \frac{T}{C_L}\left(\frac{\partial \tau}{\partial T}\right)_L = \frac{AT}{C_L}\left\{\frac{L}{L_0} - \left(\frac{L_0}{L}\right)^2\right\}. \qquad (11.5.9)$$

Hence for an axial deformation, $L_0 \to L$, starting at temperature T_0, show that the final temperature, T, is given by

$$\ln\left(\frac{T}{T_0}\right) = \frac{A}{2L_0 L C_L}(L^3 - 3L_0^2 L + 2L_0^3). \qquad (11.5.10)$$

Confirm that $T > T_0$ in both stretching and compression.

11.5.3. A rubber cylinder undergoes a reversible isothermal axial deformation, $L_0 \to L$, at temperature T_0. Demonstrate that the work done, W, and the heat input, Q, are given by

$$W = -Q = \frac{AT_0}{2L_0L} (L^3 - 3L_0^2 L + 2L_0^3). \qquad (11.5.11)$$

11.5.4. (a) The equation of state for rubber (11.5.1) does not take account of the thermal expansivity of rubber in the unstressed state. Nevertheless, the equation correctly indicates that the linear thermal expansivity, λ, of stretched rubber is negative, where

$$\lambda = \frac{1}{L} \left(\frac{\partial L}{\partial T} \right)_\tau. \qquad (11.5.12)$$

Establish that λ is negative when $L > L_0$ by first showing that

$$\lambda = \frac{1}{T} \left(\frac{L_0^3 - L^3}{2L_0^3 + L^3} \right). \qquad (11.5.13)$$

(b) Obtain $(\partial S/\partial \tau)_T$. Show that S decreases both in stretching and compression at a given temperature.

11.5.5. The thermal expansion of rubber in the unstressed state can be included in the equation of state, (11.5.1), by substituting $(1 + \lambda(T - T_0))L_0$ in place of L_0. Here λ is the linear expansivity measured at $\tau = 0$. A typical value for λ is $2 \times 10^{-4} \mathrm{K}^{-1}$ at $T_0 = 293$ K.

(a) Show that U is now a weak function of L and that to first order in λT,

$$U = U_0 + C_L(T - T_0) + \lambda A T^2 \left(\frac{L^2}{2L_0} - \frac{2L_0^2}{L} + \frac{3L_0}{2} \right). \qquad (11.5.14)$$

(b) Use the approach taken in question 11.5.2. to determine $(\partial T/\partial L)_S$. To first order in λT show that

$$\left(\frac{\partial T}{\partial L} \right)_S = \frac{AT}{C_L} \left\{ \frac{L}{L_0} - \left(\frac{L_0}{L} \right)^2 - \lambda T \left[\frac{L}{L_0} + 2 \left(\frac{L_0}{L} \right)^2 \right] \right\}. \qquad (11.5.15)$$

(c) Establish the following approximate expression for $\Delta T = T - T_0$ in an isentropic stretching process, $L_0 \to L$, by setting $T = T_0$ on the right-hand side of (11.5.15) and integrating with respect to L:

$$\Delta T = \frac{AT_0L_0}{C_L} \left\{ \frac{(L^2 - L_0^2)(1 - \lambda T_0)}{2L_0^2} + \frac{(L_0 - L)(1 + 2\lambda T_0)}{L} \right\}. \qquad (11.5.16)$$

(d) Consider an adiabatic stretching process, $L_0 \to (1 + \varepsilon)L_0$, starting at $T = T_0$. Assume $\varepsilon \ll 1$. Show that $\Delta T = 0$ when, to first order,

$$\varepsilon = \frac{2\lambda T_0}{1 + \lambda T_0}. \tag{11.5.17}$$

In question 11.5.2 we showed that the temperature of rubber increases during isentropic stretching processes. Equation (11.5.16), which takes account of the thermal expansivity, shows that for small extensions the temperature will decrease instead. Use the values for T_0 and λ given above to evaluate the fractional extension for which ΔT remains negative.

11.5.6. The thermal properties of a metallic wire are represented by the heat capacity at constant length, C_L, eqn (11.5.2), the coefficient of linear expansion, λ, eqn (11.5.12), and the isothermal Young's modulus, E_T, which is given by (7.4.4). Here τ is the tension of the wire. In the following assume that C_L, λ, and E_T are constant and let A be the cross-sectional area of the wire.

(a) By expressing τ as a function of T and L show that

$$d\tau = -\lambda A E_T dT + \frac{A E_T}{L} dL. \tag{11.5.18}$$

(b) Given $\tau = 0$ at $T = T_0$ and $L = L_0$, show that

$$\tau = -\lambda A E_T (T - T_0) + \frac{A E_T}{L_0} (L - L_0), \tag{11.5.19}$$

provided $(L - L_0)/L_0 \ll 1$. Obtain the Helmholtz free energy:

$$F = U_0 - TS_0 + (T - T_0)C_L - TC_L \ln\left(\frac{T}{T_0}\right)$$

$$- \lambda A E_T (T - T_0)(L - L_0) + \frac{A E_T}{2L_0}(L - L_0)^2. \tag{11.5.20}$$

(c) Determine U and S as functions of T and L. Show that when the length of the wire is changed isothermally, $L \to L(1 + \varepsilon)$, where $\varepsilon \ll 1$, the change in tension is associated with the corresponding change in U rather than S. (This result is in contrast to vulcanized rubber.)

(d) Show that in an isentropic process,

$$\left(\frac{\partial T}{\partial L}\right)_S = -\frac{\lambda A E_T T}{C_L}, \tag{11.5.21}$$

and in an isothermal process, the heat input, Q, is given by

$$\left(\frac{dQ}{dL}\right)_T = \lambda A E_T T. \tag{11.5.22}$$

11.6 Magnetic systems

We have previously considered two forms for the energy fundamental relation of a magnetizable system (Section 8.2). The correct form depends on how we choose to identify the field, B_0, in the absence of the medium. When the field source is a current-carrying coil, the applied fields, $B_{0\Phi}$ and B_{0J}, correspond to fixed flux-turns, Φ, and fixed coil current, J, respectively. For a linear isotropic body in which the demagnetizing field may be ignored, the fields are related by

$$B_{0\Phi} = B_{0J}(1 + \gamma_m \chi_m) \tag{11.6.1}$$

when the source is a long solenoid. Here γ_m is a geometric factor determined by the degree to which the body fills the solenoid (see eqns (8.2.25), (8.2.27)). When the volume of the system is small compared with the solenoid, $\gamma_m \to 0$; for a system which fills the solenoid, $\gamma_m = 1$.

The magnetic susceptibility, χ_m, was taken to be constant in the processes considered in Section 8.2. Here we shall consider the temperature dependence of χ_m explicitly, by introducing the isothermal susceptibility, χ_{mT},

$$\chi_{mT} = \left(\frac{\partial M}{\partial H_i}\right)_T, \tag{11.6.2}$$

where H_i is the magnetic intensity in the medium. For a linear medium, χ_{mT} is independent of H_i and

$$M = \chi_{mT} H_i. \tag{11.6.3}$$

In Section 8.2 we overlooked the distinction between H_i and the magnetic intensity in the absence of the medium given the same coil current. Denote this latter quantity by H_e. Suppose the body is an ellipsoid of rotation, the axis of rotation lying parallel to the axis of the current carrying solenoid. In this situation the magnetic intensity measured in the body is given by

$$H_i = H_e - \gamma_d M, \tag{11.6.4}$$

where γ_d is the demagnetizing factor for the axial direction (Robinson 1973). For a sphere $\gamma_d = \frac{1}{3}$; in the limit of an elongated needle-shaped body, $\gamma_d \to 0$. Hence, for a linear medium,

$$H_i = \frac{H_e}{1 + \gamma_d \chi_{mT}}. \tag{11.6.5}$$

Thus the effect of including the demagnetizing field is achieved by replacing χ_{mT} by $\chi_{mT}/(1 + \gamma_d\chi_{mT})$. In particular, (11.6.1) becomes

$$B_{0\Phi} = B_{0J}\left[1 + \frac{\gamma_m\chi_{mT}}{1 + \gamma_d\chi_{mT}}\right], \tag{11.6.6}$$

and since $B_{0J} = \mu_0 H_e$, we have from (11.6.3), (11.6.5), and (11.6.6),

$$M = \chi_{mT}H_i = \frac{\chi_{mT}B_{0J}}{\mu_0(1 + \gamma_d\chi_{mT})} \tag{11.6.7a}$$

$$= \frac{\chi_{mT}B_{0\Phi}}{\mu_0(1 + (\gamma_m + \gamma_d)\chi_{mT})}. \tag{11.6.7b}$$

Thus we require only an expression for χ_{mT} in terms of T, V, and N to obtain the equation of state linking M and $B_{0\Phi}$ or B_{0J}.

To illustrate we shall consider the *Curie–Weiss law* (Rosenberg 1988). When the temperature of a ferromagnetic body is rather greater than the Curie temperature, the body behaves as a paramagnetic medium for which

$$\chi_{mT} = \frac{c}{T - T_p}. \tag{11.6.8}$$

Here c is the Curie constant and T_p is the paramagnetic Curie temperature, which is chosen to fit measured data. T_p is approximately the same as the ferromagnetic transition temperature, T_c. In iron, for instance, $T_c = 1043$ K, and $T_p = 1093$ K. The Curie–Weiss equation becomes inapplicable as T approaches T_c, where χ_{mT} exhibits divergent behaviour. The form of this divergence (Heller 1967; also see Section 12.5) is $\chi_{mT} \sim (T - T_c)^{-\gamma}$ where $\gamma \approx 1.33$. This behaviour is related to the onset of spontaneous ordering within the body due to internal interactions.

The Curie–Weiss equation is also applicable to dilute paramagnetic salts, in which case T_p may be of the order of a few millikelvin, and to antiferromagnetic materials, in which case T_p is negative. For example, in MnO, $T_p = -680$ K whereas the Néel temperature, the transition temperature above which the material is paramagnetic, is 116 K. When the temperature of interest is much greater than $|T_p|$ the expression reduces to the *Curie law:* $\chi_{mT} = c/T$.

We may now write down the equations of state for a paramagnetic body which conforms to the Curie–Weiss law; from eqns (11.6.7) and (11.6.8) we have:

$$M = \frac{B_{0J}c}{\mu_0(T - T_p + \gamma_d c)}, \tag{11.6.9}$$

$$M = \frac{B_{0\Phi}c}{\mu_0(T - T_p + (\gamma_m + \gamma_d)c)}. \tag{11.6.10}$$

These have the same form as the Curie–Weiss law if we replace T_p by an effective paramagnetic Curie temperature, T'_p, given by

$$T'_p = T_p - \gamma_d c \qquad (11.6.11)$$

in the B_{0J} representation, and

$$T'_p = T_p - (\gamma_m + \gamma_d)c \qquad (11.6.12)$$

in the $B_{0\Phi}$ representation.

We are now in a position to establish a fundamental relation for a paramagnetic body; we shall restrict our attention to the representation in which T, V, N, and B_{0J} are the independent variables. There is no accepted symbol for this potential which involves both the magnetic Gibbs function, G_J, and the non-magnetic Helmholtz function, F. We shall use the symbol Γ and, for brevity, we shall denote the magnetic field as B. First we write down the Gibbs equation for Γ by integrating (10.3.26) over the volume and including the volume work term, $-P\,\delta V$:

$$\delta\Gamma = -S\,\delta T - P\,\delta V - \int_V M\,\delta B\,dV + \mu\,\delta N. \qquad (11.6.13)$$

Following (11.1.2) we write Γ as the sum of a magnetic work part and a field-free part:

$$\Gamma(T, V, B, N) = \Gamma(T, V, 0, N) + \Gamma(T, V, B, N) - \Gamma(T, V, 0, N) \qquad (11.6.14)$$

$$= \Gamma(T, V, 0, N) + \int_0^B \left(\frac{\partial\Gamma}{\partial B'}\right)_{T,V,N} dB'. \qquad (11.6.15)$$

Now, $\Gamma(T, V, 0, N) = F(T, V, N)$ is the Helmholtz free energy for the system when $B = 0$. Also given that the field is uniform and that the system is homogeneous and linear,

$$\left(\frac{\partial\Gamma}{\partial B'}\right)_{T,V,N} = -VM, \qquad (11.6.16)$$

where M is given by (11.6.7a). Hence,

$$\Gamma(T, V, B, N) = F(T, V, N) - \frac{VB^2\chi_{mT}}{2\mu_0(1 + \gamma_d\chi_{mT})} \qquad (11.6.17)$$

$$= F(T, V, N) - \frac{VB^2 c}{2\mu_0(T - T'_p)}, \qquad (11.6.18)$$

T'_p being given by (11.6.11). To proceed we shall assume that c is a function of N/V and we show this explicitly by introducing c_c defined by

$$c = \frac{Nc_c}{V}. \qquad (11.6.19)$$

Note that T_p' is also a function of c, and hence of N/V, but we shall take this effect to be of second order and neglect it. Thus (11.6.18) yields the following expression for the magnetic fundamental relation:

$$\Gamma(T, V, B, N) = F(T, V, N) - \frac{NB^2 c_c}{2\mu_0(T - T_p')}. \tag{11.6.20}$$

This equation allows us to illustrate the magnetocaloric effect in a dilute paramagnetic solid. Since $S = -(\partial\Gamma/\partial T)_{V,B,N}$ we obtain

$$S(T, V, B, N) = S(T, V, 0, N) - \frac{NB^2 c_c}{2\mu_0(T - T_p')^2}, \tag{11.6.21}$$

where $S(T, V, 0, N)$ is the entropy when $B = 0$. Consider an isentropic process in which the magnetic field changes, $B_a \to B_b$. To obtain an expression for the corresponding temperatures, T_a and T_b, we need an expression for $S(T, V, 0, N)$ as a function of T. For simplicity we shall assume here that the molar heat capacity at constant V is small enough that $\Delta S = S(T_b, V, 0, N) - S(T_a, V, 0, N)$ can be ignored in this process. Accordingly we find

$$\frac{T_b - T_p'}{T_a - T_p'} = \frac{B_b}{B_a}. \tag{11.6.22}$$

Thus as the final magnetic field is reduced to zero, the final temperature of the system approaches T_p'. This illustrates the principle of cooling by *adiabatic demagnetization* (for experimental background see Zemansky and Dittman (1981)), an important refrigeration process in low temperature physics. There is also a potential for magnetic refrigeration at higher temperatures. In that application an important material property is the change in S due to a given field B at a fixed temperature. The magnitude of this term,

$$\frac{VB^2}{2\mu_0} \frac{\partial}{\partial T} \left(\frac{\chi_{mT}}{1 + \gamma_d \chi_{mT}} \right)_{V,B,N}, \tag{11.6.23}$$

peaks near the Curie temperature. Consequently low-hysteresis ferromagnetic materials, whose Curie temperature is in the range 250 K–300 K, are of special interest for magnetic refrigeration in large-scale applications. Certain composites of Gd meet these conditions. Since χ_{mT} may be much greater than unity in such materials, the distinction between the B_{0J} and $B_{0\Phi}$ representations is significant. By contrast, in cryogenic cooling it is normal to use dilute paramagnetic materials because this helps to ensure that T_p' is small. In these applications, $\chi_{mT} \ll 1$, and we may then take $B_{0J} \approx B_{0\Phi}$.

Questions

11.6.1. A system of weakly interacting spin-$\frac{1}{2}$ particles is subject to a magnetic field B. The particles have two quantum states of energy $\pm\mu_B B$, where μ_B is the Bohr magneton. When the system geometry is appropriate, the local magnetic field, B, and the field produced by the source in the absence of the system, B_0, are equivalent. For this system we shall use the $B_{0\Phi}$ representation (Section 8.2) so that the Helmholtz free energy per unit volume, f_Φ, is naturally a function of $B_{0\Phi} = B$. This can be expressed as

$$f_\Phi = -nRT\ln\left(e^{\varepsilon/RT} + e^{-\varepsilon/RT}\right). \tag{11.6.24}$$

Here $f_\Phi = F_\Phi/V$, $n = N/V$, and

$$\varepsilon = \tilde{N}_A \mu_B B. \tag{11.6.25}$$

(a) Show that:

(i)
$$u_\Phi = -n\varepsilon\left(\frac{e^{\varepsilon/RT} - e^{-\varepsilon/RT}}{e^{\varepsilon/RT} + e^{-\varepsilon/RT}}\right), \tag{11.6.26}$$

(ii)
$$u_\Phi = Ts + \mu n. \tag{11.6.27}$$

(This is the Euler form. Note the magnetic term does not appear. Although S, B, and N are the natural variables for U_Φ, B is not extensive. See question 9.5.4.)

(b) Figure 11.3 shows loci for states for a given field, B, in (s, T) coordinates. Confirm that these have the correct form. The process $a \to b \to c$ shown represents isothermal magnetization followed by adiabatic demagnetization, resulting in a reduction of system temperature.

(c) Obtain an expression for the molar heat capacity at constant magnetic field, defined by

$$c_B = \frac{T}{n}\left(\frac{\partial s}{\partial T}\right)_B.$$

Show that c_B is a peaked function of T, having a maximum at $T \sim 0.9\ \varepsilon/R$, and that when $T \gg \varepsilon/R$,

$$c_B = R\left(\frac{\varepsilon}{RT}\right)^2. \tag{11.6.28}$$

This peak is the Schottky hump illustrated in Fig. 10.3. To reconcile (11.6.28) with (10.2.37) replace E in (10.2.37) by 2ε.

(d) Obtain an expression for the magnetic moment per unit volume, M, and show that the magnetic enthalpy is

$$h_\Phi = u_\Phi + MB = 0.$$

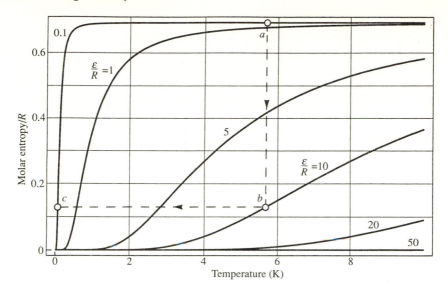

Fig. 11.3 Entropy–temperature representation for states of constant applied field B for a system of spin-$\frac{1}{2}$ particles; ε is given by (11.6.25). In the process $a \rightarrow b$ the field is increased keeping T constant by contact with some external reservoir, in the process $b \rightarrow c$ the field is reduced reversibly whilst maintaining the system adiabatically isolated.

Notice that in the Φ representation B corresponds to V for a fluid system, and M to P. Since $u_\Phi = -MB$, the system may be treated as an ideal quantum gas in the sense of eqn (11.3.12); the corresponding Grüneisen constant is $\gamma_g = -1$. It follows from (11.3.13) that T/B is constant in an isentropic process. We revisit this result again in part (h).

(e) Given that $M = \chi_{mT} B / \mu_0$, where μ_0 is the vacuum permeability, show that, when $\varepsilon \ll RT$,

$$\chi_{mT} = \frac{nc_c}{T}, \qquad (11.6.29)$$

where

$$c_c = \frac{\mu_0 (\mu_B \tilde{N}_A)^2}{R}. \qquad (11.6.30)$$

This is an example of the Curie law; $c = nc_c$ is the Curie constant.

(f) Show that the maximum value of M, obtained by taking $T \rightarrow 0$, is

$$M_{\max} = \mu_B \tilde{N}_A n. \qquad (11.6.31)$$

Let $m = M/M_{max}$; show that

$$s = -nR \left\{ \left(\frac{1-m}{2} \right) \ln \left(\frac{1-m}{2} \right) + \left(\frac{1+m}{2} \right) \ln \left(\frac{1+m}{2} \right) \right\}.$$
$$(11.6.32)$$

(g) Show that the magnetic Gibbs function, $g_\Phi = g_\Phi(T, M, n)$
$= u_\Phi - Ts + BM$, is

$$g_\Phi = nRT \left\{ \left(\frac{M_{max} - M}{2M_{max}} \right) \ln \left(\frac{M_{max} - M}{2M_{max}} \right) \right.$$

$$\left. + \left(\frac{M_{max} + M}{2M_{max}} \right) \ln \left(\frac{M_{max} + M}{2M_{max}} \right) \right\}, \quad (11.6.33)$$

and confirm that

$$B = \left(\frac{\partial g_\Phi}{\partial M} \right)_{T,n}. \quad (11.6.34)$$

(h) From (f) it is evident that s can be expressed as a function of M and n alone. In this respect a system comprising weakly interacting spins is unusual. For instance, because the magnetic enthalpy $h_\Phi(s, M) = 0$, the molar heat capacity at constant magnetization is

$$c_M = \frac{1}{n} \left(\frac{\partial h_\Phi}{\partial T} \right)_{M,n} = \frac{T}{n} \left(\frac{\partial s}{\partial T} \right)_{M,n} = 0, \quad (11.6.35)$$

and in an isentropic process (for example, adiabatic demagnetization) in which the magnetic field is changed, $B_a \to B_b$, then $M_a = M_b$. For this process show that

$$\left(\frac{\partial T}{\partial B} \right)_{s,n} = - \left(\frac{\partial M}{\partial s} \right)_{B,n} = \frac{2\mu_B \tilde{N}_A}{R \ln \left(\frac{1+m}{1-m} \right)} = \frac{T}{B}, \quad (11.6.36)$$

and hence that

$$\frac{T_a}{T_b} = \frac{B_a}{B_b}. \quad (11.6.37)$$

This last result follows from the fact that s is a function of B/T, a result we anticipated in (d). It is peculiar to a system of spins in which the magnetic interaction alone contributes to the internal energy, u_Φ.

(i) For an isentropic process $a \to b$, show that the work done on the medium per unit volume is

$$w_{ab} = M_a(B_a - B_b). \quad (11.6.38)$$

11.6.2. A particular paramagnetic system is subject to an external magnetic field, B, defined by the current J in the source circuit (Section 8.2). The demagnetizing effect is negligible. Various representations for the fundamental relation of the system are given in eqns (9.5.43), (9.5.47), (10.3.46), and (10.3.47). Since the energy, entropy, heat input, and work input will be determined for a fixed unit volume they will be denoted by the lower case symbols, u, s, q, and w.

(a) Suppose the magnetization changes $M_a \to M_b$ in a process carried out reversibly at constant T. Show that the work per unit volume, w_e, by the current source, due to the medium is

$$w_e = \frac{\mu_0 T}{2nc_c} (M_b^2 - M_a^2) = \frac{nc_c}{2\mu_0 T} (B_b^2 - B_a^2). \quad (11.6.39)$$

Show that the heat input satisfies the first law relation, $q + w_e = \Delta u_J$.

(b) In a heating process performed quasistatically at constant field B, show that the molar heat capacity is given by

$$c_B = \frac{T}{n} \left(\frac{\partial s}{\partial T} \right)_{B,n} = \frac{1}{T^2} \left(c_0 + \frac{c_c B^2}{\mu_0} \right), \quad (11.6.40)$$

and that for a process $a \to b$, given B is constant, the heat input is

$$q = n \left(c_0 + \frac{c_c B^2}{\mu_0} \right) \left(\frac{1}{T_a} - \frac{1}{T_b} \right). \quad (11.6.41)$$

Show that the work done by the current source which maintains B constant in this process is

$$w_e = \frac{nc_c B^2}{\mu_0} \left(\frac{1}{T_b} - \frac{1}{T_a} \right), \quad (11.6.42)$$

and that $w_e + q = \Delta u_J$.

(c) Consider a heating process, $a \to b$, performed quasistatically at constant magnetization, M. Obtain an expression for the corresponding molar heat capacity, c_M, and show that the heat input is

$$q = nc_0 \left(\frac{1}{T_a} - \frac{1}{T_b} \right). \quad (11.6.43)$$

Establish the general result:

$$c_B - c_M = \frac{T\mu_0}{n\chi_{mT}} \left(\frac{\partial M}{\partial T} \right)_B^2. \quad (11.6.44)$$

By evaluating both sides, confirm that (11.6.44) holds for this particular system.

For a heating process, $a \to b$, performed quasistatically at constant magnetization, M, show that the magnetic field, B, changes by

$$\Delta B = \frac{\mu_0 M (T_b - T_a)}{n c_c}. \tag{11.6.45}$$

Demonstrate that $w_e = 0$ in this process. Also show that the current source of the field coil nevertheless does some work.

(d) The system is moved relative to the source coil, the initial and final magnetic fields at the system being B_a and B_b, respectively. The process is carried out quasistatically, at constant temperature, T, and the current, J, in the source coil remains constant. Show that the mechanical work done on the system, w_m, the heat input, q, and the work done by the electrical source, w_e, are given by

$$w_m = \frac{n c_c}{2 \mu_0 T} (B_a^2 - B_b^2) = q, \tag{11.6.46}$$

$$w_e = \frac{n c_c}{\mu_0 T} (B_b^2 - B_a^2). \tag{11.6.47}$$

Confirm that

$$w_e + w_m + q = \Delta u_J.$$

(e) Consider an isentropic process in which the field changes, $B_a \to B_b$. Assume the system remains stationary and let T_a and T_b be the initial and final temperatures for the system. Show that

$$\frac{T_b}{T_a} = \left(\frac{\mu_0 c_0 + c_c B_b^2}{\mu_0 c_0 + c_c B_a^2} \right)^{1/2}, \tag{11.6.48}$$

and that the work done by the current source is

$$w_e = n c_0 \left(\frac{1}{T_a} - \frac{1}{T_b} \right). \tag{11.6.49}$$

Demonstrate that the change in magnetization is given by

$$M_b^2 - M_a^2 = \frac{n^2 c_0 c_c^2}{\mu_0 T_a^2} \left(\frac{B_b^2 - B_a^2}{\mu_0 c_0 + c_c B_b^2} \right). \tag{11.6.50}$$

The influence of the magnetic field on the temperature is the *magnetocaloric effect*. At room temperature the heat capacity of paramagnetic materials due to lattice vibration is so large that the magnetocaloric effect is insignificant, but at low temperatures (a few kelvin or less) this contribution to the heat capacity must approach zero (by the third law) and for this reason the magnetocaloric effect is then an effective means for cooling. To illustrate, consider the limit in which B_b is small but B_a is large, in the sense that

$$c_c B_b^2 \ll \mu_0 c_0, \qquad c_c B_a^2 \gg \mu_0 c_0.$$

For this situation show that $T_b \ll T_a$ and $M_b \ll M_a$. Notice that if $c_0 = 0$, then $T_b/T_a = B_b/B_a$ and $M_b = M_a$.

(f) Suppose that the process of adiabatic demagnetization described in (e) is carried out by moving the system relative to the source coil whilst keeping the current J in the source coil constant. Show that the mechanical work, w_m, and electrical work, w_e, of the current source are given by

$$w_m = nc_0 \left(\frac{1}{T_a} - \frac{1}{T_b} \right) + \frac{nc_c}{\mu_0} \left(\frac{B_a^2}{T_a} - \frac{B_b^2}{T_b} \right), \quad (11.6.51)$$

$$w_e = \frac{nc_c}{\mu_0} \left(\frac{B_b^2}{T_b} - \frac{B_a^2}{T_a} \right), \qquad (11.6.52)$$

and confirm that $w_e + w_m + q = \Delta u_J$.

11.6.3. A long uniform metallic rod of volume V lies coaxially within a current-carrying solenoid. The rod is cooled whilst the current in the solenoid is maintained at a fixed value by a current source. At a certain temperature the rod, a type-I superconductor (Rosenberg 1988), becomes superconducting. All magnetic flux is then excluded from the body of the rod (the *Meissner effect*). The electrical resistance of the material also vanishes and all currents are then confined to a thin layer at the surface of the rod. In effect the rod exhibits perfect diamagnetism, and we may set $\chi_{mT} = -1$. In the following assume that demagnetizing phenomena may be disregarded and that the rod remains in the superconducting state.

(a) Let $\Gamma(T, V, B, N)$ be the potential identified by the Gibbs equation (11.6.13). Show that

$$\Gamma(T, V, B, N) = \Gamma(T, V, 0, N) + \frac{VB^2}{2\mu_0}, \quad (11.6.53)$$

where B is the field in the solenoid in the absence of the rod, given the same coil current.

(b) Show that the molar heat capacity at constant V and B is a function of T and $v = V/N$ only.

(c) The solenoid field is increased from 0 to B at constant temperature. Let W_{mag} be the work done by the current source which is attributable to the rod. Show that

$$W_{mag} = -\frac{VB^2}{2\mu_0}. \qquad (11.6.54)$$

(d) The rod is moved into the coil from a field free region whilst T and the coil current are constant. Show that the mechanical work done in this process is

$$W_{mech} = \frac{VB^2}{2\mu_0} = \Delta\Gamma \qquad (11.6.55)$$

and that the work done by the current source is

$$W_{elect} = -\frac{VB^2}{\mu_0}. \qquad (11.6.56)$$

(e) In the process described in (d) show that the rod will tend to be expelled from the solenoid. Describe what would happen if the coil flux were constant instead.

12

Transformations of matter

12.1 Phase changes

Introduction

The term *phase* is well illustrated by the familiar phases of water: ice, liquid water, and steam. More formally a phase of a system is a homogeneous state, identified by such thermodynamic properties as the density or chemical composition. Single-phase systems are simple systems in the sense used in Section 9.2. Thus a single-phase fluid may be represented by the extensive variables $U, V, N_1, \ldots, N_r$. On the other hand, when these state parameters are varied it is found that the corresponding systems are not always homogeneous. On varying U say, keeping $V, N_1, \ldots, N_r$ fixed, we may find that the equilibrium state changes from a single phase to a state consisting of a mixture of two or more phases. This process is called a *phase transition*.

Figure 12.1 illustrates the onset of the liquid–vapour phase transition when the temperature of a fixed volume and mass of water is varied near the anomalous point at 4°C. The coordinates are $\log(P)$ and u. Below 4°C, or $u = 20 \, \text{kJ kg}^{-1}$, the equilibrium pressure of the liquid decreases as the temperature increases. When the pressure reaches the saturated vapour pressure some of the liquid changes into the vapour phase; as the temperature is increased further the liquid begins to expand again and the system reverts back to the liquid phase. The figure illustrates how the gradient of the P–u locus at constant v undergoes sharp changes at these points. This suggests that the phase change is associated with an underlying discontinuity in the equilibrium state functions.

We shall treat a multi-phase system as a composite system consisting of several simple systems. In a phase change process the extensive parameters for one subsystem, (mole-numbers, volume, entropy) increase at the expense of the others. Since this process is driven by the instabilities of the individual phases, we expect the equations of state for the phases to display these instabilities. We have seen these intrinsic instabilities already in the van der Waals gas (Section 11.2). It is therefore important to distinguish between the equations of state for an individual phase and the equations which represent the system as a whole.

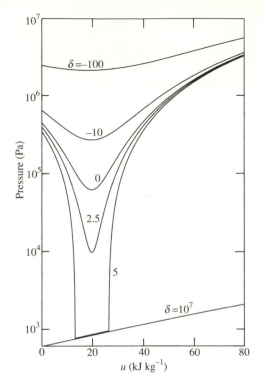

Fig. 12.1 Isochores for water near the anomalous point at 4°C. The specific volume is related to δ by $v = 10^{-3} + \delta \times 10^{-8}$ m^3 kg^{-1}. The isochore corresponding to $\delta = 5$ exhibits a partial liquid–vapour phase change when the vapour pressure equals the saturated vapour pressure. The isochore for $\delta = 10^7$ lies entirely within the mixed two-phase region.

The Gibbs phase rule

Consider a fluid system in which two phases of a single-component substance are in equilibrium. The interface between the phases is non-restrictive with respect to energy, volume, and matter transfer. Thus, denoting the intensive variables for the phases by T, P, μ and T', P', μ' we can express the equilibrium condition by

$$T = T, \qquad P = P', \qquad \mu = \mu'. \tag{12.1.1}$$

Evidently the phases must be distinguished by a property other than T, P, and μ, such as the molar volume, the molar heat capacity, the compressibility, or perhaps another property such as a symmetry. Note that the pressure equality holds strictly only when the common surface of the phases is without curvature, and we shall assume this condition is observed.

We shall now determine the number of independent thermodynamic variables for this system. First, we know that a single-component homogeneous system has three independent state coordinates. For instance, if we choose the energy representation for the fundamental relation, the independent variables would be S, V, and N. Equivalently by using the Helmholtz or Gibbs potential we would use T, V, and N, or T, P, and N as the natural state coordinates. Whatever representation is selected, at least one variable must be extensive to define the size of the system. As a consequence the three intensive variables, T, P, and μ, are not independent. Indeed, we know already that they are related by the Gibbs–Duhem relation (9.5.37). For a single-component system this can be written in terms of the molar entropy, s, and molar volume, v, as

$$s\,dT - v\,dP + d\mu = 0. \tag{12.1.2}$$

Thus, given that N is fixed for a single-phase, single-component system, there are three intensive variables, two of which can be varied independently.

For two phases we must use at least one extensive variable for each phase, and, by eqn (12.1.1), we require only three intensive variables. However, only one of the three can be independent because the intensive variables are now constrained by two Gibb–Duhem relations (12.1.2), one for each phase.

Thus a single-component system in a state consisting of two phases in equilibrium can have only one independent intensive variable. We may choose to use the temperature as the independent variable, so in a liquid–vapour or solid–vapour transition, the pressure, called the saturated vapour pressure, P_s, must be a unique function of T:

$$P_s = P_s(T). \tag{12.1.3}$$

Guggenheim (1967) obtained an empirical relation for $P_s(T)$ for the liquid–vapour transition of gases which comply with the principle of corresponding states (question 11.2.3). This equation is expressed in terms of the critical pressure, P_c, and critical temperature, T_c:

$$\ln\left(\frac{P_s}{P_c}\right) \approx 5.3\left(1 - \frac{T_c}{T}\right). \tag{12.1.4}$$

When there are three phases in equilibrium in a single-component system there can be no independent intensive variables, since T, P, and μ will be constrained by three Gibbs–Duhem equations. Such a state, called a triple point, therefore has uniquely defined intensive parameters. It is for this reason that the triple point of water has been adopted as the primary standard in defining the Kelvin temperature.

Consider the general case of a system having r independent components coexisting in p different phases. We assume they are not chemically reactive. For this system there will be $r + 2$ intensive variables, corresponding to $T, P, \mu_1, \mu_2, \ldots, \mu_r$. In equilibrium the intensive parameters are common to all phases, but they are not independent since there are p Gibbs–Duhem relations, one for each phase. Hence the number of independent intensive variables is

$$i = r + 2 - p. \qquad (12.1.5)$$

This is the *Gibbs phase rule*.

Notice that the phase rule restricts the total number of phases that can coexist in a system of r components to $r + 2$. Thus we may not have more than three coexisting phases in a single-component system. Apart from this, the phase rule places no restriction on the amount of each phase; the total amounts of the phases are independent thermodynamic variables for a multi-phase system.

In addition, the phase rule provides useful clues on the effect of particular changes to a system. Consider, for example, a mixture of ice, water, and water vapour at the triple point. Since the system has no degree of freedom, the temperature is well defined. But if we admit a second component, such as another gas, the system will then have two components with $r + 2 - p = 2 + 2 - 3 = 1$ degree of freedom. Thus the temperature and pressure are no longer uniquely defined. For this reason the purity of water used in a triple point standard temperature cell is important in precision thermometry.

The Clapeyron equation

Consider a single-component fluid system in a two-phase state. An explicit form for the coexistence relationship (12.1.3) cannot be established from purely thermodynamic considerations alone. However, we can obtain some properties of the function $P_s(T)$ from the Gibbs–Duhem relation. Suppose the system undergoes an infinitesimal change of state, $a \rightarrow b$, along the coexistence line, as in Fig. 12.2. For the phase labelled 1, the Gibbs–Duhem relation for this state change is

$$\delta\mu_s = -s_1\,\delta T + v_1\,\delta P_s. \qquad (12.1.6)$$

Here P_s denotes the saturated vapour pressure, μ_s the corresponding chemical potential, and T is taken to be an independent parameter. In phase 2 we have the same values for T, P_s, and μ_s and hence the corresponding equation is

$$\delta\mu_s = -s_2\,\delta T + v_2\,\delta P_s. \qquad (12.1.7)$$

On eliminating $\delta\mu_s$ we obtain an expression for the derivative of $P_s(T)$,

$$\frac{dP_s}{dT} = \frac{s_2 - s_1}{v_2 - v_1}. \qquad (12.1.8)$$

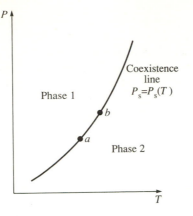

Fig. 12.2 Two-phase coexistence line for a single-component system.

Now $s_2 - s_1 = L/T$, where L is the latent heat per mole for the phase change process. Hence (12.1.8) can be written

$$\frac{\mathrm{d}P_s}{\mathrm{d}T} = \frac{L}{T(v_2 - v_1)}, \tag{12.1.9}$$

which is the *Clapeyron equation*. For a liquid–vapour phase change process we denote the liquid and vapour phases by 1 and 2, respectively. Provided the state is not close to the critical point, v_1 will be small compared to v_2 and we may use the ideal gas approximation for v_2. Thus, $v_2 = RT/P_s$ and we obtain the *Clausius–Clapeyron equation*:

$$\frac{\mathrm{d}P_s}{\mathrm{d}T} = \frac{LP_s}{RT^2}. \tag{12.1.10}$$

The normal–superconducting transition

Equations analogous to the Clapeyron equation (12.1.9) may be established for other phase change phenomena. The normal–superconducting transition in type-I superconducting metals is a particular example. The properties of these metals are briefly reviewed in question (11.6.3).

We may treat the transition as a phase change in a single-component system, but unlike the fluid system considered above there are now four intensive variables. In this case the corresponding form of the Gibbs phase rule restricts the number of independent intensive variables to two, rather than just one.

We shall adopt the same assumptions used in question (11.6.3): the body is a long uniform rod of type-I superconductor located coaxially in a long current-carrying solenoid. The applied field B is that produced in the solenoid in the absence of the body, given the same coil current.

The Clapeyron equation for the normal–superconducting transition may be derived on the basis of the chemical potential, as in the fluid case. But the rod is properly regarded as a complete entity. It has surface currents, which are responsible for its magnetization, and these are a property of the rod as a whole. We shall therefore consider the phase change in terms of the thermodynamic potentials for the complete rod. First, the energy fundamental relation is given by (9.5.18),

$$U_J = U_J(S, V, I, N)$$

where $I = MV$ is the magnetic moment of the rod. Let G be the thermodynamic potential having the natural variables T, P, B, and N; this may be regarded as the Gibbs free energy with respect to both P and B. (We drop the subscript J for simplicity.) Then,

$$\delta G = -S\,\delta T + V\,\delta P - I\,\delta B + \mu\,\delta N. \tag{12.1.11}$$

Since the rod behaves as a perfectly diamagnetic body in the superconducting state, $\chi_{mT} = -1$. The equation of state then follows from $I = MV$ and $M = \chi_{mT} B/\mu_0$:

$$I = -\frac{VB}{\mu_0}. \tag{12.1.12}$$

It follows that G is an increasing function of B in the superconducting phase. Thus if B increases at constant T, P, and N, a point is reached at which $G_s > G_n$, (where the subscripts s and n denote the superconducting and normal phases, respectively). The stable state of the rod is then the normal state and the system may undergo a phase transition. Actually the sharpness of the transition depends on the preparation of the material. The field at which the transition occurs, called the *critical field*, B_c, is a function of T and P. Assume that P is constant and consider δG evaluated on the coexistence line for two states of the rod, one a normal state, the other a superconducting state. The field is critical for both states and so $\delta G_s = \delta G_n$. Hence, from (12.1.11),

$$-S_s\,\delta T + V_s\,\delta P - I_s\,\delta B_c = -S_n\,\delta T + V_n\,\delta P - I_n\,\delta B_c, \tag{12.1.13}$$

and so the critical field must satisfy

$$\left(\frac{\partial B_c}{\partial T}\right)_P = \frac{S_s - S_n}{I_n - I_s}. \tag{12.1.14}$$

Now I_n can be ignored compared with I_s, which is given by (12.1.12). Hence we find

$$\left(\frac{\partial B_c}{\partial T}\right)_P = \frac{\mu_0(S_s - S_n)}{V_s B_c}, \tag{12.1.15}$$

or, rearranging,

$$S_s - S_n = \frac{V_s}{2\mu_0}\left(\frac{\partial(B_c^2)}{\partial T}\right)_P.$$

(12.1.16)

An indicative approximation for the function $B_c(T)$, given that P is constant, is the parabolic relation,

$$B_c(T) = B_{c0}\left\{1 - \left(\frac{T}{T_c}\right)^2\right\},$$

(12.1.17)

where B_{c0} is the limiting value of B_c as $T \to 0$ and T_c, the *critical temperature*, is the highest temperature at which the metal exhibits superconductivity (see Fig. 12.3). Because $(\partial B_c/\partial T)_P$ is negative, $S_s < S_n$, and hence the rod requires an input of heat in the superconducting $\to$ nominal transition.

Questions

12.1.1. (a) Show that the Guggenheim relation (12.1.4) is consistent with the Clausius–Clapeyron equation provided the molar latent heat of evaporation is given by

$$L = 5.3RT_c.$$

(12.1.18)

(b) The critical pressure of many non-polar elements and compounds lies in the range 2–10 MPa. Use (12.1.4) to show that

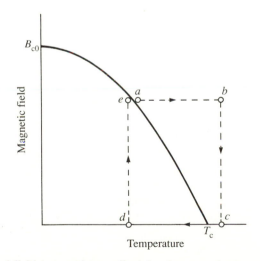

Fig. 12.3 Critical field (or coexistence line) for the normal-superconducting transition in a type-I superconductor. The states *a* and *e*, which are adjacent to the critical field, correspond to the normal and superconducting phases, respectively.

the corresponding normal boiling temperature at one atmosphere, T_B, is

$$T_B = 0.59T_c \pm 0.06T_c, \tag{12.1.19}$$

and hence obtain *Trouton's rule*,

$$L \approx 9RT_B. \tag{12.1.20}$$

12.1.2. It is intended to calibrate a thermometer to an accuracy of 0.01 K
by using the temperature of saturated steam as a known temperature. Determine the permissible error in the measurement of
the steam pressure. Take atmospheric pressure to be 10^5 Pa. The
latent heat of vaporization of water at this pressure is 2.26
MJ kg^{-1}.

12.1.3. For a skater weighing 650 N the area of contact with the ice is
$10^{-5} m^2$. Given the temperature of the ice is $-4°C$, determine
whether or not the ice under the skate will melt. The latent heat
of fusion of ice is 330 kJ kg^{-1}; the density of ice and water is
920 kg m^{-3} and 1000 kg m^{-3}, respectively.

12.1.4. Consider a rod of type-I superconducting metal.
(a) Show that $S_n = S_s$ at $T = T_c$.
(b) Use the third law to show that $(\partial B_c/\partial T)_P \to 0$ as $T \to 0$.
(c) Demonstrate that the parabolic model (12.1.17) satisfies
these conditions.
(d) Use (12.1.16) to show that the molar heat capacities, c_s and
c_n, evaluated at constant field in the superconducting and
normal states respectively, are related by

$$c_s - c_n = \frac{Tv_s}{2\mu_0} \left(\frac{\partial^2 (B_c^2)}{\partial T^2} \right)_P. \tag{12.1.21}$$

where v_s is the molar volume in the superconducting state.
Here the volume change with temperature is ignored.
(e) In the classification of Ehrenfest a phase transition in which
the latent heat is non-zero is said to be a first-order transition, while one having no latent heat but a discontinuity in
the molar heat capacity is of second order. Show that the
normal–superconducting transition is of second order when
$B = 0$, but is otherwise of first order.
(f) Given that (12.1.17) holds, show that at T_c,

$$c_s - c_n = \frac{4v_s (B_{c0})^2}{\mu_0 T_c}. \tag{12.1.22}$$

12.1.5. A rod of metallic superconductor (Section 12.1) is in the normal
state at the boundary of the superconducting region. By cooling

the rod at constant applied field, B, the process labeled $a \to e$ in Fig. 12.3, the rod becomes superconducting at essentially the same temperature. The field source is a solenoid connected to a current source.

(a) Show that the heat input to the rod in the direct process $a \to e$ is

$$Q_{ae} = \frac{T_a V}{2\mu_0} \left[\frac{\partial (B_a)^2}{\partial T} \right]_P, \qquad (12.1.23)$$

where V is the volume of the rod.

(b) For the direct process $a \to e$ show that the work done by the current source which maintains the applied field B constant is

$$W_{ae} = -\frac{V(B_a)^2}{\mu_0}. \qquad (12.1.24)$$

(c) Consider the process $a \to b \to c \to d \to e$ shown in Fig. 12 3. Use equation (12.1.21) to show that the heat input to the system in the processes $a \to b$ and $c \to d$ are related by

$$Q_{ab} = -Q_{cd} + \frac{T_a V}{2\mu_0} \left[\frac{\partial (B_a)^2}{\partial T} \right]_P - \frac{V(B_a)^2}{2\mu_0}. \qquad (12.1.25)$$

(d) Show that the work done by the current source for the solenoid in the process $d \to e$ is

$$W_{de} = -\frac{V(B_a)^2}{2\mu_0}, \qquad (12.1.26)$$

and confirm that

$$Q_{ab} + Q_{cd} + W_{de} = Q_{ae} + W_{ae}. \qquad (12.1.27)$$

12.1.6. The magnetic susceptibility of oxygen, one of a few gases which exhibit significant paramagnetic properties, is given by the Curie law:

$$\chi_{mT} = \frac{nc_c}{T}. \qquad (12.1.28)$$

Here $c_c = 1.23 \times 10^{-5} \, \text{m}^3 \, \text{K} \, \text{mol}^{-1}$ and $n = N/V$ is the molar concentration. Let G denote the thermodynamic potential having the natural variables T, P, B, and N; take B to be the applied field defined by the source coil current.

(a) Show that

$$G(T, P, B, N) = G(T, P, N) - \frac{NB^2 c_c}{2\mu_0 T} = N\mu, \qquad (12.1.29)$$

where μ is the chemical potential of the gas in the field B and $G(T, P, N)$ is the Gibbs free energy in the field-free state.

(b) Suppose a region of homogeneous magnetic field is occupied by a paramagnetic gas which is in unrestricted equilibrium with the same gas in a field-free region. Assume $G(T, P, N)$ is given by the ideal gas equation (10.3.44), suitably adapted for one component. Use the equilibrium condition,

$$\mu(T, P, B) = \mu(T, P_0, 0) \qquad (12.1.30)$$

to establish that the pressure, P, of the gas in the presence of the field is related to the pressure, P_0, in the field-free region by

$$P = P_0 \exp\left(\frac{c_c B^2}{2\mu_0 R T^2}\right). \qquad (12.1.31)$$

(c) Calculate the pressure enhancement $\delta P / P_0$ in oxygen at $T = 300\,\text{K}$ in a magnetic field of 10 tesla.

12.2 Stability of phases

Basic stability conditions

We shall consider the stability conditions for a homogeneous single–component fluid system, Σ, in a state characterized by U, V, and N. The entropy is given by the fundamental relation $S = S(U, V, N)$. From the maximum entropy principle we know that Σ will be in a stable state provided there is no other state having greater entropy for the same values of U, V and N. Suppose, for instance we are able to divide Σ into two subsystems, Σ_a and Σ_b, such that,

$$S(U_a, V_a, N_a) + S(U_b, V_b, N_b) > S(U, V, N)$$

where $U = U_a + U_b$, $V = V_a + V_b$, and $N = N_a + N_b$. In that case the original state will be unstable so that the system will tend to separate into subsystems in order to maximize $S(U_a, V_a, N_a) + S(U_b, V_b, N_b)$. That process would represent a phase change and, in general, the partitioned system would not be homogeneous. Thus, in order for the fluid to be stable against such an internal process it is necessary that

$$S(U_a + U_b, V_a + V_b, N_a + N_b) \geqslant S(U_a, V_a, N_a) + S(U_b, V_b, N_b), \qquad (12.2.1)$$

for all U_a, V_a, N_a and U_b, V_b, N_b.

According to the minimum energy principle (Section 9.7) we expect there will be a corresponding condition in the energy representation. This condition can be established formally (Galgani and Scotti 1968) from (12.2.1).

Consider a partitioning of Σ into subsystems Σ_a and Σ_b such that the entropy of Σ, $S = S_a + S_b$, as well as V and N, is constant. The energy of Σ may not be the same in the partitioned as in the homogeneous unpartitioned state. In the homogeneous state the energy, U_h, is given by

$$U_h = U(S, V, N) = U(S_a + S_b, V_a + V_b, N_a + N_b), \qquad (12.2.2)$$

and in the partitioned state the energy is $U_a + U_b$ where

$$U_a = U(S_a, V_a, N_a), \qquad U_b = U(S_b, V_b, N_b). \qquad (12.2.3)$$

Since the entropy of the homogeneous and inhomogeneous states of Σ is the same,

$$S(U_h, V_a + V_b, N_a + N_b) = S(U_a, V_a, N_a) + S(U_b, V_b, N_b) \qquad (12.2.4)$$

and it then follows from (12.2.1) that

$$S(U_h, V_a + V_b, N_a + N_b) \leqslant S(U_a + U_b, V_a + V_b, N_a + N_b). \qquad (12.2.5)$$

Hence, since S is a monotonic increasing function of U at constant V and N, we conclude that $U_h \leqslant U_a + U_b$, or

$$U(S_a + S_b, V_a + V_b, N_a + N_b) \leqslant U(S_a, V_a, N_a) + U(S_b, V_b, N_b). \qquad (12.2.6)$$

Thus Σ is stable against internal processes leading to inhomogeneity provided it has no partitioned state of lower energy at constant entropy. This condition is equivalent to (12.2.1) but we shall find it is more convenient to use.

Curvature conditions

To illustrate (12.2.6) suppose $V_a = V_b = V$, $N_a = N_b = N$ and consider the stability condition for Σ under a transfer of entropy between Σ_a and Σ_b. Let $S_a = S + \delta S$, $S_b = S - \delta S$; Σ will be stable under this perturbation provided:

$$U(2S, 2V, 2N) \leqslant U(S + \delta S, V, N) + U(S - \delta S, V, N),$$

or

$$U(S, V, N) \leqslant \tfrac{1}{2}\{U(S + \delta S, V, N) + U(S - \delta S, V, N)\}. \qquad (12.2.7)$$

This equation has a simple geometric interpretation. The right-hand side represents U at the midpoint of a straight line in the U–S plane joining the points with coordinates $(U(S + \delta S), \ S + \delta S)$ and $(U(S - \delta S), \ S - \delta S)$, as shown in Fig. 12.4. The left-hand side is the corresponding equilibrium value of U. Thus any chord connecting two points on the locus of equilibrium states in the U–S plane must lie above the locus. Such functions are said to be *convex* (meaning 'convex-down'). If the chord lies below the curve, the function is *concave*. Thus, for the system to be stable U must be a convex function of S at constant V and N. It also follows from (12.2.6)

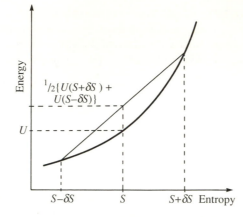

Fig. 12.4 Illustrating the curvature condition, eqn (12.2.7), that U must be a convex-down function of S when a single-phase system is stable.

that U must be a convex function of V and N and this condition must hold for large as well as small variations, δS, δV, and δN.

The implications of the convex condition are far-reaching. Suppose Fig. 12.5 represents the fundamental relation of a system obtained on the basis of a particular model, $U = U_m(S, V, N)$. The equation represents states which are locally stable along the curves *abc* and *efg*, since these are clearly convex. Nevertheless the entire region *bcdef* lies above the chord

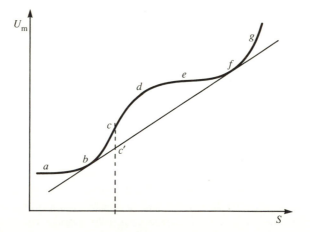

Fig. 12.5 Example of an unstable single-phase model fundamental relation, $U_m(S)$. Single-phase states lying along the locus *b–c–d–e–f* cannot be equilibrium states. States between *b* and *c* and between *e* and *f* are locally stable but globally unstable; they are metastable states. States between *c* and *e* are locally unstable.

drawn between b and f, and hence equation (12.2.6) is not satisfied. Assume the model is valid for the states b and f. Then for states with $S_b < S < S_f$, the system may achieve a state of lower energy at constant S by separating into two subsystems, one having the same density as state b, the other having the same density as state f. For example, if we attempt to place the system in state c the system would tend — given the entropy is constant — to adopt the state c' instead. This process would represent a phase transition.

Local stability criteria

Assume that U is well behaved in the thermodynamic domain of interest so that we may expand the terms on the right-hand side of (12.2.7) as a Taylor series in δS. The first term yields

$$U(S + \delta S, V, N) = U(S, V, N) + \frac{\partial U}{\partial S}\delta S + \frac{1}{2}\frac{\partial^2 U}{\partial S^2}\delta S^2 + \frac{1}{6}\frac{\partial^3 U}{\partial S^3}\delta S^3 + \cdots$$

$$(12.2.8)$$

and on substituting in (12.2.7) we find, to second order,

$$\frac{\partial^2 U}{\partial S^2}\delta S^2 \geqslant 0 \qquad (12.2.9)$$

or,

$$\left(\frac{\partial^2 U}{\partial S^2}\right)_{V,N} \geqslant 0. \qquad (12.2.10)$$

Note that this is a necessary, but not sufficient, condition for the stability of a homogeneous system. It can be expressed in terms of directly measurable quantities using

$$\frac{\partial^2 U}{\partial S^2} = \left(\frac{\partial T}{\partial S}\right)_{V,N} = \frac{T}{Nc_v},$$

and hence,

$$c_v \geqslant 0. \qquad (12.2.11)$$

Other stability criteria may be obtained from (12.2.6) by considering variations in V and N, but for the moment it is useful to take a different approach. We showed in Section 10.2 that the equilibrium state of a system, Σ in thermal contact with a heat reservoir minimizes the free energy, F, at the reservoir temperature. Indeed, we may replace ΔU, for the composite system of the reservoir and Σ, by ΔF. Here ΔU is evaluated at constant entropy and ΔF is determined at constant temperature. Hence, for a system held at constant T, (12.2.6) may be expressed in terms of F as

$$F(T, V_a + V_b, N_a + N_b) \leqslant F(T, V_a, N_a) + F(T, V_b, N_b). \quad (12.2.12)$$

On considering changes in V and N separately we obtain the following conditions for stability:

$$\left(\frac{\partial^2 F}{\partial V^2}\right)_{T,N} \geqslant 0, \quad (12.2.13)$$

and

$$\left(\frac{\partial^2 F}{\partial N^2}\right)_{T,V} \geqslant 0. \quad (12.2.14)$$

Since $\partial F / \partial V = -P$, it also follows that

$$\kappa_T = -\frac{1}{V}\left(\frac{\partial V}{\partial P}\right)_{T,N} \geqslant 0. \quad (12.2.15)$$

Stability criteria involving the temperature dependence of F follow from the energy condition, (12.2.10). In terms of T this condition becomes

$$\left(\frac{\partial S}{\partial T}\right)_{V,N} \geqslant 0, \quad (12.2.16)$$

and since $S = -\partial F / \partial T$, we obtain

$$\left(\frac{\partial^2 F}{\partial T^2}\right)_{V,N} \leqslant 0. \quad (12.2.17)$$

Evidently F must be a convex-down function in the F–V and F–N planes, but a concave-down function in the F–T plane. It is also clear that the stability criteria for different potentials are closely related. We shall examine these relationships in a more general context in Section 12.3.

The physical interpretation of the requirements $c_v \geqslant 0$ and $\kappa_T \geqslant 0$ is expressed by *Le Châtelier's principle*: the response of a system in equilibrium to an external perturbation is such as to reduce the perturbation. For example, suppose two simple systems of fixed volume, Σ_1 and Σ_2, are in equilibrium with respect to heat exchange. We perturb the state of Σ_1 by increasing its energy a small amount. Because $c_v \geqslant 0$, the temperature of Σ_1 will increase. In the subsequent process energy will be transferred to Σ_2 and the temperature difference will be reduced. If, on the other hand, c_v were negative, the temperature difference would have the opposite sign. The composite system would be unstable because further energy would then be transferred to Σ_1 from Σ_2 and the perturbation would lead to an ongoing process. Since such a process would be contrary to Le Châtelier's principle, c_v cannot be negative for a system in equilibrium.

12.3   General stability criteria

It is evident from Section 12.2 that the conditions for stability in a homogeneous system can be expressed in a number of ways. However, the relationships between the different criteria are not obvious. Accordingly we shall re-examine the stability conditions in a general way in this section. The results obtained here are useful subsequently, but not essential, so you may move directly to Section 12.4 if you wish, without loss of continuity.

We shall use $X_1, X_2, \ldots, X_t$ to denote the extensive variables of a system, and $P_1, P_2, \ldots, P_t$ to denote the corresponding intensive variables. In this notation equation (12.2.7) may be easily generalized to include displacements in all extensive variables. Now if $\delta X_1/X_1 = \delta X_2/X_2 = \cdots = \delta X_t/X_t$, then the generalized form of (12.2.7) reduces to a trivial equality because $U(X_1, X_2, \ldots, X_t)$ is first-order homogeneous, in which case the equality condition has no special significance for the stability of the system. Accordingly we shall choose to keep one extensive variable, X_t say, fixed. We shall define the corresponding specific quantities (or generalized densities), by

$$x_i = X_i/X_t \qquad 1 \leqslant i \leqslant t-1, \qquad (12.3.1)$$

$$u = U/X_t, \qquad (12.3.2)$$

$$u(x_1, x_2, \ldots, x_{t-1}) = \frac{1}{X_t} U(X_1, X_2, \ldots, X_t), \qquad (12.3.3)$$

and the associated Legendre transforms of u are given by expressions such as

$$\psi(P_1, P_2, \ldots, P_r) = u - \sum_{i=1}^{r} x_i P_i. \qquad (12.3.4)$$

Note that $\psi(P_1, P_2, \ldots, P_r)$ is a function of $P_1, P_2, \ldots, P_r, x_{r+1}, \ldots, x_{t-1}$. For simplicity we do not specify the variables x_i in the arguments of ψ; instead x_i will be implied whenever its conjugate variable P_i is omitted. In this notation equation (12.2.7) can be generalized:

$$u(x_1, x_2, \ldots, x_{t-1}) \leqslant \tfrac{1}{2} \{ u(x_1 + \delta x_1, x_2 + \delta x_2, \ldots, x_{t-1} + \delta x_{t-1})$$
$$+ u(x_1 - \delta x_1, x_2 - \delta x_2, \ldots, x_{t-1} - \delta x_{t-1}) \}. \qquad (12.3.5)$$

Given that the function u is well behaved, we may express the terms on the right-hand side as a Taylor series:

$$u(x_1 + \delta x_1, x_2 + \delta x_2, \ldots, x_{t-1} + \delta x_{t-1}) = u(x_1, x_2, \ldots, x_{t-1}) + \delta u$$
$$+ \tfrac{1}{2}\delta^2 u + \tfrac{1}{6}\delta^3 u + \cdots . \qquad (12.3.6)$$

Expressions for $\delta u, \delta^2 u, \ldots,$ are given in Appendix D. On substituting in

(12.3.5) the odd orders cancel, yielding the condition $\delta^2 u \geqslant 0$. More explicitly, if we use the notation

$$u_{mn} = \frac{\partial^2 u}{\partial x_m \partial x_n},$$ (12.3.7)

then

$$\delta^2 u = \sum_{m=1}^{t-1} \sum_{n=1}^{t-1} u_{mn} \, \delta x_m \, \delta x_n \geqslant 0,$$ (12.3.8)

for all deviations in the densities $\delta x_m \, \delta x_n$.

Since $u_{mn} = u_{nm}$ are real quantities the matrix $[u_{mn}]$ is a real symmetric matrix. In addition a matrix which satisfies (12.3.8) is said to be *non-negative*. All diagonal elements of a non-negative matrix must be positive or zero since (12.3.8) must hold in the special case that only one of the deviations, δx_m, is non-zero. However, the condition on the diagonal elements is not sufficient to ensure that $\delta^2 u \geqslant 0$. Instead, sufficient criteria may be established by selecting new thermodynamic coordinates so that $\delta^2 u$ is expressed as the sum of squared deviations. This transformation is equivalent to diagonalizing the matrix $[u_{mn}]$ by choosing new thermodynamic variables. Now there are many ways to do this, but because $[u_{mn}]$ is real and symmetric the transformation can always be achieved by means of a linear non-singular transformation (Korn and Korn 1968) of the type

$$\delta x_m = \sum_{i=1}^{t-1} a_{mi} \, \delta \bar{x}_i.$$ (12.3.9)

Here $\delta \bar{x}_i$ denotes a deviation in the transformed variables $\bar{x}_1, \ldots, \bar{x}_{t-1}$ and $[a_{mi}]$ is the matrix of a transformation which diagonalizes $[u_{mn}]$. Thus, on substituting for δx_m and δx_n in (12.3.8), we express $\delta^2 u$ as

$$\delta^2 u = \sum_{i=1}^{t-1} \bar{u}_{ii} (\delta \bar{x}_i)^2,$$ (12.3.10)

where $[\bar{u}_{ii}]$ is a diagonal matrix whose elements are given by

$$\bar{u}_{ij} = \sum_{m=1}^{t-1} \sum_{n=1}^{t-1} a_{mi} u_{mn} a_{nj}.$$ (12.3.11)

When $\delta^2 u$ is expressed in the form (12.3.10) the condition $\delta^2 u \geqslant 0$ is equivalent to the requirement that $\bar{u}_{ii} \geqslant 0$ for all i between 1 and $t-1$.

It is useful to distinguish between states for which $\delta^2 u > 0$ for *all* displacements δx_m and those for which $\delta^2 u = 0$ for *some* displacements. The first category are normal stable states in that the system will tend to revert to its original state after any displacement δx_m. Here we assume the global stability requirements are satisfied (see Fig. 12.5). In this situation the matrix $[\bar{u}_{ij}]$ cannot have any vanishing diagonal elements. Now the number of positive and zero diagonal elements of the matrix $[\bar{u}_{ij}]$ is the same for every transformation which diagonalizes a non-negative matrix $[u_{mn}]$. This is the Jacobi–Sylvester law of inertia (Korn and Korn 1968). Thus for normal stability it is sufficient to establish that the diagonalized matrix $[\bar{u}_{ij}]$ has no zero elements in one diagonal form only.

If, on the other hand, $\delta^2 u = 0$ for one or more displacements then $[\bar{u}_{ij}]$ must have one or more zero elements on the diagonal. Such states are found on the boundary between stable and unstable regions. The classification of the state as stable, unstable, or critically stable (see Section 12.5) then depends on the higher-order differentials. For the moment we shall put these matters to one side and restrict the discussion to normal stable states. With this limitation we have

$$u_{mm} = \left(\frac{\partial P_m}{\partial x_m}\right)_{x_1, x_2, \ldots, x_{t-1}} > 0 \qquad \text{for } 1 \leqslant m < t - 1 \qquad (12.3.12)$$

is a necessary condition to ensure that $\delta^2 u > 0$.

To establish sufficient stability criteria we shall now set up a linear transformation which expresses $\delta^2 u$ in a diagonal form, as in (12.3.10). We follow a step-by-step procedure (Tisza 1966), beginning by writing (12.3.8) in a way that isolates the terms involving δx_1:

$$\delta^2 u = u_{11}(\delta x_1)^2 + 2\sum_{n=2}^{t-1} u_{1n}\, \delta x_1\, \delta x_n + \sum_{m=2}^{t-1} \sum_{n=2}^{t-1} u_{mn}\, \delta x_m\, \delta x_n. \qquad (12.3.13)$$

Now in the representation with independent variables $x_1, \ldots, x_{t-1}$, the corresponding variation in P_1 is given by

$$\delta P_1 = \sum_{n=1}^{t-1} \frac{\partial P_1}{\partial x_n}\, \delta x_n = \sum_{n=1}^{t-1} u_{1n}\, \delta x_n, \qquad (12.3.14)$$

so we have

$$u_{11}(\delta x_1)^2 = \frac{(\delta P_1)^2}{u_{11}} - 2\sum_{n=2}^{t-1} u_{1n}\, \delta x_1\, \delta x_n - \sum_{m=2}^{t-1} \sum_{n=2}^{t-1} \frac{u_{1m} u_{1n}}{u_{11}}\, \delta x_m\, \delta x_n,$$

and hence, substituting in (12.3.13),

$$\delta^2 u = \frac{(\delta P_1)^2}{u_{11}} + \sum_{m=2}^{t-1} \sum_{n=2}^{t-1} \left\{\frac{u_{mn} u_{11} - u_{1m} u_{1n}}{u_{11}}\right\} \delta x_m\, \delta x_n. \qquad (12.3.15)$$

Now let $u_m = \partial u/\partial x_m$ and re-express the term in brackets using Jacobians

$$\frac{u_{mn} u_{11} - u_{1m} u_{1n}}{u_{11}} = \frac{\partial(u_n, u_1)}{\partial(x_m, x_1)} \left(\frac{\partial x_1}{\partial u_1}\right)_{x_m} = \frac{\partial(u_n, u_1)}{\partial(x_m, x_1)} \frac{\partial(x_m, x_1)}{\partial(x_m, u_1)}$$

$$= \left(\frac{\partial u_n}{\partial x_m}\right)_{u_1} = \left(\frac{\partial P_n}{\partial x_m}\right)_{P_1} = \psi_{mn}(P_1), \qquad (12.3.16)$$

and hence from (12.3.15)

$$\delta^2 u = \frac{(\delta P_1)^2}{u_{11}} + \sum_{m=2}^{t-1} \sum_{n=2}^{t-1} \psi_{mn}(P_1)\, \delta x_m\, \delta x_n. \qquad (12.3.17)$$

Clearly to ensure $\delta^2 u > 0$ we must have $\psi_{mm}(P_1) > 0$ for $2 \leqslant m \leqslant t-1$.

This process of completing the square may be continued. Subject to the condition $\delta P_1 = 0$, we have the following equation corresponding to (12.3.14):

$$\delta P_2 = \sum_{n=2}^{t-1} \psi_{2n}(P_1)\, \delta x_n, \qquad (12.3.18)$$

and on repeating the above procedure we obtain

$$\delta^2 u = \frac{(\delta P_1)^2}{u_{11}} + \frac{(\delta P_2)^2}{\psi_{22}(P_1)} + \sum_{m=3}^{t-1} \sum_{n=3}^{t-1} \psi_{mn}(P_1, P_2)\, \delta x_m\, \delta x_n$$

$$= \sum_{m=1}^{t-2} \frac{(\delta P_m)^2}{\psi_{mm}(P_1, P_2, \ldots, P_{m-1})} + \psi_{t-1,t-1}(P_1, \ldots, P_{t-2})(\delta x_{t-1})^2$$

$$(12.3.19)$$

where we have defined $\psi_{11} = u_{11}$. Notice that δP_m is the displacement in P_m subject to $P_1, P_2, \ldots, P_{m-1}, x_{m+1}, \ldots, x_{t-1}$ being fixed. Thus each δP_m refers to a different set of constraints and so its physical interpretation must reflect this. We see that (12.3.19) is the diagonal form we sought. It follows that

$$\psi_{mm}(P_1, P_2, \ldots, P_{m-1}) = \left(\frac{\partial P_m}{\partial x_m} \right)_{P_1, \ldots, P_{m-1}, x_{m+1}, \ldots, x_{t-1}} > 0, \qquad 1 \leqslant m \leqslant t-1,$$

is necessary and sufficient for $\delta^2 u > 0$. Also by inversion we have an equivalent condition involving P_m as the independent variable instead of x_m:

$$\left(\frac{\partial x_m}{\partial P_m} \right)_{P_1, \ldots, P_{m-1}, x_{m+1}, \ldots, x_{t-1}} > 0, \qquad 1 \leqslant m \leqslant t-1.$$

Thus by simply reordering the $P_1, P_2, \ldots, P_m$ and the $x_{m+1}, x_{m+2}, \ldots, x_{t-1}$ we conclude that

$$\frac{\partial x_m}{\partial P_m} > 0 \quad \text{and} \quad \frac{\partial P_m}{\partial x_m} > 0, \qquad (12.3.20)$$

for any representation in which P_m or x_m, respectively, are among the natural variables. These conditions may be written in terms of the corresponding potentials,

$$\frac{\partial^2 \psi}{\partial P_m^2} < 0 \quad \text{and} \quad \frac{\partial^2 \psi}{\partial x_m^2} > 0, \qquad (12.3.21)$$

where ψ is any potential density having P_m or x_m as a natural variable. In geometric terms these equations show that, for a normal stable state, ψ must be a concave-down function of its intensive parameters, P_m, and a convex-down function of the densities, x_m. These conditions represent a generalized expression of Le Châtelier's principle. We shall reserve consideration of the case in which $\partial^2 \psi / \partial x_m^2$ vanishes until Section 12.5.

We may also establish relations between the derivatives, $\partial P_m / \partial x_m$, evaluated in different representations. Consider a representation in which P_n and x_m are natural variables; the other variables need not be specified explicitly:

$$\left(\frac{\partial P_m}{\partial x_m}\right)_{P_n} = \frac{\partial(P_m, P_n)}{\partial(x_m, P_n)} = \frac{\partial(P_m, P_n)}{\partial(x_m, x_n)} \frac{\partial(x_m, x_n)}{\partial(x_m, P_n)}$$

$$= \left\{ \left(\frac{\partial P_m}{\partial x_m}\right)_{x_n} \left(\frac{\partial P_n}{\partial x_n}\right)_{x_m} - \left(\frac{\partial P_n}{\partial x_m}\right)_{x_n} \left(\frac{\partial P_m}{\partial x_n}\right)_{x_m} \right\} \left(\frac{\partial x_n}{\partial P_n}\right)_{x_m}$$

$$= \left(\frac{\partial P_m}{\partial x_m}\right)_{x_n} - \left\{ \left(\frac{\partial P_n}{\partial x_m}\right)_{x_n} \right\}^2 \left(\frac{\partial x_n}{\partial P_n}\right)_{x_m}. \qquad (12.3.22)$$

But from (12.3.20) we know $(\partial x_n / \partial P_n)_{x_m} > 0$. Hence it follows from (12.3.22) that

$$\left(\frac{\partial P_m}{\partial x_m}\right)_{x_n} \geqslant \left(\frac{\partial P_m}{\partial x_m}\right)_{P_n} > 0. \qquad (12.3.23)$$

By continuation we obtain an ordering of the derivatives:

$$\left(\frac{\partial P_m}{\partial x_m}\right)_{x_1, x_2, \ldots, x_{t-1}} \geqslant \left(\frac{\partial P_m}{\partial x_m}\right)_{P_1, x_2, \ldots, x_{t-1}} \geqslant \left(\frac{\partial P_m}{\partial x_m}\right)_{P_1, P_2, \ldots, x_{t-1}} \geqslant \cdots > 0. \quad (12.3.24)$$

Here x_m and P_m are not in the list of fixed variables. This result, which represents the principle of *Le Châtelier–Braun*, may be interpreted in the following way: when a system in equilibrium is subject to an external perturbation any indirect response which occurs will take place in a way which reduces the effect of the perturbation. To illustrate, consider a gas contained in a cylinder fitted with a movable piston. If the specific volume changes by δv the change in pressure will depend on whether the temperature or the entropy is constant. Since the variable conjugate to v is $-P$, eqn (12.3.23) yields

$$\left(\frac{\partial(-P)}{\partial v}\right)_s \geqslant \left(\frac{\partial(-P)}{\partial v}\right)_T > 0.$$

Thus when the volume decreases, such that $v \to v - \delta v$, the corresponding pressure change must satisfy

$$(\delta P)_s \geqslant (\delta P)_T > 0. \qquad (12.3.25)$$

Now in the isothermal volume change there will be an indirect process: heat will be transferred from the gas to its surroundings. Equation (12.3.25) indicates that the heat transfer will take place in such a way that the pressure increase (the direct response of the system) is less than it would have been if the process had been executed adiabatically.

Questions

12.3.1. Parts of a single-component fluid system undergo small departures from equilibrium due to an external perturbation. The entropy of the whole system and its material content remain constant.

(a) Show that the system will be stable with respect to these changes provided

$$\delta^2 u = u_{ss}(\delta s)^2 + 2u_{sv}\,\delta s\,\delta v + u_{vv}(\delta v)^2 > 0. \quad (12.3.26)$$

Here u, s, and v are the molar energy, entropy and volume and u_{ss}, u_{sv}, u_{vv} denote second-order derivatives of u with respect to s and v.

(b) Show that $\delta^2 u$ can be expressed as

$$\delta^2 u = \frac{1}{u_{vv}}\left\{ (u_{vv}\,\delta v + u_{sv}\,\delta s)^2 + (u_{ss}u_{vv} - u_{sv}u_{sv})\,(\delta s)^2 \right\}$$

$$= v\kappa_s(\delta P)^2 + \frac{T}{c_{\mathrm{p}}}\,(\delta s)^2. \quad (12.3.27)$$

where δP is the change in pressure at constant s and δs is the local deviation in s at constant P. Note (12.3.27) is a particular case of (12.3.19).

(c) Show that $\delta^2 u$ can also be expressed as

$$\delta^2 u = \frac{1}{u_{ss}}\left\{ (u_{ss}\,\delta s + u_{vs}\,\delta v)^2 + (u_{vv}u_{ss} - u_{vs}u_{vs})\,(\delta v)^2 \right\}$$

$$= \frac{c_v}{T}\,(\delta T)^2 + \frac{1}{v\kappa_T}\,(\delta v)^2, \quad (12.3.28)$$

where δT and δv are the local displacements in T and v at constant v and T, respectively.

(d) Give criteria for the system to be stable in terms of c_v, c_p, κ_T, and κ_s.

12.3.2. (a) Apply the Jacobi–Sylvester law of inertia to (12.3.27) and (12.3.28) to show that, if c_v is finite at the critical point of a single-component fluid system, then $c_p \to \infty$ at the critical point and κ_s must remain finite. Show that α must also diverge. (Recollect eqns (10.4.17) and (10.4.34).)

(b)   Show that the divergence of c_p and α follows directly from (10.4.17), (10.4.32), and (10.4.34), without invoking the Jacobi–Sylvester law of inertia, provided we can assert that $(\partial P/\partial T)_v$ is finite at the critical point. (This must hold on the critical isochore below the critical temperature because $(\partial P/\partial T)_v$ is then the gradient of the liquid–vapour coexistence line, eqn (12.1.3).)

(c)   It has been found that c_v diverges weakly at the critical point in fluids. Show that κ_s must also diverge.

12.3.3. (a)   The stability condition for a system characterized by the densities $x_1, x_2, \ldots, x_{t-1}$ is given by eqn (12.3.8). If we avoid critically stable states by excluding the equality condition, we have

$$\delta^2 u = \sum_{m=1}^{t-1} \sum_{n=1}^{t-1} u_{mn} \, \delta x_m \, \delta x_n > 0. \qquad (12.3.29)$$

Show that (12.3.29) may also be written as

$$\delta^2 u = \sum_{m=1}^{t-1} \delta P_m \, \delta x_m > 0. \qquad (12.3.30)$$

(b)   By using a representation in which $P_1, P_2, \ldots, P_r, x_{r+1}, \ldots, x_{t-1}$ are the independent variables, express δx_m in terms of $\delta P_1, \delta P_2, \ldots, \delta P_r, \delta x_{r+1}, \ldots, \delta x_{t-1}$ for $m \leqslant r$. Similarly for $n \geqslant r+1$ express δP_n in terms of the same set of displacements.

(c)   Assume (12.3.30) holds in the representation $P_1, P_2, \ldots, P_r, x_{r+1}, \ldots, x_{t-1}$. Use the results of (b) to show that $\delta^2 u$ can be expressed as the sum of two quadratic forms,

$$\delta^2 u = \sum_{m=1}^{r} \sum_{m'=1}^{r} \frac{\partial x_m}{\partial P_{m'}} \delta P_m \, \delta P_{m'} + \sum_{n=r+1}^{t-1} \sum_{n'=r+1}^{t-1} \frac{\partial P_n}{\partial x_{n'}} \delta x_n \, \delta x_{n'},$$
$$(12.3.31)$$

and hence obtain eqn (12.3.20).

12.3.4. (a)   In the notation of eqn (12.3.22) demonstrate that

$$\left(\frac{\partial x_m}{\partial P_m}\right)_{x_n} = \left(\frac{\partial x_m}{\partial P_m}\right)_{P_n} - \left\{\left(\frac{\partial x_n}{\partial P_m}\right)_{P_n}\right\}^2 \left(\frac{\partial P_n}{\partial x_n}\right)_{P_m} \qquad (12.3.32)$$

$$= \left(\frac{\partial x_m}{\partial P_m}\right)_{P_n} - \left\{\left(\frac{\partial P_n}{\partial P_m}\right)_{x_n}\right\}^2 \left(\frac{\partial x_n}{\partial P_n}\right)_{P_m}. \qquad (12.3.33)$$

(b)   Show that the principle of Le Châtelier–Braun may be expressed as

$$\left(\frac{\partial x_m}{\partial P_m}\right)_{P_1, P_2, \ldots, P_{t-1}} \geqslant \left(\frac{\partial x_m}{\partial P_m}\right)_{x_1, P_2, \ldots, P_{t-1}}$$

$$\geqslant \left(\frac{\partial x_m}{\partial P_m}\right)_{x_1, x_2, \ldots, P_{t-1}} \geqslant \cdots > 0, \qquad (12.3.34)$$

where the list of fixed variables excludes P_m and x_m.

(c) Confirm that equations (10.4.17), (10.4.35), and (11.6.44) are particular examples of (12.3.32) and that (11.5.8) is an example of (12.3.33).

12.4   Illustrative phase change phenomena

Fluids

The Helmholtz free energy for a fluid system in equilibrium with a heat reservoir must satisfy (12.2.12):

$$F(T, V_a + V_b, N_a + N_b) \leqslant F(T, V_a, N_a) + F(T, V_b, N_b). \quad (12.4.1)$$

It is useful to introduce a parameter λ $(0 \leqslant \lambda \leqslant 1)$ and define

$$V_\lambda = (1 - \lambda)V_a + \lambda V_b, \qquad N_\lambda = (1 - \lambda)N_a + \lambda N_b. \quad (12.4.2)$$

Then (12.4.1) implies

$$F(T, V_\lambda, N_\lambda) \leqslant (1 - \lambda)F_a + \lambda F_b \qquad (12.4.3)$$

where $F_a = F(T, V_a, N_a)$ and $F_b = F(T, V_b, N_b)$. In the particular case where $N_a = N_b = N$ and $V_a \neq V_b$, eqn (12.4.3) is a more specific condition than (12.4.1) since trivial homogeneous state changes are excluded.

The geometrical interpretation of (12.4.3) in this situation is illustrated in Fig. 12.6. As we vary λ the left-hand side represents a segment of the isotherm between the states (F_a, V_a) and (F_b, V_b) in the F–V plane; the right-hand side represents the chord connecting these two points. Thus a chord drawn between any two points must lie above the equilibrium line. If a model equation of state does not satisfy this condition, the system will generally undergo a phase change process.

As an illustration of this condition we shall use the Peng–Robinson equation of state, which represents real fluid properties rather better than the van der Waals equation (see question 12.4.1 for details). Figure 12.7 shows the isotherms of argon calculated from the Peng–Robinson equation together with isotherms determined by measurement (Michels *et al.* 1958). The figure shows that the equation fails in the mixed-phase region in the same way as the van der Waals equation, but elsewhere the Peng–Robinson equation is seen to offer a reasonable model for determining the properties of the liquid and vapour states.

Another view of a Peng–Robinson isotherm is presented in Fig. 12.8,

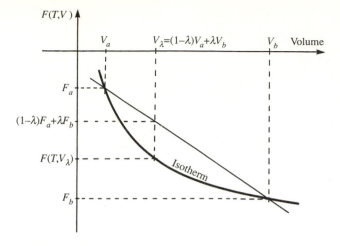

Fig. 12.6   Illustrating the inequality (12.4.3), which requires that the Helmholtz free energy should be a convex function of its extensive natural variables when T is constant.

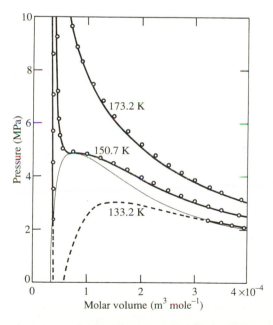

Fig. 12.7   Isotherms of argon calculated using the Peng–Robinson equation. The points (o) are derived from the measurements of Michels *et al.* (1958). The saturated liquid and vapour states are represented by the inverted bell-shaped solid line. The non-equilibrium states of the Peng–Robinson equation within the saturation envelope are represented by the dashed curve.

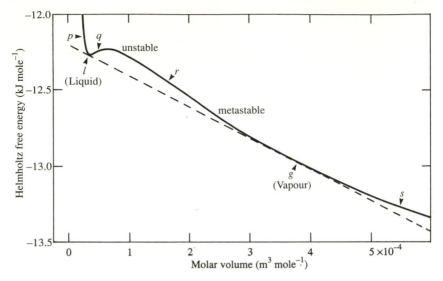

Fig. 12.8 Isotherm of argon at 130 K (solid line) in v–f coordinates, calculated from the Peng–Robinson equation of state. States lying on the solid line between l and g are either metastable or unstable. The equilibrium states lie on the double tangent (dashed line), which represents a mixture of the liquid and vapour states given by eqn (12.4.4).

which shows the molar free energy, f, plotted as a function of molar volume. The requirement for the chord to be above the equilibrium state in the f–v plane fails for $v_l \leqslant v \leqslant v_g$. Thus the states of the single-phase model are unstable between the points labelled l and g, corresponding to the saturated liquid and saturated vapour states, respectively. Instead, the equilibrium states are inhomogeneous mixtures of liquid and vapour, the composition being determined by the parameter λ,

$$f_e(\lambda) = (1 - \lambda) f(T, v_l) + \lambda f(T, v_g). \qquad (12.4.4)$$

Here f_e denotes the free energy of the equilibrium state for 1 mole and λ, which is the fraction of the mixture in the gas phase, is given by

$$\lambda = \frac{v - v_l}{v_g - v_l}. \qquad (12.4.5)$$

Although the equilibrium state is a heterogeneous state when $v_l \leqslant v \leqslant v_g$, nevertheless the states in the intervals l–q and g–r are locally stable. These states fail the global stability condition, (12.4.3), but the isotherm is locally convex. Such states are said to be metastable. Under suitable conditions systems may be prepared in such states. However, where the convexity condition is violated, in the region q–r for instance, the model is locally unstable and such states are not accessible. In a reversible phase change process a system may not pass through the metastable states since the process would then involve non-quasistatic steps whenever the system relaxes from the metastable single-phase state to the equilibrium mixed-phase state of lower free energy.

The tangent construction shown in Fig. 12.8 may be used to establish the molar volume for the saturated liquid and vapour states, v_l and v_g. Alternatively, we may solve the equilibrium equations

$$\mu(T, v_l) = \mu(T, v_g) \qquad (12.4.6a)$$

and

$$P(T, v_l) = P(T, v_g) = P_s. \qquad (12.4.6b)$$

Here P_s is the equilibrium pressure in the saturated state. To illustrate this solution, Fig. 12.9 show the locus of points $\mu(T, v)$, $P(T, v)$ as v is varied from the liquid to the vapour state, the arrows indicating the direction of increasing volume. The figure has been obtained using the Peng–Robinson equation for argon. The branches for which μ is a decreasing function of v are either stable or metastable, (since $(\partial \mu / \partial v)_T = -1/\kappa_T$ and κ_T must be positive). The rising branch in Fig. 12.9, q–r, is unstable. The point of intersection, $l = g$, represents the equilibrium saturated states and the branches l–q and g–r are metastable. However, the model equation provides no indication as to how much of each metastable branch is physically realizable.

The conditions for equilibrium (12.4.6a, b) allow us to revisit Maxwell's equal area rule (see Section 11.2). Recollect that we are concerned with state functions obtained from a model equation of state. The equilibrium states of the system are the states of the model which satisfy the criteria for equilibrium, such as (12.4.1). First we express $\mu(T, v)$ in terms of the molar free energy, f, in order to introduce v as the independent variable:

$$\mu(T, v_g) = f(T, v_g) + P_s v_g, \qquad (12.4.7)$$

$$\mu(T, v_l) = f(T, v_l) + P_s v_l, \qquad (12.4.8)$$

and so from (12.4.6a, b) we have

$$f(T, v_g) + P_s v_g - f(T, v_l) - P_s v_l = \int_{v_l}^{v_g} \left(\frac{\partial f(T, v')}{\partial v'} \right)_T dv' + P_s (v_g - v_l) = 0.$$

Hence,

$$\int_{v_l}^{v_g} \{ P(T, v') - P_s \} \, dv' = 0, \qquad (12.4.9)$$

where $P(T, v')$ represents the model equation of state and P_s is the two-phase saturated state pressure. This equation expresses Maxwell's rule that

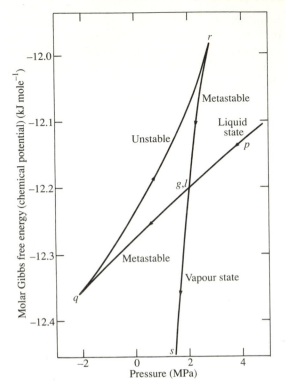

Fig. 12.9 Chemical potential of argon at 130 K showing the locus of states as the molar volume v is changed, calculated using the Peng–Robinson equation of state. The arrows indicate the direction of increasing v. The states p, l, q, r, g, and s correspond to the states with the same labels as in Fig. 12.8.

the area of the lobes cut off the model isotherm by the equilibrium line are equal, as illustrated in Fig. 11.1. The logic of using the unphysical part of the equation of state to determine the equilibrium pressure, P_s, is not in question, since the rule concerns the properties of the model equation of state, not the equilibrium state function. All we assume is that the model equation is well behaved in the domain of interest and that it is applicable in the neighbourhood of the states (T, v_g) and (T, v_l). If, for instance, we piece together a model equation of state which fits measurements in the single-phase, saturated, and metastable regions but is otherwise discontinuous in the two-phase region, the equal area rule would not be applicable. In reaching this conclusion we have made no assumption about the temperature dependence of the molar heat capacity at constant volume, c_v.

Superfluid helium

At atmospheric pressure, helium undergoes a phase change at 4.2 K from a gas to a normal liquid, called He-I. At 2.18 K it undergoes a further phase change from He-I to an abnormal form, He-II. This phase change is unusual. First, it is not a first-order phase change because the latent heat is zero. Instead, the molar heat capacity at the saturated pressure varies in a characteristically anomalous way (Fig. 12.10) which suggests the name, *λ-transition*, for the He-I to He-II phase change.

Many properties of He-II may be represented by a two-fluid model in which the liquid is regarded as a mixture of the normal and superfliud components. The existence of the superfluid form (see, for instance, Riedi (1988)) is related to the very light mass of the He atom, the fact that it has very weak atom–atom interactions, and to the integer spin of the dominant isotope, ^{4}He. The properties of the superfluid include: (i) it has zero entropy; and (ii) it exhibits no viscosity. Consequently a narrow flow passage which can restrict the flow of the normal component will nevertheless transmit the superfluid. Such a passage, called a superfluid leak, is therefore a barrier to entropy flow but not to mass flow.

This is unusual, since a wall which is non-restrictive to matter normally transmits heat also. Two normal systems separated by a non-restrictive wall are in equilibrium when both the chemical potentials and the temperatures are equal (see questions 9.6.2 and 9.7.2). By contrast, when two vessels of He-II are connected by a superfluid leak, only the chemical potentials need to be equal for equilibrium. We may therefore have two vessels of He-II in equilibrium with respect to matter flow whilst being maintained at different temperatures. Under this condition $\delta\mu = \mu_b - \mu_a = 0$, where μ_a and μ_b are the chemical potentials for He-II in the two vessels. The corresponding Gibbs–Duhem relation (9.5.37) becomes, $s\,\delta T - v\,\delta P = 0$,

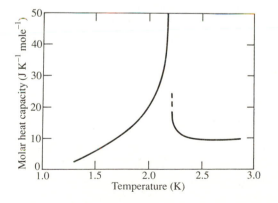

Fig. 12.10 Molar heat capacity of liquid helium at saturation, illustrating the characteristic λ-shape of the He-I, He-II phase transition.

where δT and δP represent the pressure and temperature difference between the two vessels. Thus we obtain

$$\left(\frac{\partial P}{\partial T}\right)_{\mu} = \frac{s}{v}. \tag{12.4.10}$$

This equation, called the *London equation*, shows that the vessel at the higher temperature will have the higher pressure due to the flow of the superfluid component through the connecting passage. If that vessel has a suitable opening a jet of liquid helium will emerge; this is the well-known fountain effect.

Binary solutions

A mixture of two liquids will exhibit characteristic phase change phenomena if the components are miscible at certain concentrations but immiscible at others. Here we shall examine this process using a simple model equation for the Gibbs free energy.

For many solutions, called *ideal binary solutions*, the Gibbs free energy may be expressed in a form similar to that for a mixture of ideal gases (10.3.44). We label the chemical components of the mixture 1 and 2 and define the mole fractions, x_i, in terms of the mole numbers, N_1 and N_2,

$$x_i = \frac{N_i}{N_1 + N_2}. \tag{12.4.11}$$

Following (10.3.44), G can be expressed as

$$G_{\text{ideal}}(T, P, N_1, N_2) = N_1\mu_1^0(T, P) + N_2\mu_2^0(T, P) + N_1 RT \ln(x_1)$$
$$+ N_2 RT \ln(x_2). \tag{12.4.12}$$

Here μ_1^0 and μ_2^0 denote the chemical potentials for the components, 1 and 2, in the unmixed form and the last two terms represent the entropy of mixing. Loosely speaking the Gibbs free energy of the mixture has this form when the molecular interactions between the different components are the same as the self interactions. The molar Gibbs potential,

$$g = \frac{G}{N_1 + N_2}, \tag{12.4.13}$$

is a convex-down function of x_1 (or x_2) taken as the independent variable, bearing in mind that

$$x_1 + x_2 = 1. \tag{12.4.14}$$

Hence the ideal binary solution is stable against a phase transition since (12.3.21) is satisfied.

The influence of other interactions between the components may be included in this model by adding further terms involving the products

$x_1 x_2$. The simplest such interaction term has the form $A(N_1 + N_2)x_1 x_2$, where A is a function of T and P. Binary solutions which may be represented in this way are *simple mixtures* (Guggenheim 1967). If, in addition, A is independent of T the system is a *regular solution*. Thus, for a simple mixture the molar free energy can be expressed as

$$g(T, P, x_1) = x_1 \mu_1^0(T, P) + (1 - x_1)\mu_2^0(T, P) + x_1 RT \ln(x_1)$$
$$+ (1 - x_1)RT \ln(1 - x_1) + Ax_1(1 - x_1). \quad (12.4.15)$$

From (12.3.21) we know the system will be stable against a separation into phases of different composition provided $(\partial^2 g/\partial^2 x_1)_T > 0$. Now $(\partial^2 g/\partial^2 x_1)_T$ is minimized when $x_1 = 0.5$, the minimum value being

$$\left(\frac{\partial^2 g}{\partial^2 x_1}\right)_T = 4RT - 2A. \quad (12.4.16)$$

Then $(\partial^2 g/\partial^2 x_1)_T = 0$ when $T = T_c$, where

$$T_c = \frac{A}{2R}. \quad (12.4.17)$$

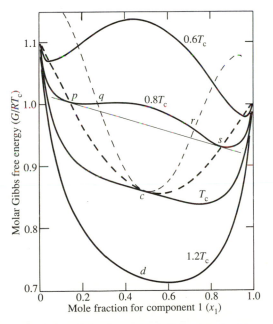

Fig. 12.11 Isotherms of the Gibbs free energy for a regular binary solution (solid lines) determined from eqn (12.4.15), taking $\mu_1^0 = RT_c$; $\mu_2^0 = 1.1 RT_c$. The thick dashed line, p–c–s, is the two-phase equilibrium coexistence line, obtained from eqn (12.4.27). The single-phase stability limit, obtained from (12.4.25) is represented by the thin dashed line q–c–r. The critical point is at c.

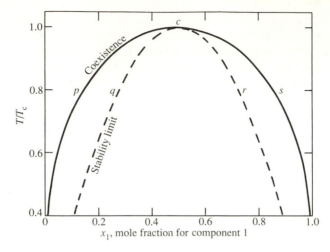

Fig. 12.12 Two-phase coexistence line (p–c–s) and the single-phase stability limit for a regular binary solution. These are also known as the binodal and spinodal lines, respectively. The points p, q, r, s, and c correspond to the labelled points in Fig. 12.11.

Thus when $A > 0$ and $T < T_c$ the isotherms of the Gibbs function will be concave at $x_1 = 0.5$, as illustrated in Fig. 12.11. The system is therefore not stable as a single phase near $x_1 = 0.5$. On the other hand, for $T > T_c$, g is convex at constant T and the single-phase system is stable at all concentrations, x_1.

Evidently T_c is the critical temperature for this system. When $T < T_c$ there is a range of concentrations for which g lies above the tangent line, such as the tangent at the points p and s in Fig. 12.11. Consequently the system separates into a two-phase mixture in this region, the individual coexisting phases having the compositions of the states p and s. This behaviour is analogous to the liquid–vapour system illustrated in Fig. 12.8. Note that not all the states on the single-phase locus connecting p ans s are locally unstable. Since the local stability condition $(\partial^2 g/\partial^2 x_1)_T > 0$ holds in the intervals p–q and r–s, these regions are metastable. Figure 12.12 shows the locus of coexistence states, and the boundary between the metastable and unstable regions, in $x_1 - T$ coordinates.

Questions

12.4.1. This is a rather long problem which requires some use of a computer. The aim is to to determine the saturation properties of a fluid by integrating its equation of state. In this example we shall use the Peng–Robinson equation, one of several empirical equations similar to the van der Waals equation. However, the

Peng–Robinson equation is rather better than the van der Waals equation for estimating state properties. The equation, as in Reid *et al.* (1986), may be written

$$P = \frac{RT}{v-b} - \frac{a}{v^2 + 2vb - b^2}\left\{1 + x\left[1 - \left(\frac{T}{T_c}\right)^{1/2}\right]\right\}^2, \quad (12.4.18)$$

where a and b are determined by the conditions (11.2.2) and (11.2.3) in terms of the critical pressure and temperature, T_c and P_c:

$$a = \frac{0.45724\,R^2 T_c^2}{P_c} \quad (12.4.19)$$

and

$$b = \frac{0.07780\,RT_c}{P_c}. \quad (12.4.20)$$

In addition,

$$x = 0.37464 + 1.54226\omega - 0.26992\omega^2 \quad (12.4.21)$$

where ω is a constant (the *acentric factor*). For simple monatomic gases, $\omega \ll 1$; for argon $\omega = 1 \times 10^{-3}$. In the following we shall use the subscript 0 to denote an ideal gas limiting state for the system.

(a) The molar heat capacity of argon at constant volume is $c_v = 1.5R$ in the ideal gas limit where $T \to T_0$, $v \to v_0$. Given the molar volume v, and temperature T, show that

$$c_v = 1.5R + \frac{x(x+1)\{I(v) - I(v_0)\}}{2(TT_c)^{1/2}} \quad (12.4.22)$$

where

$$I(v) = -\int \frac{a\,dv}{v^2 + 2vb - b^2} = \frac{a}{2\sqrt{2}}\ln\left\{\frac{v + (1 + 2\sqrt{2})b}{v + (1 - 2\sqrt{2})b}\right\}, \quad (12.4.23)$$

assuming $v + (1 - 2\sqrt{2})b > 0$.

(b) Use (11.1.13) to show that the molar free energy is given by

$$f(T, v) = u_0 - Ts_0 + 1.5R(T - T_0)$$

$$- 1.5RT\ln\left(\frac{T}{T_0}\right) - RT\ln\left(\frac{v-b}{v_0-b}\right)$$

$$+ \{I(v) - I(v_0)\}\left\{1 + x\left[1 - \left(\frac{T}{T_c}\right)^{1/2}\right]\right\}^2 \quad (12.4.24)$$

and obtain an expression for $s(T, v)$.

(c) The molar mass of argon is $3.995 \times 10^{-2}\,\mathrm{kg\,mol^{-1}}$ and the critical parameters are (Michels *et al.* 1958)

$$P_c = 4.898\,\mathrm{MPa}, \qquad T_c = 150.87\,\mathrm{K},$$

$$v_c = 74.56 \times 10^{-6}\,\mathrm{m^3\,mol^{-1}}.$$

Select a suitable state, in the ideal gas region and evaluate the corresponding values of u_0 and s_0 for argon using the fundamental relation for a monatomic ideal gas, (9.3.1).

Obtain an expression for the chemical potential for argon, $\mu(T, v)$, on the basis of the Peng–Robinson model. Plot μ as a function of P, using v as an independent variable, for a constant value of T between $100\,\mathrm{K}$ and $140\,\mathrm{K}$. Hence determine the saturated vapour pressure of argon at this temperature. Compare your value with the data in Table 12.1.

(d) Devise an iterative routine to obtain numerical values for v_g, v_l, and P_s by solving equations (12.4.6a) and (12.4.6b), given the temperature, T. Determine the saturated liquid and vapour entropy. Table 12.1 shows results for argon obtained by a procedure of this kind. The triple point is $83.8\,\mathrm{K}$.

(e) Confirm, by numerical calculation, that the saturated vapour pressure of argon obtained from the Peng–Robinson equa-

Table 12.1 Saturated state properties of argon calculated using the Peng–Robinson equation of state. The subscripts l and g denote the liquid and vapour phases, respectively. In comparison with tabulated data (Perry and Chilton 1973), the error in P_s is less than 1%; v_l is too small by up to 15% below $140\,\mathrm{K}$, and too large by nearly 10% at the critical point; v_g is within 2% below $140\,\mathrm{K}$ but is too large by almost 10% at the critical point; s_l and s_g are within 1% of tabulated data below $140\,\mathrm{K}$ and are in error by 2% at the critical point.

T	P_s	v_l	v_g	u_l	u_g	s_l	s_g
(K)	(MPa)	$(10^{-5}\,\mathrm{m^3\,mol^{-1}})$		$(\mathrm{J\,mol^{-1}})$		$(\mathrm{J\,mol^{-1}\,K^{-1}})$	
83.8	0.069	2.46	986	-4813	1025	53.5	131.3
90	0.134	2.53	539	-4552	1085	56.5	127.1
100	0.323	2.68	238	-4114	1160	61.1	121.5
110	0.667	2.86	120	-3646	1200	65.6	116.7
120	1.22	3.11	65.9	-3135	1190	70.1	112.5
130	2.03	3.50	37.9	-2554	1103	74.8	108.3
140	3.19	4.17	21.6	-1841	875	80.2	103.5
150	4.74	6.39	10.0	-637	149	89.1	95.4

tion of state is in reasonable agreement with the Guggenheim relation (12.1.4).

12.4.2. (a) Given that $A > 0$ in eqn (12.4.15), obtain the critical temperature of a simple binary mixture, $T_c = A/2R$, by assuming $(\partial^3 g/\partial^3 x_1)_T = 0$, $(\partial^2 g/\partial^2 x_1)_T = 0$ at the critical point. Show that the boundary between the regions of local stability and instability is given by

$$x_1 = \frac{1 \pm \sqrt{1 - T/T_c}}{2}. \qquad (12.4.25)$$

(b) The equation for the coexistence line in a simple mixture (p–c–s in Figs 12.11 and 12.12) may be obtained as follows. First note that the equation for g, (12.4.15), can be expressed as a linear term plus a term $g_s(P, T, x_1)$ which is symmetrical about $x_1 = 0.5$:

$$g = \mu_2^0(P, T) + x_1(\mu_1^0(P, T) - \mu_2^0(P, T)) + g_s(P, T, x_1). \qquad (12.4.26)$$

It follows that the double tangent which defines the points p and s must have zero gradient in x_1–g_s coordinates. Hence show that the states on the coexistence locus must satisfy

$$\ln\left(\frac{x_1}{1 - x_1}\right) + \frac{2T_c}{T}(1 - 2x_1) = 0. \qquad (12.4.27)$$

The solutions of this equation may be obtained numerically. Show that if x_1 is on the coexistence locus, then $1 - x_1$ is also.

(c) Establish that the chemical potential for component 1 in the mixture is given by

$$\mu_1(T, P, x_1) = \mu_1^0(P, T) + RT\ln(x_1) + 2RT_c(1 - x_1)^2. \qquad (12.4.28)$$

Let x_1 and x_1' denote the mole fractions for component 1 in coexisting phases when $T < T_c$. By evaluating μ_1 and μ_2 demonstrate that

$$\mu_1(T, P, x_1) = \mu_1(T, P, x_1') \qquad \text{and}$$
$$\mu_2(T, P, x_1) = \mu_2(T, P, x_1'). \qquad (12.4.29)$$

12.4.3. According to the Gibbs phase rule the state of a single-component system in the form of a two-phase mixture can be represented by the temperature together with the amounts of the two phases. Yet in Appendix B the state of a mixture of water and steam is represented by global state variables, such as U, V, and N, as if the

system consisted of a single phase. Explain how these two representations are linked.

12.5   Critical phenomena

The critical states of a system lie between the regions of phase stability and
instability a rather special way. Consider, for example, the critical point,
c, of a simple binary mixture shown in Fig. 12.11. For $T < T_c$ the line
q–c–r is the boundary of the locally stable single-phase region. Along this
line $(\partial^2 g/\partial x_1^2)_T = 0$. When $T > T_c$ the isotherms are convex-down every-
where and the system is stable, globally and locally. The critical point is
the state at which both the limiting stability line and the coexistence line
are tangential with the critical isotherm. Thus the condition $(\partial^2 g/\partial x_1^2)_T = 0$
holds at the critical point and also in neighbouring states on the critical
isotherm. Hence the critical state is identified by the conditions

$$\left(\frac{\partial^2 g}{\partial x_1^2}\right)_T = 0, \qquad \left(\frac{\partial^3 g}{\partial x_1^3}\right)_T = 0. \tag{12.5.1}$$

The equations we used to identify the critical state of a van der Waals gas,
(11.2.2) and (11.2.3), may be expressed similarly,

$$\left(\frac{\partial^2 f}{\partial v^2}\right)_T = 0, \qquad \left(\frac{\partial^3 f}{\partial v^3}\right)_T = 0, \tag{12.5.2}$$

and the generalization to other systems is straightforward.

   These conditions indicate that the isotherm of the thermodynamic potential must be flattened in the region of the critical point, as Fig. 12.11 shows.
Consequently, in the event the system departs from equilibrium the change
in the potential which would normally restore the system to its initial state
is greatly reduced at the critical point. To illustrate, suppose there is a local
departure from uniform concentration in a binary mixture. The concentration elsewhere must make a compensating deviation. This situation is like
that shown in Fig. 12.6, where the state of the non-uniform system lies on
the chord joining the two states of deviation. By analogy we can understand
that if $g(x_1)$ is a convex function (such as at state d in Fig. 12.11) the free
energy of the system in the non-uniform state will exceed that of the uniform state. The system will revert to its initial state accordingly. However,
the restoring potential will be reduced if the curvature of $g(x_1)$ is small, as
at the critical state. Thus we expect that fluctuations, which occur in all
thermodynamic systems, will be greatly enhanced. In addition, the spatial
scale over which the fluctuations will exhibit correlated behaviour will
increase as the critical point is approached. In liquid–vapour transitions,
such density fluctuations are the cause of *critical opalescence*, a dramatic
enhancement in the scattering of light by the fluid at the critical point.

The critical point fluctuations suggest that the simple thermodynamic state representation of the critical state is unsatisfactory and incomplete. Indeed the theory of critical phenomena has been an active field of research since the time of van der Waals, especially since the 1950s when it became clear that the thermodynamic potentials are singular at the critical point. This is not surprising since the system may pass from a stable to an unstable state at the critical point without passing through metastable states (see Fig. 12.11). In addition, the response functions of the system, such as κ_T, c_p, and α, diverge at the critical point; we have seen this already from the van der Waals model (questions 11.2.5 and 11.2.7). But the van der Waals equation fails to account correctly for the form of these divergences. This may seem to be a detail, which could perhaps be related to some incidental aspect of the van der Waals equation of state. However, this difficulty actually derives from the underlying statistical model, rather than the van der Waals equation itself, and so the problem is significant. Other properties derived from the van der Waals equation exhibit similar descrepancies at the critical point; for later reference we list some of them now.

First consider the isothermal compressibility, κ_T. In the neighbourhood of the critical point we shall replace T and v by the variables $\hat{T}$ and $\hat{v}$ defined by

$$\hat{T} = \frac{T}{T_c} - 1 \quad \text{and} \quad \hat{v} = \frac{v}{v_c} - 1, \tag{12.5.3}$$

where T_c and v_c denote the critical temperature and volume; we shall use this notation for other variables also. For $T > T_c$ the temperature variation of κ_T along the critical isochore of a van der Waals gas may be obtained from eqns (11.2.21) and (11.2.22). Setting $\hat{v} = v_r - 1 = 0$, we obtain

$$\kappa_T \sim |\hat{T}|^{-1}. \tag{12.5.4}$$

Here we have used $|\hat{T}|$ since this limiting form also holds on the van der Waals liquid–vapour coexistence line for $T < T_c$ (see question 12.5.2). Measurements, on the other hand, yield:

$$\kappa_T \sim |\hat{T}|^{-\gamma}, \tag{12.5.5}$$

where γ is a *critical point exponent*, and $\gamma = 1.23$, for many fluids. In the theory of critical phenomena there are a number of other such exponents (for introductory treatments see Stanley (1971), Waldram (1985), or Rowlinson (1988)), of which we shall consider three, α, β, and δ which are defined as follows:

(1)   molar heat capacity at constant volume on the critical isochore:

$$c_v \sim |\hat{T}|^{-\alpha}; \tag{12.5.6}$$

(2)   saturated liquid–vapour density difference:

$$\rho_l - \rho_g \sim (-\hat{T})^\beta, \tag{12.5.7}$$

where $\rho = 1/v$;

(3)   pressure–density relation on the critical isotherm:

$$|\hat{P}| \sim |\hat{\rho}|^\delta. \tag{12.5.8}$$

For a van der Waals gas we find (see question 12.5.1) $\beta = 0.5$, $\delta = 3$, whereas the measured values for pure fluids are $\beta = 0.33$, $\delta = 4$–5. In addition, c_v has a discontinuity of $4.5R$ at the critical point for states confined to the critical isochore of a van der Waals gas, but no divergence (question 12.5.3). Hence $\alpha = 0$, whereas it is now accepted that $\alpha \approx 0.11$, which indicates a weak divergence in c_v at the critical point.

Clearly the measured values for the critical exponents are inconsistent with the van der Waals model. As mentioned, this difficulty concerns not just the van der Waals equation, but the statistical assumptions which it implies. In such theories, called mean field theories, the interactions between atoms are represented by an average field due to all other particles; the fluctuating part of the interatomic field is disregarded. The mean field theories are applicable only when the interatomic forces have extremely long range, a condition which is not observed in normal fluids.

Similar difficulties arise in other critical phenomena, such as the ferro-magnetic–paramagnetic transition in zero magnetic field. Yet the corresponding critical exponents for the ferromagnetic transition are approximately the same as those for fluids. This suggests that the physical processes which determine the critical exponents have a universal character, being essentially the same in systems which are otherwise dissimilar. In addition, it can be established on quite general grounds that the critical exponents are linked by equations of constraint, such as the Rushbrooke inequality, $\alpha + 2\beta + \gamma \geqslant 2$.

The first theory to offer a unified account of these features of the critical point was a phenomenological proposal, called the *scaling hypothesis*, put forward by Widom and others in 1965. According to this suggestion the thermodynamic potentials near the critical point may be expressed as the sum of two functions, one of which is singular. To illustrate the scaling hypothesis we shall consider the molar Helmholtz function for a fluid, $f(T, v)$. According to the hypothesis, f can be expressed as

$$f(T, v) = -s_c T - P_c v + f_s(\hat{T}, \hat{v}) \tag{12.5.9}$$

in the single-phase region in the neighbourhood of the critical point. The scaling hypothesis also requires that f_s should be a singular function satisfying the relation

$$f_s(\lambda^a \hat{T}, \lambda^b \hat{v}) = \lambda f_s(\hat{T}, \hat{v}), \tag{12.5.10}$$

for all λ. Here a and b are constants, called scaling parameters. Such a function is a *generalized homogeneous function*. Equation (12.5.10) expresses a scaling rule: under a scale change near the critical point in which $\hat{T} \to \lambda^a \hat{T}$, $\hat{v} \to \lambda^b \hat{v}$, then $f_s \to \lambda f_s$. Thus the scaling hypothesis asserts that the form of f_s is preserved in a scale change provided $\hat{T}$ and $\hat{v}$ change according to the scaling rule. Thus the form of f_s is closely proscribed by the scaling hypothesis and, in turn, this fixes the scaling relations for other properties of the system at the critical point, including the critical point indices.

As an example consider the molar heat capacity at constant volume on the critical isochore at $T > T_c$. Here $\hat{v} = 0$ and $\hat{T} > 0$. First take the temperature derivative of $f(T, v)$. From (12.5.3) and (12.5.9) we obtain

$$s = s_c - \frac{\partial f_s}{\partial T} = s_c - \frac{1}{T_c} \frac{\partial f_s}{\partial \hat{T}}, \tag{12.5.11}$$

and hence

$$\hat{s} = \frac{s - s_c}{s_c} = \frac{-1}{s_c T_c} \frac{\partial f_s}{\partial \hat{T}}. \tag{12.5.12}$$

Now in the neighbourhood of the critical point,

$$c_v = -T_c \frac{\partial^2 f}{\partial T^2} = s_c \frac{\partial \hat{s}}{\partial \hat{T}}, \tag{12.5.13}$$

and the scaling relation for $\hat{s}$ is obtained by differentiating (12.5.10) with respect to T. The left-hand side yields

$$\frac{1}{T_c} \frac{\partial f_s(\lambda^a \hat{T}, \lambda^b \hat{v})}{\partial \hat{T}} = \frac{\lambda^a}{T_c} \frac{\partial f_s(\lambda^a \hat{T}, \lambda^b \hat{v})}{\partial (\lambda^a \hat{T})} = -\lambda^a \hat{s}(\lambda^a \hat{T}, \lambda^b \hat{v}), \tag{12.5.14}$$

and the right-hand side is simply $-\lambda \hat{s}(\hat{T}, \hat{v})$. Hence,

$$\lambda^a \hat{s}(\lambda^a \hat{T}, \lambda^b \hat{v}) = \lambda \hat{s}(\hat{T}, \hat{v}). \tag{12.5.15}$$

From (12.5.13) we see that the corresponding relation for c_v is obtained by differentiating (12.5.15) with respect to $\hat{T}$ again:

$$\lambda^{2a} c_v(\lambda^a \hat{T}, \lambda^b \hat{v}) = \lambda c_v(\hat{T}, \hat{v}). \tag{12.5.16}$$

On the critical isochore $\hat{v} = 0$, and since (12.5.16) must hold for all λ we may set $\lambda^a \hat{T} = 1$. On rearranging, we find

$$c_v(\hat{T}, \hat{v}) = \hat{T}^{-\alpha} c_v(1, 0),$$

or

$$c_v(\hat{T}, \hat{v}) \sim \hat{T}^{-\alpha} \tag{12.5.17}$$

where

$$\alpha = 2 - 1/a. \tag{12.5.18}$$

The scaling relation for $\hat{P}$ is obtained from (12.5.9) using

$$P = -\frac{\partial f}{\partial v} = P_c - \frac{1}{v_c}\frac{\partial f_s}{\partial \hat{v}}, \qquad (12.5.19)$$

yielding

$$\lambda^b \hat{P}(\lambda^a \hat{T}, \lambda^b \hat{v}) = \lambda \hat{P}(\hat{T}, \hat{v}). \qquad (12.5.20)$$

On the critical isotherm $\hat{T} = 0$. Hence putting $\lambda^b|\hat{v}| = 1$ we obtain

$$|\hat{P}| \sim |\hat{v}|^\delta \qquad (12.5.21)$$

where

$$\delta = 1/b - 1. \qquad (12.5.22)$$

Note that (12.5.21) is equivalent to (12.5.8) since $\hat{v} \approx -\hat{\rho}$ provided $|\hat{v}| \ll 1$.

The isothermal compressibility on the critical isochore, given $\hat{T} > 0$, is obtained from $\partial^2 f_s/\partial \hat{v}^2$ by setting $\hat{v} = 0$ and $\lambda^a \hat{T} = 1$. We obtain

$$\kappa_T \sim \hat{T}^{-\gamma}, \qquad (12.5.23)$$

where

$$\gamma = \frac{1 - 2b}{a}. \qquad (12.5.24)$$

To complete the link between the scaling parameters and the critical indices we require an expression for β involving the ρ-$\hat{T}$ relation in the two-phase region. From the Gibbs phase rule (Section 12.1) we know that there is only one intensive degree of freedom for states lying on the saturation boundary of the single-phase region. So for these states, which we shall identify by the subscript s, it must be possible to express $\hat{P}_s$ as a function of $\hat{T}_s$. Since (12.5.20) also holds, it follows that $\hat{v}_s \to \lambda^b \hat{v}_s$ when $\hat{T}_s \to \lambda^a \hat{T}_s$. Hence, expressing $\hat{v}_s$, as a function of $\hat{T}_s$, we must have

$$\hat{v}_s(\lambda^a \hat{T}_s) = \lambda^b \hat{v}_s(\hat{T}_s), \qquad (12.5.25)$$

and, setting $\lambda^a \hat{T}_s = -1$, (since $\hat{T}_s < 0$) we find

$$\hat{v}_s \sim (-\hat{T}_s)^{b/a}. \qquad (12.5.26)$$

Since $\hat{v} \approx -\hat{\rho}$, we obtain

$$\rho_l - \rho_g \sim (-\hat{T})^\beta \qquad (12.5.27)$$

on the liquid–vapour coexistence line, where

$$\beta = b/a. \qquad (12.5.28)$$

These equations show how the critical indices α, β, γ, and δ are related to the scaling constants, a and b. Clearly the values of the scaling constants are subject to mutual constraints since there are only two scaling constants. One may simply show, for example,

$$\beta(\delta - 1) = \gamma, \tag{12.5.29}$$

$$\alpha + \beta(\delta + 1) = 2, \tag{12.5.30}$$

$$\alpha + 2\beta + \gamma = 2. \tag{12.5.31}$$

These relations were originally established as inequalities on the basis of thermodynamic arguments (Stanley 1971).

Despite its success, the scaling hypothesis initially drew criticism because it lacked a fundamental basis. The hypothesis was seen to be heuristically reasonable but the theory was incomplete and no means existed to calculate the scaling constants. This problem was solved in 1971 by Wilson who established a statistical foundation for the scaling hypothesis and methods for calculating the critical exponents. For a descriptive introduction see Waldram (1985); Reichl (1980) provides a more advanced treatment.

Questions

12.5.1. (a) Use (11.2.13) to obtain μ as a function of T and v for a van der Waals gas:

$$\mu(T, v) = u_0 - Ts_0 + c_v \left\{ T - T_0 - T \ln\left(\frac{T}{T_0}\right) \right\}$$

$$- a\left(\frac{2}{v} - \frac{1}{v_0}\right) + RT\left\{\frac{v}{v-b} - \ln\left(\frac{v-b}{v_0-b}\right)\right\}. \tag{12.5.32}$$

(b) Given that $|\hat{v}| \ll 1$, show that

$$\hat{\rho} = -\hat{v}. \tag{12.5.33}$$

(c) Use (11.2.14) and (12.5.32) together with the definitions of $\hat{T}$ and $\hat{v}$, eqn (12.5.3), to show that the liquid–vapour phase equilibrium condition, $\mu(T, v_l) = \mu(T, v_g)$, can be expressed as

$$\phi(\hat{T}, \hat{v}_l) = \phi(\hat{T}, \hat{v}_g), \tag{12.5.34}$$

where

$$\phi(\hat{T}, \hat{v}) = \ln(3\hat{v} + 2) + \frac{9}{4(\hat{v} + 1)(\hat{T} + 1)} - \frac{1}{3\hat{v} + 2}. \tag{12.5.35}$$

Here the subscripts l and g refer to the saturated liquid and vapour states, respectively. By expanding $\phi(\hat{T}, \hat{v})$ as a power series in $\hat{T}$ and $\hat{v}$ (to terms of third order) establish that $\hat{v}_l$ and $\hat{v}_g$ are related by

$$4(\hat{T} - \hat{T}^2)(1 - \hat{v}_l - \hat{v}_g) + \hat{v}_l^2 + \hat{v}_l\hat{v}_g + \hat{v}_g^2 = 0. \quad (12.5.36)$$

(d)   Express the van der Waals equation of state (11.2.17) near the critical point in terms of $\hat{P}$, $\hat{T}$ and $\hat{v}$. To third order in $\hat{v}$ show that the reduced pressure is given by

$$P_r = 1 + 4\hat{T} - 6\hat{T}(\hat{v} - 1.5\hat{v}^2) - 1.5\hat{v}^3. \quad (12.5.37)$$

Hence show that the equilibrium condition, $P(T, v_l) = P(T, v_g)$, can be expressed as

$$4\hat{T}\{1 - 1.5(\hat{v}_l + \hat{v}_g)\} + \hat{v}_l^2 + \hat{v}_l\hat{v}_g + \hat{v}_g^2 = 0. \quad (12.5.38)$$

(e)   Use (12.5.36) and (12.5.38) to obtain equations for the volume of saturated liquid and vapour near the critical point of a van der Waals gas:

$$\hat{v}_l = \hat{T} - 2\sqrt{(-\hat{T})} \quad (12.5.39)$$

$$\hat{v}_g = \hat{T} + 2\sqrt{(-\hat{T})} \quad (12.5.40)$$

and hence

$$\rho_l - \rho_g = 4\sqrt{(-\hat{T})}, \quad (12.5.41)$$

$$\rho_l + \rho_g = -2\hat{T} \quad (12.5.42)$$

Equation (12.5.41) shows that the critical exponent defined by (12.5.7) is $\beta = 0.5$, for a van der Waals gas. Equation (12.5.42) shows that the average of the coexisting liquid and vapour densities is a linear decreasing function of the temperature near the critical point of a van der Waals gas. This result, which is consistent with measurements, is the *law of rectilinear diameters*. It provides a means for determining the critical density.

(f)   Show that the reduced pressure on the critical isotherm of a van der Waals gas is given by

$$P_r = 1 - 1.5\hat{v}^3. \quad (12.5.43)$$

Hence the critical exponent defined by (12.5.8) is $\delta = 3$ for a van der Waals gas.

12.5.2.   (a)   The reduced compressibility of a van der Waals gas, κ_{Tr}, defined by (11.2.20), is given in (11.2.22). Show that

$$\kappa_{Tr} = \frac{1}{6\hat{T}}, \quad (12.5.44)$$

on the critical isochore.

(b)   Equation (12.5.44) is not applicable when $T < T_c$, since the van der Waals equation does not represent equilibrium states in the mixed-phase region. Instead, the approach to the critical point may be made along the saturated gas phase line. For this case show that

$$\kappa_{Tr} = \frac{1}{12(-\hat{T})}. \qquad (12.5.45)$$

(Hint: use (11.2.22) and (12.5.40) and retain only the lowest order in $(-\hat{T})$.) Equations (12.5.44) and (12.5.45) show that the critical exponent defined by (12.5.5) is $\gamma = 1$, for a van der Waals gas.

12.5.3. For states on the critical isochore the molar heat capacity of a van der Waals gas at constant volume exhibits a characteristic jump at the critical point. The existence of this jump may be demonstrated in the following way. Throughout, we assume that $T \leqslant T_c$, and that c_v is independent of T in the single-phase region for a van der Waals gas.

(a) Let λ be the mole fraction of the fluid in the gas phase for a state on the critical isochore near the critical point. Use (12.5.41) and (12.5.42) to show

$$\lambda = 0.5 + 0.25\sqrt{(-\hat{T})}. \qquad (12.5.46)$$

(b) Show that the molar entropy in the single-phase region in the neighbourhood of the critical point can be written

$$s(T, \hat{v}) = s_c + c_v \ln\left(\frac{T}{T_c}\right) + R\ln(1 + 1.5\hat{v}) \qquad (12.5.47)$$

where $\hat{v}$ is defined by (12.5.3) and s_c is the molar entropy at $T = T_c$, $v = v_c$. (Hint: you might like to begin with eqn (11.2.13).)

(c) On the critical isochore in the mixed-phase region the entropy is given by

$$s(T) = (1 - \lambda)s_l + \lambda s_g, \qquad (12.5.48)$$

where the subscripts l and g refer to the saturated liquid and vapour states, respectively. By expressing $\hat{v}_l$ and $\hat{v}_g$ as functions of $\hat{T}$ obtain an expression for $s(T)$ correct to first order in $(-\hat{T})$ on the critical isochore for $T \leqslant T_c$. You should find that

$$s(T) = s_c + c_v\hat{T} + \frac{9}{2}R\hat{T}. \qquad (12.5.49)$$

Hence show that the molar heat capacity on the critical isochore of a van der Waals gas has a discontinuity of $4.5R$ at the critical point.

12.5.4. A ferromagnetic body exhibits a ferromagnetic response below the
Curie temperature, T_c; above T_c the body is paramagnetic. In
zero magnetic field this transition is a critical point transition
which has the following critical exponents:

(1) $c_B \sim \hat{T}^{-\alpha}$ (molar heat capacity at constant B; $B = 0$,
                                    $\hat{T} > 0$);

(2) $M \sim (-\hat{T})^{\beta}$ (magnetization; $B = 0$, $\hat{T} < 0$);

(3) $\chi \sim \hat{T}^{-\gamma}$ (isothermal magnetic susceptibility; $B = 0$,
                                    $\hat{T} > 0$);

(4) $|B| \sim |M|^{\delta}$ (field on the critical isotherm; $\hat{T} = 0$).

Assume B is the applied field due to a long solenoid carrying a
given coil current. The Gibbs free energy per unit volume, g, is
a function of T and B, as in (10.3.25). According to the scaling
hypothesis g can be expressed as

$$g(T, B) = -s_c T + g_s(\hat{T}, B), \qquad (12.5.50)$$

in the neighbourhood of the critical point. Here g_s is a generalized homogeneous function which has the scaling relation

$$g_s(\lambda^p \hat{T}, \lambda^q B) = \lambda g_s(\hat{T}, B). \qquad (12.5.51)$$

Show that the critical indices are related to the scaling constants,
p and q, by eqns (12.5.18), (12.5.22), (12,5.24), and (12.5.28) provided we substitute $a = p$ and $b = 1 - q$. Confirm that eqns
(12.5.29)–(12.5.31) are correct.

12.6 Chemical reactions

Gibbs free energy for a mixture

The equilibrium state for a chemically reactive system may be obtained by
much the same methods as those we used for fluids undergoing phase
changes (Section 12.4). Here we shall consider gas phase reactions, but the
results are also applicable to many solutions. To be specific we shall first
consider a reaction in equilibrium at constant pressure. For this situation
the equilibrium condition is that the Gibbs free energy, G, of the reacting
mixture should be a minimum at fixed P and T. We shall therefore begin
by establishing suitable expressions for G.

For each chemical species, labelled i, we may obtain the molar Gibbs
function g_i, when it is in an isolated form, using (11.1.17). Recollect that
g_i is the same as the chemical potential in this situation:

$$g_i = \mu_i^0(T, P) = h_i(T_0, P_0) - T s_i(T_0, P_0) + \int_{T_0}^{T} c_{pi}(T', P_0) \, dT'$$

$$- T \int_{T_0}^{T} \frac{c_{pi}(T', P_0)}{T'} \, dT' + \int_{P_0}^{P} v_i(T, P') \, dP', \quad (12.6.1)$$

where μ_i^0 denotes the chemical potential of component i in isolation. If the components are permitted to mix, without reacting, the Gibbs free energy of the mixture may be expressed in a similar form:

$$G = H(T_0, P_0) - TS(T_0, P_0) + N \int_{T_0}^{T} c_p(T', P_0) \, dT'$$

$$- NT \int_{T_0}^{T} \frac{c_p(T', P_0)}{T'} \, dT' + \int_{P_0}^{P} V(T, P') \, dP', \qquad (12.6.2)$$

where H, S, c_p, and V represent properties of the mixture. Now, many systems exhibit the ideal gas properties,

$$H = \sum_{i=1}^{r} N_i h_i \qquad (12.6.3)$$

$$V = \sum_{i=1}^{r} N_i v_i, \qquad (12.6.4)$$

and since $c_p = \frac{1}{N} \left(\frac{\partial H}{\partial T} \right)_{P,N}$, we have

$$c_p = \sum_{i=1}^{r} x_i c_{pi}, \qquad (12.6.5)$$

where $x_i = N_i/N$, $N = \Sigma N_i$. However, for the mixture $S \neq \Sigma N_i s_i$ because the entropy of mixing must also be included. Assuming that the ideal gas mixing expression (9.3.20) holds, we have

$$S = \sum_{i=1}^{r} N_i(s_i - R \ln(x_i)), \qquad (12.6.6)$$

and hence for the mixture,

$$G(P, T, N_1, \ldots, N_r) = \sum_{i=1}^{r} N_i(\mu_i^0(T, P) + RT \ln(x_i)). \quad (12.6.7)$$

In writing this expression we assume the reaction process can be inhibited so that the amounts of the components, $N_1, \ldots, N_r$, may be treated as independent parameters. This is simply a device which allows us to treat the process of mixing separately from the reaction process.

We may express the pressure dependence of G explicitly using (12.6.1). This applies to the pure component i (that is, in the unmixed form):

$$\mu_i^0(T, P) = \mu_i^0(T, P_0) + \int_{P_0}^{P} v_i(T, P') \, dP', \qquad (12.6.8)$$

where v_i is the molar volume for species i alone. Thus the integral may be evaluated using the ideal gas relation, $Pv_i = RT$,

$$\mu_i^0(T, P) = \mu_i^0(T, P_0) + RT \ln \left(\frac{P}{P_0} \right), \tag{12.6.9}$$

where $\mu_i^0(T, P_0)$, the chemical potential for species i in the unmixed form at a standard pressure, P_0, is given by

$$\mu_i^0(T, P_0) = h_i(T_0, P_0) - T s_i(T_0, P_0) + \int_{T_0}^{T} c_{pi}(T', P_0) \, dT'$$

$$- T \int_{T_0}^{T} \frac{c_{pi}(T', P_0)}{T'} \, dT'. \tag{12.6.10}$$

In the special case of a monatomic ideal gas $\mu_i^0(T, P_0)$ may be calculated from (10.3.44). More general expressions may be found in standard texts on chemical thermodynamics.

From (12.6.7) and (12.6.9) we obtain the chemical potential for component i in the mixture,

$$\mu_i(T, P, x_i) = \mu_i^0(T, P_0) + RT \ln \left(\frac{P x_i}{P_0} \right). \tag{12.6.11}$$

and the Gibbs free energy of the mixture,

$$G(P, T, N_1, \ldots, N_r) = \sum_{i=1}^{r} N_i \mu_i(T, P, x_i)$$

$$= \sum_{i=1}^{r} N_i \left\{ \mu_i^0(T, P_0) + RT \ln \left(\frac{P x_i}{P_0} \right) \right\}. \tag{12.6.12}$$

Chemical equilibrium

Let us first consider a simple reaction involving two reactants, A_1 and A_2, and one product species, A_3, which are linked by the chemical equation

$$A_1 + A_2 \leftrightarrow A_3. \tag{12.6.13}$$

A molecular recombination–dissociation process is an example of such a reaction. Suppose there is initially one mole each of the reactants, A_1 and A_2; subsequently there are λ moles of A_3, where λ is called the *degree of reaction*. There will then be $1 - \lambda$ mole each of A_1 and A_2, and the mole fractions will be given by

$$x_1 = x_2 = \frac{1 - \lambda}{2 - \lambda}, \qquad x_3 = \frac{\lambda}{2 - \lambda}. \tag{12.6.14}$$

From (12.6.12) we may obtain G as a function of T, P, and λ; we take $P = P_0$ for simplicity.

$$G(T, P_0, \lambda) = \mu_1^0(T, P_0) + \mu_2^0(T, P_0) + \lambda \Delta G^0(T, P_0)$$

$$+ RT \left\{ 2(1 - \lambda) \ln \left(\frac{1 - \lambda}{2 - \lambda} \right) + \lambda \ln \left(\frac{\lambda}{2 - \lambda} \right) \right\} \tag{12.6.15}$$

where
$$\Delta G^0(T, P_0) = \mu_3^0(T, P_0) - \mu_1^0(T, P_0) - \mu_2^0(T, P_0), \quad (12.6.16)$$

is called the *standard free energy* for the reaction. Figure 12.13 illustrates (12.6.15), showing how G depends on λ for different values of ΔG^0. Since G is a minimum with respect to λ at equilibrium, it is evident that the reaction (12.6.13) moves to the left when $\Delta G^0 > 0$, and to the right when $\Delta G^0 < 0$. Even when $\Delta G^0 = 0$ the reaction proceeds to some extent, due to the entropy of mixing in a constant pressure process. Clearly ΔG^0 is a primary parameter in determining the degree of reaction at equilibrium.

Of course the equilibrium degree of reaction may be obtained by minimizing an explicit expression for G for a particular reaction, such as (12.6.15), but it is normally more convenient to formulate the problem in the following way. Consider a chemical process having m reactants and n products:

$$a_1 A_1 + a_2 A_2 + \cdots + a_m A_m \leftrightarrow a_{m+1} A_{m+1} + a_{m+2} A_{m+2} + \cdots + a_r A_r$$
$$(12.6.17)$$

where $r = m + n$. Here the coefficients $a_1, a_2, \ldots, a_r$, are positive integers, or ratios of integers, and the A_i denote whole atoms or molecules of the participating elements or compounds. We define the *stoichiometric coefficients* for the reaction, v_i, by

$$v_i = -a_i \qquad \text{for } i = 1, 2, \ldots, m, \qquad (12.6.18a)$$

$$v_i = a_i \qquad \text{for } i = m+1, m+2, \ldots, r. \qquad (12.6.18b)$$

The reaction may now be written as

$$\sum_{i=1}^{r} v_i A_i \leftrightarrow 0. \qquad (12.6.19)$$

This equation constrains the changes in the mole numbers of the reactants and products in the reaction process. This restriction, which derives from the requirement that the total amount of each element is fixed, may be expressed more conveniently in terms of the degree of reaction, λ. Suppose the reaction proceeds such that $N_i \rightarrow N_i + \delta N_i$. Then $\delta N_i \propto v_i$, and hence $\delta N_i / v_i$ is a constant. By analogy with the degree of reaction, λ, introduced for the simple reaction (12.6.13), we have

$$\delta N_i = v_i \delta \lambda, \qquad i = 1, 2, \ldots, r. \qquad (12.6.20)$$

Thus when $\delta \lambda > 0$, then $\delta N_i < 0$ for the reactants, and $\delta N_i > 0$ for the products, and the reaction (12.6.17) proceeds from left to right. The associated change in the Gibbs free energy of the reacting mixture, given that T and P are constant, is

$$\delta G = \sum_{i=1}^{r} \left(\frac{\partial G}{\partial N_i} \right)_{T,P} \delta N_i = \sum_{i=1}^{r} \mu_i v_i \delta \lambda, \qquad (12.6.21)$$

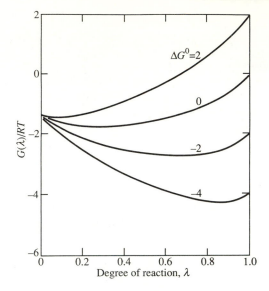

Fig. 12.13 Gibbs free energy, $G(\lambda)$, for the reaction (12.6.13) represented as a function of the degree of reaction, λ, at constant T and P. $G(\lambda)$ is a minimum at equilibrium. The figure shows how the equilibrium state depends on ΔG^0, the standard free energy for the reaction. The figure has been obtained from (12.6.15) taking $\mu_1^0 = \mu_2^0 = 0$.

or

$$\left(\frac{\partial G}{\partial \lambda}\right)_{T,P} = \sum_{i=1}^{r} \mu_i \nu_i. \tag{12.6.22}$$

Thus the condition for equilibrium, $\delta G = 0$, is simply expressed as

$$\sum_{i=1}^{r} \mu_i \nu_i = 0. \tag{12.6.23}$$

Here μ_i, the chemical potential for species i in the reacting mixture at equilibrium, is given by (12.6.11). Hence,

$$RT \sum_{i=1}^{r} \nu_i \ln\left(\frac{Px_i}{P_0}\right) = -\sum_{i=1}^{r} \nu_i \mu_i^0(T, P_0). \tag{12.6.24}$$

The right-hand side depends on the temperature only and it is convenient to express it in terms of the standard free energy for the reaction, ΔG^0, which we define as in (12.6.16),

$$\Delta G^0 = \Delta G^0(T, P_0) = \sum_{i=1}^{r} \nu_i \mu_i^0(T, P_0). \tag{12.6.25}$$

Physically ΔG^0 represents the change in G in a process which begins with a_i moles of each reactant A_i, $i = 1, \ldots, m$, all separated in stable states at pressure P_0 and temperature T. The final state for the process is the end state of the reaction, having a_i moles of the product species, A_i, $i = m+1, \ldots, r$, also in isolated forms at the standard pressure P_0, and at temperature T.

Now let

$$K_p = \exp\left(\frac{-\Delta G^0}{RT}\right). \tag{12.6.26}$$

Then (12.6.24) can be written as

$$\sum_{i=1}^{r} v_i \ln\left(\frac{Px_i}{P_0}\right) = \ln(K_p) \tag{12.6.27}$$

or

$$\prod_{i=1}^{r} \left(\frac{Px_i}{P_0}\right)^{v_i} = K_p, \tag{12.6.28}$$

which is the well-known *law of mass action*. For our purposes the partial pressure of the component i, P_i, is given by

$$P_i = \frac{Px_i}{P_0}. \tag{12.6.29}$$

Since the reference pressure, P_0, is normally taken to be 1 standard atmosphere, P_i is expressed in units of standard atmospheres and (12.6.28) becomes

$$\prod_{i=1}^{r} (P_i)^{v_i} = K_p. \tag{12.6.30}$$

Alternatively we may express the equilibrium condition in terms of the mole-fractions, x_i, by defining

$$K_x = \left(\frac{P_0}{P}\right)^{\Sigma v_i} K_p, \tag{12.6.31}$$

so that (12.6.28) becomes

$$\prod_{i=1}^{r} (x_i)^{v_i} = K_x. \tag{12.6.32}$$

In order to calculate the reaction constant K_p the standard free energy for the reaction, ΔG^0, is required. Equation (12.6.25) expresses ΔG^0 in terms of the corresponding chemical potentials, $\mu_i^0(T, P_0)$, for the reactants and products in isolated forms. This suggests that we shall require the values of $\mu_i^0(T, P_0)$) for every element and compound involved in a reaction in order to determine ΔG^0. However, the absolute values for the chemical potentials are not needed in ordinary chemical reactions because the number of atoms of each element is constant. To illustrate, suppose the reactants in (12.6.13) are compounds of the elements A_5, A_6, A_7, A_8 such that

$$A_5 + A_6 \leftrightarrow A_1 \quad \text{and} \quad A_7 + A_8 \leftrightarrow A_2. \tag{12.6.33}$$

The corresponding standard free energy for these reactions is called the *free energy of formation*, ΔG_f^0. Thus for A_1 and A_2,

$$\Delta G_{f1}^0 = \mu_1^0 - \mu_5^0 - \mu_6^0, \qquad \Delta G_{f2}^0 = \mu_2^0 - \mu_7^0 - \mu_8^0, \tag{12.6.34}$$

dropping the parameters T and P_0 for brevity. Hence for the reaction $A_2 + A_2 \leftrightarrow A_3$, the standard free energy is given by

$$\Delta G^0 = \mu_3^0 - \mu_1^0 - \mu_2^0$$
$$= \Delta G_{f3}^0 - \Delta G_{f1}^0 - \Delta G_{f2}^0, \tag{12.6.35}$$

where

$$\Delta G_{f3}^0 = \mu_3^0 - \mu_5^0 - \mu_6^0 - \mu_7^0 - \mu_8^0$$

is the free energy of formation for the compound A_3 from its elements. Equation (12.6.35) shows that we do not require the chemical potentials of the individual elements to determine the standard free energy for a chemical reaction. It is sufficient to know the free energy of formation of the compounds involved, information which is widely tabulated. The free energy of formation of the elements is zero by definition.

For reactions at temperatures other than the standard (normally 25°C) we require information of the temperature dependence of K_p. This may be obtained from (12.6.26).

$$\frac{d \ln (K_p)}{dT} = -\frac{1}{R} \frac{\partial}{\partial T} \left(\frac{\Delta G^0}{T} \right), \tag{12.6.36}$$

where we have taken P_0 to be constant. Recollect that ΔG^0 is the change in the free energy for a chemical system between two equilibrium states. Hence ΔG^0 has all the thermodynamic properties of the Gibbs function. In particular, from (10.3.57) we obtain

$$H = -T^2 \frac{\partial}{\partial T} \left(\frac{G}{T} \right)_{P,N} \tag{12.6.37}$$

and hence

$$\frac{d \ln (K_p)}{dT} = \frac{\Delta H^0}{RT^2}, \tag{12.6.38}$$

where $\Delta H^0 = \Sigma v_i h_i^0$ is the enthalpy change of the system corresponding to ΔG^0. Equation (12.6.38) is the *van't Hoff equation*.

Particle reactions

High-energy reactions which involve particle and nuclear transformations are important in the theory of stellar evolution and in the theory of the early universe. In such processes the energy of the particles may be relativistic and consequently the thermodynamic potentials must include the mass energy of the system. The normal chemical assumption, that the mass and mole numbers of individual chemical elements are fixed, may not hold under these conditions. Instead, the process invariants include such quantities as the total number of baryons and the total number of leptons of a particular kind.

In order to illustrate these processes we begin with the grand potential for a gas of weakly interacting fermions. This fundamental relation, which may be obtained in integral form by standard statistical methods (see for example, Landsberg (1961), Reif (1965), and Landau and Lifshitz (1980)), is

$$\Psi(T, V, \mu) = -\frac{4\pi g V k T}{h^3} \int_0^\infty p^2 \ln\left\{1 + \exp\left[\frac{\mu - E(p)}{kT}\right]\right\} dp, \quad (12.6.39)$$

where

$$E(p) = (p^2 c^2 + m^2 c^4)^{1/2}, \quad (12.6.40)$$

is the relativistic energy of a particle of rest mass m and momentum p. Here c is the speed of light, h is Planck's constant, and k is the Boltzmann constant. In addition, $g = 2s + 1$, where s is the intrinsic spin of the particle. Note that g must be even for fermions.

In this expression μ, which includes the rest mass energy of the particle, is the chemical potential per particle (not per mole). Hence, on differentiating Ψ with respect to μ we obtain $-\tilde{N}$, the number of particles in the system, not the number of moles. It turns out to be more convenient to determine $\tilde{N}$ before carrying out the integral over momentum, p. Thus we have

$$\tilde{N} = \frac{4\pi g V}{h^3} \int_0^\infty \frac{p^2 \, dp}{1 + \exp\left(\dfrac{E(p) - \mu}{kT}\right)}. \quad (12.6.41)$$

The integral can be evaluated under certain limiting and simplifying assumptions. First consider the limit, known as a *cold degenerate fermion gas*, for which

$$\mu - mc^2 \gg kT. \quad (12.6.42)$$

Under this condition the denominator of the integrand has two possible values. Let p_0 be defined by

$$p_0^2 c^2 + m^2 c^4 = \mu^2. \quad (12.6.43)$$

When $p > p_0$ then $(E(p) - \mu)/kT \gg 1$, in which case the integrand is negligible; alternatively when $p < p_0$, $(E(p) - \mu)/kT \ll -1$ and the denominator is then unity. Hence under the condition (12.6.42) we may use the approximation

$$\tilde{N} = \frac{4\pi g V}{h^3} \int_0^{p_0} p^2 \, dp = \frac{4\pi g V p_0^3}{3h^3}. \tag{12.6.44}$$

It is convenient to introduce a dimensionless parameter, x, defined by

$$x = \frac{p_0}{mc}, \tag{12.6.45}$$

which is a measure of the extent to which particles of momentum p_0 are relativistic. Then,

$$\tilde{N} = \frac{4\pi g V}{3} \left(\frac{xmc}{h} \right)^3, \tag{12.6.46}$$

and

$$\mu^2 = m^2 c^4 (1 + x^2). \tag{12.6.47}$$

The cold degenerate fermion gas is a useful model for the equilibrium reaction involving neutron beta decay and its inverse. We shall briefly review this reaction, which is relevant in the theory of dwarf stars. The equations for the process are,

$$e^- + p \rightarrow n + \nu_e \tag{12.6.48a}$$

$$n \rightarrow e^- + p + \bar{\nu}_e \tag{12.6.48b}$$

where n, p, e^-, ν_e, $\bar{\nu}_e$ denote the neutron, proton, electron, the electron neutrino, and its antiparticle respectively. We simplify the model by not including the Coulomb interaction between the electrons and protons (see Shapiro and Teukolsky (1983)). In addition we assume that the neutrinos escape freely and hence that the neutrino density will be negligible. Under this restriction the reaction equilibrium condition (12.6.23) can be expressed as

$$\mu_n = \mu_e + \mu_p. \tag{12.6.49}$$

An additional constraint is that the numbers of electrons and protons should be equal to ensure charge neutrality:

$$\tilde{N}_e = \tilde{N}_p. \tag{12.6.50}$$

Hence from (12.6.46) we have

$$m_e x_e = m_p x_p, \tag{12.6.51}$$

where m_e and m_p denote the rest masses of the electron and proton, respectively. Thus from (12.6.47) and (12.6.49),

$$m_n(1 + x_n^2)^{1/2} = m_e(1 + x_e^2)^{1/2} + m_p(1 + x_p^2)^{1/2}. \qquad (12.6.52)$$

In the extreme relativistic region, where N/V is large and $x \gg 1$ for all particles, this relation takes the simplified form

$$m_n x_n = m_e x_e + m_p x_p = 2m_p x_p. \qquad (12.6.53)$$

Since $m_n \approx m_p$ we have $x_n \approx 2x_p$ and hence from (12.6.46),

$$\tilde{N}_n/\tilde{N}_p \approx 8. \qquad (12.6.54)$$

This is an extreme condition. For comparison, we shall determine x_e and x_p at the threshold for inverse beta decay, (12.6.48a), which occurs when $\mu_n = \mu_e + \mu_p$ but $\tilde{N}_n \approx 0$. Assuming $x_p \ll 1$ and setting $x_n = 0$ we obtain from (12.6.52),

$$m_n - m_p = m_e(1 + x_e^2)^{1/2}, \qquad (12.6.55)$$

or

$$x_e^2 = \frac{\delta^2}{m_e^2} - 1, \qquad (12.6.56)$$

where $\delta = m_n - m_p$. Now $\delta = 2.53 m_e$ and $m_p = 1836 m_e$ (Appendix A), yielding

$$x_p \approx \frac{(\delta^2 - m_e^2)^{1/2}}{m_p} \approx 1.27 \times 10^{-3}, \qquad (12.6.57a)$$

and

$$x_e \approx 2.32. \qquad (12.6.57b)$$

Thus our assumption $x_p \ll 1$ is justified; while the electrons are relativistic at the threshold, the neutrons and protons are not.

  The integrals arising from the fundamental relation (12.6.39) may also be evaluated in the special case that $\mu = 0$. This limit arises when we consider equilibrium under electron–positron pair production:

$$e^- + e^+ \leftrightarrow 2\nu. \qquad (12.6.58)$$

Here ν denotes a photon. In this reaction the chemical potentials for e^- and e^+ will be the same, being in equal numbers, but the chemical potential for the photon gas is zero (see (9.3.3)). Thus, for the electron and positron, $\mu_{e^+} = \mu_{e^-} = 0$. To be specific we evaluate (12.6.39) in the limit $kT \gg mc^2$. Referring to the integrand we see that

$$\exp\left[\frac{\mu - E(p)}{kT}\right] \approx 1 \qquad (12.6.59)$$

unless $p \gg mc$. We may therefore approximate the integral by setting $E(p) = pc$. In fact, by numerical integration, one may show that the error due to this substitution is less than 0.5% when $kT = 4mc^2$. Hence, for $\mu \approx 0$, $kT \geqslant 4mc^2$, (that is, $T \geqslant 2.4 \times 10^{10} \text{ K}$ for electrons) we have

$$\Psi(T, V, \mu) = -\frac{4\pi g V k T}{h^3} \int_0^\infty p^2 \ln\left\{1 + \exp\left(\frac{\mu - pc}{kT}\right)\right\} dp. \quad (12.6.60)$$

For an electron, $g = 2$; hence on setting $\mu = 0$ and rearranging (Abramowitz and Stegun 1970; Wang and Guo 1989) we obtain

$$\Psi(T, V, \mu) = -\frac{8\pi V}{h^3} \frac{kT^4}{c} \int_0^\infty \frac{z^3 \, dz}{1 + \exp z}, \quad (12.6.61)$$

$$= -\frac{7}{6} \frac{\sigma}{c} V T^4, \quad (12.6.62)$$

where σ is the Stefan–Boltzmann constant, (9.3.3a). Hence, from (10.3.67) we have the energy

$$U = -T^2 \left(\frac{\partial(\Psi/T)}{\partial T}\right)_{V, \mu/T} = \frac{7}{2} \frac{\sigma}{c} V T^4. \quad (12.6.63)$$

By comparison with (10.2.26) we see that the electron and positron gases have energy which is 7/8 that for the black-body radiation field. Note that the electron and positron density depends on T, since $\tilde{N}_{e^-}$ and $\tilde{N}_{e^+}$ are determined by (12.6.39). In fact, using $\tilde{N} = \partial\Psi/\partial\mu$ and setting $\mu = 0$, the integral may be expressed in terms of a Riemann ζ-function (Abramowitz and Stegun 1970; Wang and Guo 1989) as

$$\tilde{N} = 12\pi\zeta(3) V \left(\frac{kT}{hc}\right)^3, \quad (12.6.64)$$

where $\zeta(3) \approx 1.20$. This example illustrates the tight coupling between radiation and matter at high temperatures, a state which characterized the early evolution of the universe. Landsberg (1978) gives an introductory account.

Questions

12.6.1.   The reaction (12.6.13) reaches equilibrium at the standard pressure, P_0. Show that the corresponding degree of reaction is given by

$$\lambda = 1 - \frac{1}{\sqrt{1 + K_p}}. \quad (12.6.65)$$

12.6.2.   When the volume of a reacting system is fixed the condition for equilibrium is most conveniently expressed in terms of the molar concentrations, c_i, and a new equilibrium constant, K_c. Here,

$$c_i = N_i / V. \quad (12.6.66)$$

Show that

$$\prod_{i=1}^r (c_i)^{\nu_i} = K_c, \quad (12.6.67)$$

where

$$K_c = \left(\frac{P_0}{RT}\right)^{\Sigma \nu_i} K_p. \tag{12.6.68}$$

12.6.3. (a) A cold n–p–e$^-$ gas is in equilibrium under β decay and inverse β decay in a state with $x_e \gg 1$ and $x_n \ll 1$, $x_p \ll 1$ (using the notation introduced above). Use (12.6.51) and (12.6.52) to demonstrate that

$$x_n \approx \left(2x_p - \frac{2\delta}{m_p}\right)^{1/2}. \tag{12.6.69}$$

(b) Show that $\tilde{N}_n/\tilde{N}_p$ is maximized when $x_p \approx 2\delta/m_p$, which is about twice the threshold value, and in this situation,

$$\frac{x_n}{x_p} \approx \left(\frac{m_p}{2\delta}\right)^{1/2}. \tag{12.6.70}$$

(c) Confirm that the assumptions made for the values of x_n, x_p, and x_e in (a) hold when $\tilde{N}_n/\tilde{N}_p$ is maximized and that

$$\tilde{N}_n/\tilde{N}_p \approx 6900. \tag{12.6.71}$$

12.6.4. (a) Evaluate the grand potential given in (12.6.39) in the limit $kT \ll mc^2$, assuming that $0 \leqslant \mu \ll mc^2$. Note that $\exp((\mu - E(p))/kT) \ll 1$ in this limit, and hence the approximation

$$\ln\left\{1 + \exp\left[\frac{\mu - E(p)}{kT}\right]\right\} = \exp\left[\frac{\mu - E(p)}{kT}\right]$$

is applicable. Under these conditions the integrand is negligible for $p > mc$ and hence $E(p)$ may be approximated using a binomial expansion. Show that

$$\Psi(T, V, \mu) = -gV\left(\frac{2\pi m}{h^2}\right)^{3/2} (kT)^{5/2} \exp\left(\frac{\mu - mc^2}{kT}\right). \tag{12.6.72}$$

Note, $\displaystyle\int_0^\infty x^2 \exp(-x^2)\, dx = \sqrt{\pi}/4$.

(b) Show that the number of electron–positron pairs in equilibrium with black-body radiation under these conditions is given by

$$\tilde{N} = 2V\left(\frac{2\pi mkT}{h^2}\right)^{3/2} \exp\left(\frac{-mc^2}{kT}\right). \tag{12.6.73}$$

(c) Taking $g = 1$ show that (12.6.72) yields (9.3.1), the fundamental relation for a gas of non-relativistic structureless particles. Confirm that the non-relativistic chemical potential is $\mu - mc^2$, where μ is the relativistic chemical potential.

Appendix A

Physical constants

The fundamental constants are taken from CODATA Bulletin No 63 Nov. 1986. Values have been rounded to five significant figures.

Atomic mass unit	a.m.u.	1.6605×10^{-27} kg
Avogadro number	$\tilde{N}_A$	6.0221×10^{23} mol^{-1}
Bohr magneton	μ_B	9.2740×10^{-24} J T^{-1}
Boltzmann constant	k	1.3807×10^{-23} J K^{-1}
Electron rest mass	m_e	9.1094×10^{-31} kg
Electron charge	e	1.6022×10^{-19} C
Gravitational constant	G	6.6726×10^{-11} m^3 kg^{-1} s^{-2}
Molar gas constant	R	8.3145 J K^{-1} mol^{-1}
Permeability of a vacuum	μ_0	$4\pi \times 10^{-7}$ H m^{-1}
Permittivity of a vacuum	$\varepsilon_0 = 1/\mu_0 c^2$	8.8542×10^{-12} F m^{-1}
Planck constant	h	6.6261×10^{-34} J s
	$h/2\pi$	1.0546×10^{-34} J s
Proton rest mass	m_p	1.6726×10^{-27} kg
Neutron–proton mass difference	$m_n - m_p$	2.3055×10^{-30} kg
Speed of light	c	2.9979×10^8 m s^{-1}
Standard atmosphere	P	1.0133×10^5 Pa
Stefan–Boltzmann constant	σ	5.6705×10^{-8} W m^{-2} K^{-4}

Appendix B

Thermodynamic properties of water and steam

The figures in this appendix, representing the basic thermodynamic properties of water and steam, have been generated using empirical relations for the thermodynamic functions of water and steam (Schmidt 1982). The coordinates have been selected for convenience in the questions set in Chapters 5, 6 and 7.

To illustrate the use of the figures, consider Fig. B.1. The curves represent different isotherms of a system composed of 1 kg of water. The states include liquid water, water vapour, and mixtures of the two phases. In this representation the y-axis is the pressure, expressed in Pa on a logarithmic scale. The x-axis is the specific internal energy, u, being the internal energy per kg of the fluid. Note that specific quantities are normally denoted by lower case symbols.

The boundaries of the single-phase regions are represented by the inverted bell-shaped curve, called the saturation envelope. The purely liquid phase lies to the left of the saturation envelope, the vapour phase to the right. The enclosed region represents mixtures of the liquid and vapour phases in equilibrium at various compositions. The separation of the fluid into two phases arises because the single-phase states which would otherwise occur are unstable.

The change of phase from liquid to vapour at constant pressure is accompanied by an increase in the specific internal energy. This is associated with an increase in the molecular potential energy as the mean separation of the molecules increases. At low pressure the volume may change by a factor of more than 10^5 in the phase change process (see Figs B.2, B.3). Consequently, the internal energy change is much larger at low, than at high, presssure. At 10 MPa, for instance, the volume changes by a factor of only 13. This pressure dependence is a consequence of the difference between the compressibility of the liquid and gas phases. As a result there is an upper limit to the pressure at which a phase change occurs in an isobaric heating process, called the critical pressure. For water the critical pressure is 22.12 MPa and the critical temperature is 374.15°C. This point is located at the apex of the saturation envelope.

Notice that the isotherms are parallel to the y-axis at low pressure in the vapour region, indicating that the internal energy depends only on the temperature, not on the pressure or volume. This is a characteristic of an ideal gas. At high pressure, however, the curvature of the isotherms shows

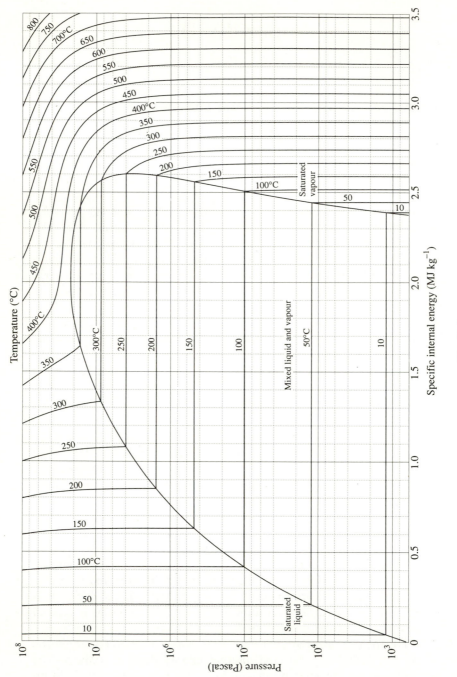

Fig. B.1

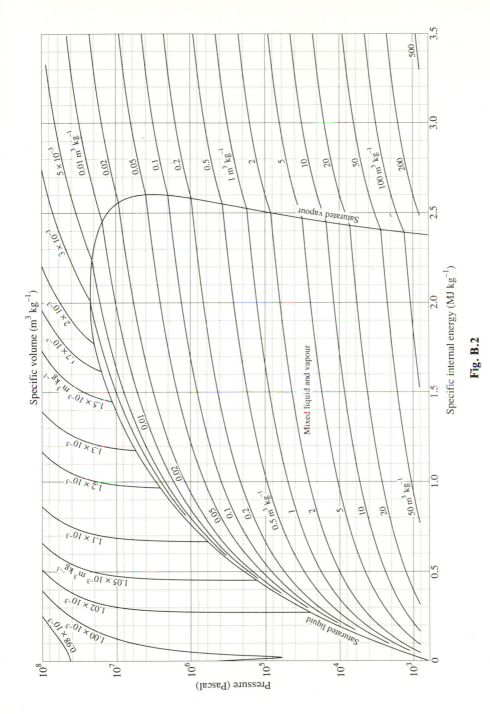

Fig. B.2

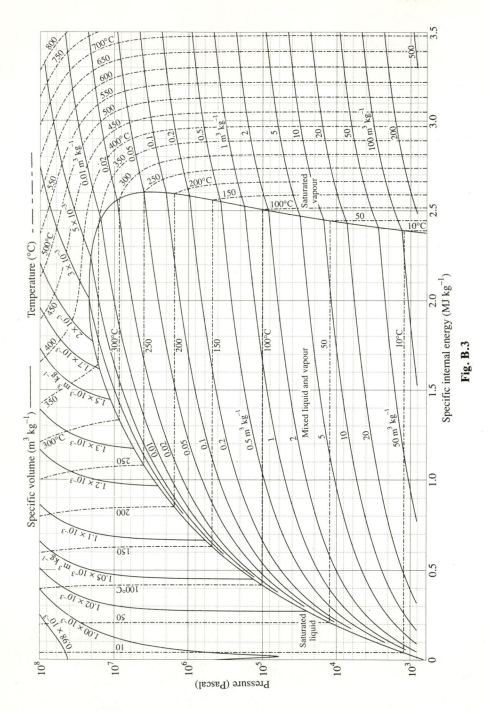

Fig. B.3

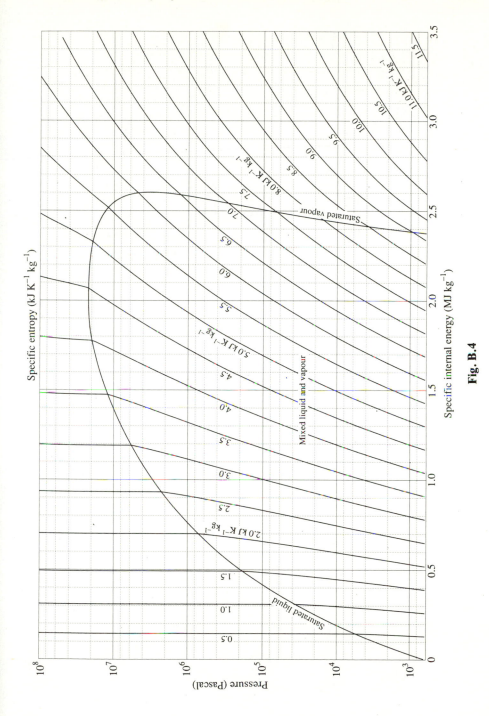

Fig. B.4

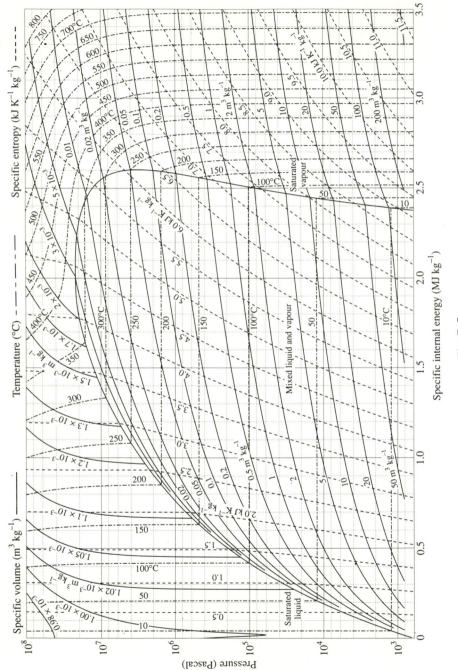

Fig. B.5

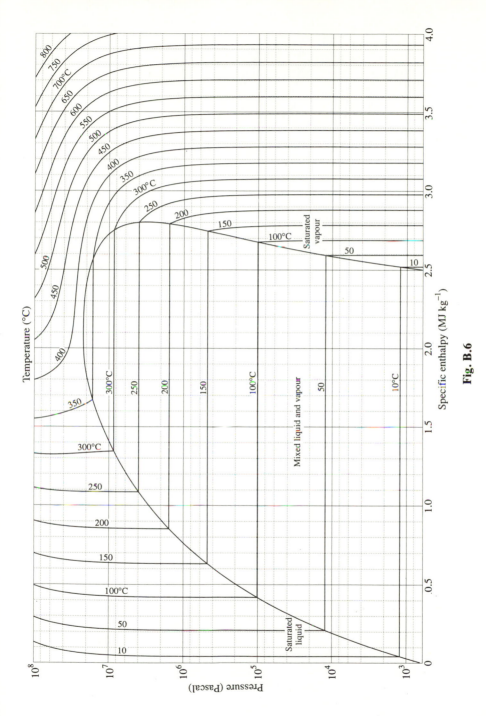

Fig. B.6

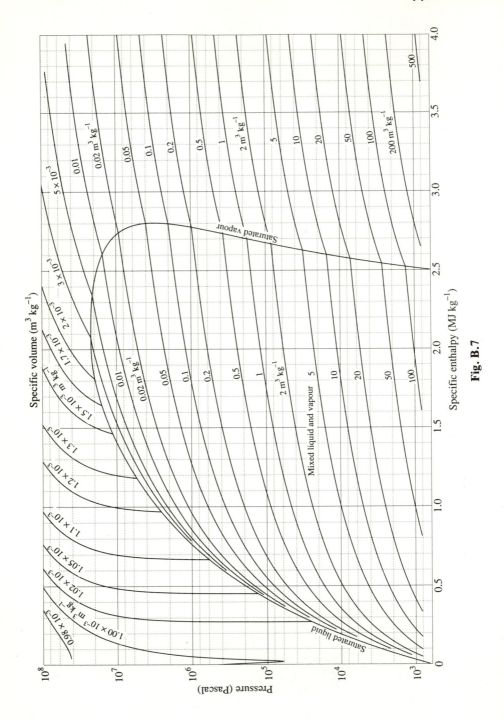

Specific volume (m³ kg⁻¹)

Specific enthalpy (MJ kg⁻¹)

Fig. B.7

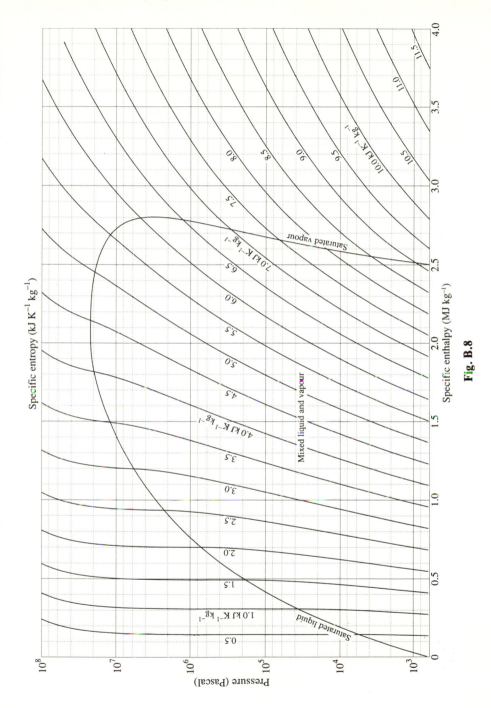

Fig. B.8

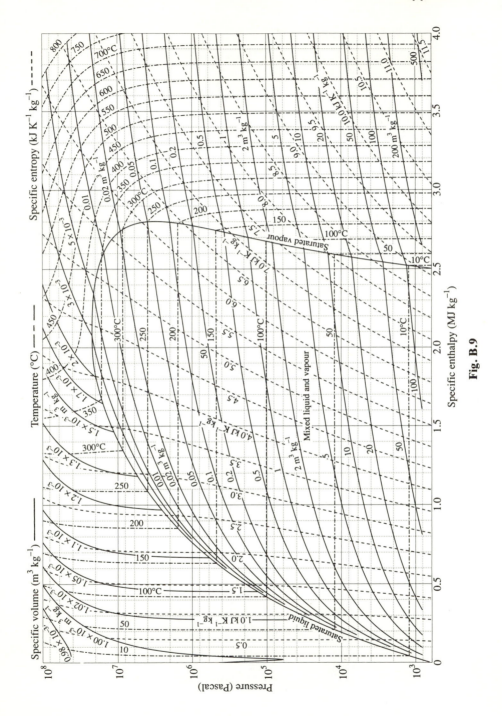

Fig. B.9

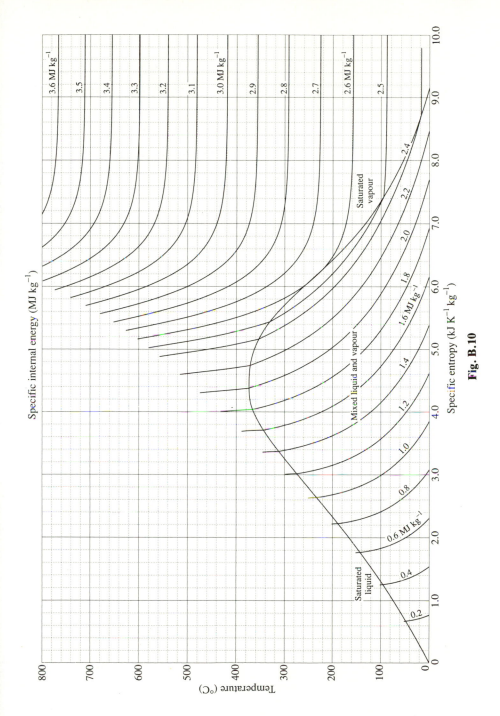

Fig. B.10

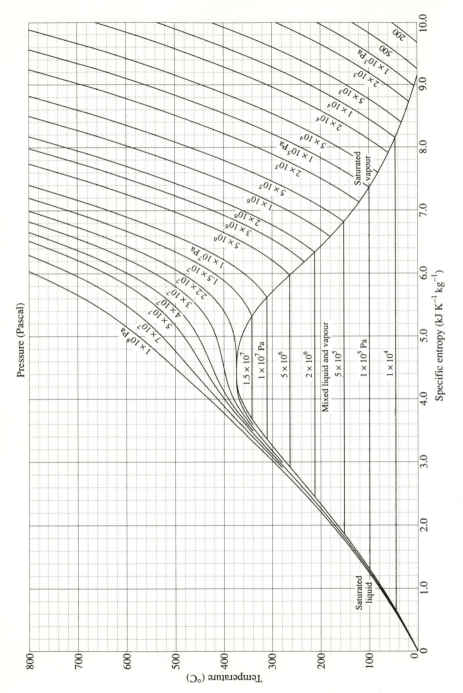

Fig. B.11

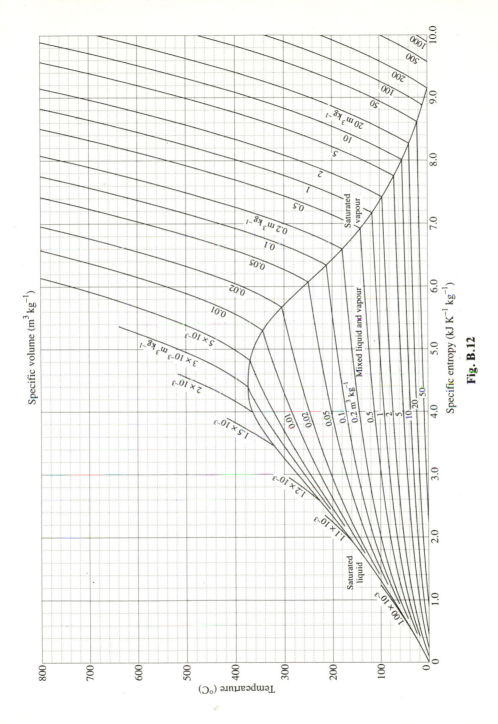

Fig. B.12

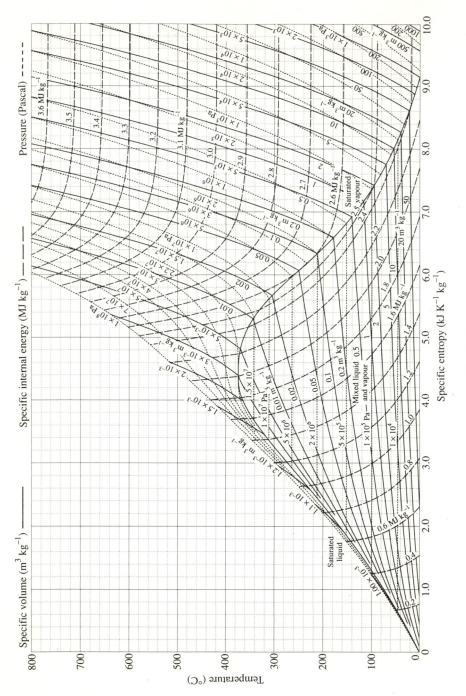

Fig. B.13

that the specific energy is less than that for an ideal gas at the same temperature. This effect is another consequence of the long-range attractive forces between molecules which become increasingly important as the inter-molecular separation is reduced at high pressure.

Figure B.2 shows a curious, but important, characteristic of water. The isochore (constant volume) locus at $v = 1.0 \times 10^{-3}\,\mathrm{m^3\,kg^{-1}}$ exhibits a sharply defined pressure minimum, corresponding to a temperature of approximately 4°C. This is due to the well-known anomalous expansivity for water at this temperature. The specific volume decreases anomalously as the temperature increases from 0–4°C given that the pressure is constant and less than about 10 MPa.

The data on water properties given in these figures are adequate for the purposes of this text. While some differences exist between these data and other compilations, such as the NBS/NRC Steam Tables (Haar *et al.* 1984), these are too small to be significant in these figures.

Appendix C

Electric and magnetic work

We establish some formal relations on electric and magnetic work which are required in Chapter 8. For brevity a number of basic results from vector calculus and electromagnetism (see, for example, Bleaney and Bleaney (1976) or Robinson (1973)) will be taken as given.

Electrical work

Let us begin with the work done by external forces when the charge of a conductor, q, is changed in the presence of a dielectric medium. We assume the conductor is surrounded by a non-conducting medium, or by free space, and we suppose q is changed by an increment δq brought from infinity. Given that the electric potential of the conductor is ϕ the work done is

$$\delta W = \phi \, \delta q. \tag{C.1}$$

We can express q in terms of the electric displacement. Let ρ be the free charge density. Then,

$$\nabla \cdot \mathbf{D} = \rho, \tag{C.2}$$

and since q is the only free charge in our system,

$$q = \int_{\text{all space}} \nabla \cdot \mathbf{D} \, dV. \tag{C.3}$$

By applying the divergence theorem the volume integral can be expressed as a surface integal over the conductor containing the charge:

$$q = \int_{\text{conductor}} \mathbf{D} \cdot d\mathbf{S}', \tag{C.4}$$

where $d\mathbf{S}'$ denotes a surface vector directed from the conductor to the medium. Hence eqn (C.1) can be written as

$$\delta W = \phi \int_{\text{conductor}} \delta \mathbf{D} \cdot d\mathbf{S}', \tag{C.5}$$

and since ϕ is constant over the surface of the conductor,

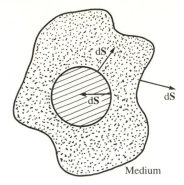

Fig. C.1 Charged conductor surrounded by a dielectric medium showing the surface elements dS and dS′.

$$\delta W = \int_{\text{conductor}} \phi \, \delta\mathbf{D} \cdot d\mathbf{S}'. \tag{C.6}$$

In this integral we may take the boundary to be the surface of the medium, where the 'medium' includes all space outside the conductor, since the product $\phi \, \delta\mathbf{D}$ will vanish faster than r^{-2} at large distances. Let dS denote on outwardly directed surface element vector of the medium. At the interface of the conductor, $d\mathbf{S} = -d\mathbf{S}'$ (Fig. C.1). Hence,

$$\delta W = -\int_{\text{medium}} \phi \, \delta\mathbf{D} \cdot d\mathbf{S}, \tag{C.7}$$

Then by using the divergence theorem again, we obtain

$$\delta W = -\int_{\text{medium}} \nabla \cdot (\phi \, \delta\mathbf{D}) \, dV. \tag{C.8}$$

Now,

$$\nabla \cdot (\phi \, \delta\mathbf{D}) = \phi \nabla \cdot \delta\mathbf{D} + (\nabla\phi) \cdot \delta\mathbf{D}, \tag{C.9}$$

and here $\nabla \cdot \delta\mathbf{D} = 0$, since there are no free charges in the medium. In addition we have

$$\mathbf{E} = -\nabla\phi - \frac{\partial \mathbf{A}}{\partial t}, \tag{C.10}$$

where $\mathbf{A}$ is the magnetic vector potential. Since the process of interest occurs quasistatically, time-varying fields can be ignored and we may use the electrostatic approximation, $\mathbf{E} = -\nabla\phi$. Hence,

$$\delta W = \phi \, \delta q = \int_{\text{medium}} \mathbf{E} \cdot \delta \mathbf{D} \, dV. \tag{C.11}$$

Similarly we can show

$$q \, \delta \phi = \int_{\text{medium}} \mathbf{D} \cdot \delta \mathbf{E} \, dV, \tag{C.12}$$

and

$$q \phi = \int_{\text{medium}} \mathbf{D} \cdot \mathbf{E} \, dV. \tag{C.13}$$

Note, again that 'medium' here means all space, excluding the conductor.

Magnetic work

Consider a conducting loop carrying a current J. We assume that the electrical resistance of the loop is negligible and we denote the magnetic flux linkage due to J by Φ. In addition we suppose the conductor is surrounded by a non-conducting medium and/or vacuum. Now a process which results in an increase in the flux, $\Phi \rightarrow \Phi + \delta\Phi$, will occur only if an external current source connected to the current loop provides the required energy input. Suppose the flux change process occurs in a time δt. The resulting induced e.m.f., $d\Phi/dt$, will oppose the current source which maintains the current J. Since $\delta\Phi = \delta t (d\Phi/dt)$, the work done by the current source is

$$\delta W = J \, \delta\Phi. \tag{C.14}$$

Notice that it is not necessary for J to change in order for work to be done; the independent variable here is Φ, not J. Now,

$$\Phi = \int_{S_c} \mathbf{B} \cdot d\mathbf{S}_c, \tag{C.15}$$

where S_c denotes a surface bounded by the conductor. We define the direction of the surface element, $d\mathbf{S}_c$, as shown in Fig. C.2. First, construct a vector element, $d\mathbf{r}$, which has the same direction as a line drawn from a point in the surface to the conductor. Second, define a line element of the conductor, $d\mathbf{l}$, which lies along the conductor and is directed in the same sense as the current J. Then $d\mathbf{S}_c$ is such that the three vectors $d\mathbf{r}$, $d\mathbf{l}$, $d\mathbf{S}_c$ form a right-handed triad, as shown.

Thus we have

$$\delta W = J \int_{S_c} \delta \mathbf{B} \cdot d\mathbf{S}_c, \tag{C.16}$$

$$= J \int_{S_c} (\nabla \times \delta \mathbf{A}) \cdot d\mathbf{S}_c, \tag{C.17}$$

where **A** is the vector potential. By Stokes' theorem we transform the surface integral into a line integral around the conductor:

$$\delta W = J \int_{\text{conductor}} \delta \mathbf{A} \cdot \mathbf{dl}. \tag{C.18}$$

We assume that the variation of **A** over the cross-section of the conductor is small enough that it can be ignored. Thus the line integral can be written as a volume integral, noting that the direction of dl is the same as the current. Let **j** be the current density in the conductor. Then

$$\delta W = \int_{\text{conductor}} \mathbf{j} \cdot \delta \mathbf{A} \, dV, \tag{C.19}$$

$$= \int_{\text{all space}} \mathbf{j} \cdot \delta \mathbf{A} \, dV. \tag{C.20}$$

Now $\mathbf{j} = \nabla \times \mathbf{H}$, ignoring any electric displacement current since the processes of interest are quasistatic. Hence,

$$\delta W = \int_{\text{all space}} (\nabla \times \mathbf{H}) \cdot \delta \mathbf{A} \, dV, \tag{C.21}$$

$$= \int_{\text{all space}} \nabla \cdot (\mathbf{H} \times \delta \mathbf{A}) \, dV + \int_{\text{all space}} \mathbf{H} \cdot \nabla \times \delta \mathbf{A} \, dV. \tag{C.22}$$

By using the divergence theorem the first term on the right-hand side can be expressed as a surface integral which will vanish as the surface extends to infinity, because the product $\mathbf{H} \times \delta \mathbf{A}$ will vanish faster than r^{-2}. In the second term we substitute $\nabla \times \delta \mathbf{A} = \delta \mathbf{B}$ and thus we have

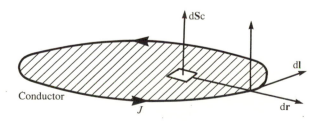

Fig. C.2 Defintion of the surface element $\mathbf{dS}_c$. The vectors **dr**, **dl** and $\mathbf{dS}_c$ form a right-handed triad.

$$\delta W = J \delta \Phi = \int\limits_{\text{all space}} \mathbf{H} \cdot \delta \mathbf{B} \, dV. \tag{C.23}$$

Similarly it may be shown that

$$\Phi \, \delta J = \int\limits_{\text{all space}} \mathbf{B} \cdot \delta \mathbf{H} \, dV, \tag{C.24}$$

and

$$J\Phi = \int\limits_{\text{all space}} \mathbf{H} \cdot \mathbf{B} \, dV. \tag{C.25}$$

Appendix D

Partial derivatives

Many thermodynamic relations are naturally expressed in terms of partial derivatives of functions of several variables. In this appendix the principal results used in this book are introduced.

Explicit functions

Let z be a continuous function of the independent variables x and y. We assume that the derivatives of z, which we define below, are also continuous. We shall consider only real functions of real variables and we shall use z to denote both the function and its value:

$$z = z(x, y). \tag{D.1}$$

Here z, is an *explicit function* of x and y; it can be represented by a surface in Cartesian coordinates, as in Fig. D.1. Consider the intersection of this surface with a plane surface parallel to the z-x plane. The plane cuts the y-axis at y. The gradient of the line of intersection is given by the partial derivative of z with respect to x, defined as

$$\frac{\partial z}{\partial x} = \left(\frac{\partial z}{\partial x}\right)_y = \lim_{\delta x \to 0} \left(\frac{z(x + \delta x, y) - z(x, y)}{\delta x}\right), \tag{D.2}$$

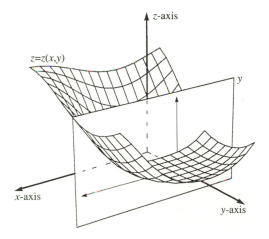

Fig. D.1   Illustrating the intersection between a plane parallel to the z-x plane and a surface $z = z(x, y)$.

where the subscript indicates that y is held constant. For brevity we shall follow the convention that it is not always necessary to display the subscript when the variables of an explicit function are held constant.

Second-order and mixed derivatives are defined by applying eqn (D.2) successively. For example,

$$\frac{\partial^2 z}{\partial x^2} = \frac{\partial}{\partial x}\left(\frac{\partial z}{\partial x}\right), \tag{D.3}$$

$$\frac{\partial^2 z}{\partial x \partial y} = \frac{\partial}{\partial x}\left(\frac{\partial z}{\partial y}\right). \tag{D.4}$$

Provided the derivatives of z are continuous, the order in which the mixed derivatives are calculated is irrelevant. This result, which is the basis for Maxwell's relations (Section 10.4), may be established by taking a Taylor expansion of $z(x + dx, y + dy)$ in two ways.

1. First expand about the point $(x, y + dy)$:

$$z(x + dx, y + dy) = z(x, y + dy) + \frac{\partial z(x, y + dy)}{\partial x}\,dx$$

$$+ \frac{1}{2!}\frac{\partial^2 z(x, y + dy)}{\partial x^2}\,(dx)^2 + \cdots. \tag{D.5}$$

Then expand $z(x, y + dy)$ about the point (x, y):

$$z(x, y + dy) = z(x, y) + \frac{\partial z(x, y)}{\partial y}\,dy + \frac{1}{2!}\frac{\partial^2 z(x, y)}{\partial y^2}\,(dy)^2 + \cdots, \tag{D.6}$$

and substituting into (D.5), we have

$$z(x + dx, y + dy) = z(x, y) + \frac{\partial z(x, y)}{\partial x}\,dx + \frac{\partial z(x, y)}{\partial y}\,dy$$

$$+ \frac{1}{2!}\left(\frac{\partial^2 z(x, y)}{\partial x^2}\,(dx)^2 + 2\frac{\partial^2 z(x, y)}{\partial x \partial y}\,dx\,dy + \frac{\partial^2 z(x, y)}{\partial y^2}\,(dy)^2\right) + \cdots. \tag{D.7}$$

2. Alternatively we may expand $z(x + dx, y + dy)$ first about the point $(x + dx, y)$, and then expand $z(x + dx, y)$ about the point (x, y). Up to second order terms, the result is the same as in (D.7), except the order of the mixed derivative is reversed. But given that z and its derivatives are continuous, the route by which we expand $z(x + dx, y + dy)$ is immaterial. We conclude that

$$\frac{\partial^2 z}{\partial x \partial y} = \frac{\partial^2 z}{\partial y \partial x}. \tag{D.8}$$

Differentials

Given increments in x and y, dx and dy, we define the differentials of z by means of the Taylor expansion, eqn (D.7):

$$z(x + dx, y + dy) - z(x, y) = dz + \frac{1}{2!}d^2z + \frac{1}{3!}d^3z + \cdots, \quad \text{(D.9)}$$

where dz, d^2z, d^3z are the first, second, ..., order differentials of z. From (D.7),

$$dz = \frac{\partial z}{\partial x}dx + \frac{\partial z}{\partial y}dy \quad \text{(D.10)}$$

$$d^2z = \frac{\partial^2 z}{\partial x^2}(dx)^2 + 2\frac{\partial^2 z}{\partial x \partial y}dx\,dy + \frac{\partial^2 z}{\partial y^2}(dy)^2 \quad \text{(D.11)}$$

$$d^3z = \frac{\partial^3 z}{\partial x^3}(dx)^3 + 3\frac{\partial^3 z}{\partial x^2 \partial y}(dx)^2\,dy + 3\frac{\partial^3 z}{\partial x \partial y^2}dx(dy)^2 + \frac{\partial^3 z}{\partial y^3}(dy)^3. \quad \text{(D.12)}$$

These equations hold whether or not x and y are independent.

Implicit functions

Instead of writing an explicit function $z = z(x, y)$ for the relation between x, y, and z, we may find it necessary to use an implicit expression such as

$$f(x, y, z) = 0. \quad \text{(D.13)}$$

Since f is an explicit function of (x, y, z), and since the differential df vanishes: $df = 0$,

$$\left(\frac{\partial f}{\partial x}\right)_{y,z}dx + \left(\frac{\partial f}{\partial y}\right)_{z,x}dy + \left(\frac{\partial f}{\partial z}\right)_{x,y}dz = 0. \quad \text{(D.14)}$$

If we now set $dz = 0$, and divide throughout by dy, we have

$$\left(\frac{\partial f}{\partial x}\right)_{y,z}\left(\frac{\partial x}{\partial y}\right)_{f,z} + \left(\frac{\partial f}{\partial y}\right)_{x,z} = 0, \quad \text{(D.15)}$$

or

$$\left(\frac{\partial x}{\partial y}\right)_{f,z} = -\frac{\left(\frac{\partial f}{\partial y}\right)_{x,z}}{\left(\frac{\partial f}{\partial x}\right)_{y,z}}. \quad \text{(D.16)}$$

This relation, called the *quotient rule*, is particularly useful. First, by swapping x and y it is clear that

$$\left(\frac{\partial y}{\partial x}\right)_{f,z} = \frac{1}{\left(\frac{\partial x}{\partial y}\right)_{f,z}}. \quad \text{(D.17)}$$

Second, by cyclic rotation of x, y and z in eqn (D.16) we obtain

$$\left(\frac{\partial y}{\partial z}\right)_{f,x} = -\frac{\left(\dfrac{\partial f}{\partial z}\right)_{x,y}}{\left(\dfrac{\partial f}{\partial y}\right)_{z,x}}, \tag{D.18}$$

$$\left(\frac{\partial z}{\partial x}\right)_{f,y} = -\frac{\left(\dfrac{\partial f}{\partial x}\right)_{y,z}}{\left(\dfrac{\partial f}{\partial z}\right)_{x,y}}, \tag{D.19}$$

and thus

$$\left(\frac{\partial x}{\partial y}\right)_{f,z} \left(\frac{\partial y}{\partial z}\right)_{f,x} \left(\frac{\partial z}{\partial x}\right)_{f,y} = -1. \tag{D.20}$$

Change of variables

Consider the explicit relation $z = z(x, y)$ and suppose that x and y can be expressed as explicit functions of the variables u and v:

$$x = x(u, v), \qquad y = y(u, v). \tag{D.21}$$

Then the differential $\mathrm{d}z$ is

$$\mathrm{d}z = \left(\frac{\partial z}{\partial x}\right)_y \mathrm{d}x + \left(\frac{\partial z}{\partial y}\right)_x \mathrm{d}y, \tag{D.22}$$

where

$$\mathrm{d}x = \left(\frac{\partial x}{\partial u}\right)_v \mathrm{d}u + \left(\frac{\partial x}{\partial v}\right)_u \mathrm{d}v, \tag{D.23}$$

and

$$\mathrm{d}y = \left(\frac{\partial y}{\partial u}\right)_v \mathrm{d}u + \left(\frac{\partial y}{\partial v}\right)_u \mathrm{d}v, \tag{D.24}$$

On substituting in (D.22) we find

$$\mathrm{d}z = \left(\frac{\partial z}{\partial u}\right)_v \mathrm{d}u + \left(\frac{\partial z}{\partial v}\right)_u \mathrm{d}v, \tag{D.25}$$

where z is treated as an explicit function of u and v, and

$$\left(\frac{\partial z}{\partial u}\right)_v = \left(\frac{\partial z}{\partial x}\right)_y \left(\frac{\partial x}{\partial u}\right)_v + \left(\frac{\partial z}{\partial y}\right)_x \left(\frac{\partial y}{\partial u}\right)_v, \tag{D.26}$$

$$\left(\frac{\partial z}{\partial v}\right)_u = \left(\frac{\partial z}{\partial x}\right)_y \left(\frac{\partial x}{\partial v}\right)_u + \left(\frac{\partial z}{\partial y}\right)_x \left(\frac{\partial y}{\partial v}\right)_u. \tag{D.27}$$

Another useful relation follows from (D.23) and (D.24) if we set $\mathrm{d}v = 0$, and divide (D.23) by (D.24):

$$\left(\frac{\partial x}{\partial y}\right)_v = \frac{\left(\frac{\partial x}{\partial u}\right)_v}{\left(\frac{\partial y}{\partial u}\right)_v} = \left(\frac{\partial x}{\partial u}\right)_v \left(\frac{\partial u}{\partial y}\right)_v. \tag{D.28}$$

This expression should be compared with eqn (D.16); note that the same variable is fixed in all derivatives in (D.28).

Finally, consider the special case in which $x = x(u, y)$ and $y = v$; from (D.27) we have

$$\left(\frac{\partial z}{\partial y}\right)_u = \left(\frac{\partial z}{\partial x}\right)_y \left(\frac{\partial x}{\partial y}\right)_u + \left(\frac{\partial z}{\partial y}\right)_x. \tag{D.29}$$

Note that on the left-hand side z is an explicit function of u and y, whereas on the right-hand side z is an explicit function of x and y, and x is an explicit function of u and y.

Jacobians

The Jacobian method offers a valuable extension of the procedures for partial derivatives. Suppose x and y are explicit functions of u and v. For simplicity we shall denote $(\partial x/\partial u)_v = x_u$, and so on, the constant parameter being implied. The Jacobian is defined by the determinant

$$\frac{\partial(x, y)}{\partial(u, v)} = \begin{vmatrix} x_u & x_v \\ y_u & y_v \end{vmatrix} = x_u y_v - y_u x_v. \tag{D.30}$$

The following properties derive directly from this definition:

$$\frac{\partial(x, y)}{\partial(x, y)} = \begin{vmatrix} x_x & x_y \\ y_x & y_y \end{vmatrix} = \begin{vmatrix} 1 & 0 \\ 0 & 1 \end{vmatrix} = 1. \tag{D.31}$$

$$\frac{\partial(x, y)}{\partial(u, v)} = -\frac{\partial(y, x)}{\partial(u, v)} = \frac{\partial(-y, x)}{\partial(u, v)}. \tag{D.32}$$

$$\frac{\partial(x, y)}{\partial(u, y)} = \begin{vmatrix} x_u & x_y \\ y_u & y_y \end{vmatrix} = \begin{vmatrix} x_u & x_y \\ 0 & 1 \end{vmatrix} = \left(\frac{\partial x}{\partial u}\right)_y. \tag{D.33}$$

In eqn (D.33) we have taken $v = y$, so $y_u = (\partial y, \partial u)_y = 0$ and $y_y = (\partial y, \partial y)_u = 1$. Now suppose that u and v are explicit functions of r and s. Consider the product

$$\frac{\partial(x, y)}{\partial(u, v)} \frac{\partial(u, v)}{\partial(r, s)} = \begin{vmatrix} x_u & x_v \\ y_u & y_v \end{vmatrix} \begin{vmatrix} u_r & u_s \\ v_r & v_s \end{vmatrix} = \begin{vmatrix} a & b \\ c & d \end{vmatrix},$$

where, for example, $a = x_u u_r + x_v v_r$, and similarly for b, c, and d. But if we regard x as an explicit function of r and s, we obtain from eqn (D.26),

$$\left(\frac{\partial x}{\partial r}\right)_s = \left(\frac{\partial x}{\partial u}\right)_v \left(\frac{\partial u}{\partial r}\right)_s + \left(\frac{\partial x}{\partial v}\right)_u \left(\frac{\partial v}{\partial r}\right)_s = x_u u_r + x_v v_r = a.$$

Thus writing $x_r = (\partial x/\partial r)_s$, and so on, we get

$$\frac{\partial(x, y)}{\partial(u, v)} \frac{\partial(u, v)}{\partial(r, s)} = \begin{vmatrix} x_r & x_s \\ y_r & y_s \end{vmatrix} = \frac{\partial(x, y)}{\partial(r, s)}. \tag{D.34}$$

Equation (D.34) may be used to demonstrate that

$$\frac{\partial(x, y)}{\partial(u, v)} \frac{\partial(p, q)}{\partial(r, s)} = \frac{\partial(x, y)}{\partial(r, s)} \frac{\partial(p, q)}{\partial(u, v)}. \tag{D.35}$$

Thus Jacobians behave rather like derivatives. This brief survey has been restricted to second-order Jacobians, but the results are applicable to higher orders as well. For example, the third-order Jacobian is defined as

$$\frac{\partial(x, y, z)}{\partial(u, v, w)} = \begin{vmatrix} x_u & x_v & x_w \\ y_u & y_v & y_w \\ z_u & z_v & z_w \end{vmatrix}, \tag{D.36}$$

and the generalization to higher orders is straightforward.

Pfaffian forms

The differential equation

$$df = X\,dx + Y\,dy + Z\,dz, \tag{D.37}$$

is known as a *Pfaffian form*. In general the integral $\int_1^2 df$ depends on the path of integration, in which case df is said to be an *inexact differential*. But the integral will be path independent if f can be expressed as a single valued function, $f = f(x, y, z)$, in which case

$$\int_1^2 df = f(x_2, y_2, z_2) - f(x_1, y_1, z_1). \tag{D.38}$$

The differential df is then said to be an *exact differential*. If we now write

$$X = \left(\frac{\partial f}{\partial x}\right)_{y, z}, \qquad Y = \left(\frac{\partial f}{\partial y}\right)_{z, x}, \qquad Z = \left(\frac{\partial f}{\partial z}\right)_{x, y}, \tag{D.39}$$

it follows from eqn (D.8) that

$$\left(\frac{\partial X}{\partial y}\right)_{x, z} = \left(\frac{\partial Y}{\partial x}\right)_{y, z}, \qquad \left(\frac{\partial Y}{\partial z}\right)_{y, x} = \left(\frac{\partial Z}{\partial y}\right)_{z, x}, \qquad \left(\frac{\partial Z}{\partial x}\right)_{z, y} = \left(\frac{\partial X}{\partial z}\right)_{x, y}. \tag{D.40}$$

These equalities represent the Maxwell relations (Section 10.4). Equation (D.40) is a necessary and sufficient condition for the Pfaffian form (D.37), to be an exact differential.

Answers to questions

Chapter 2

2.1.1	4.14×10^{-23} Pa
2.1.2	(a) 19.7 m (b) rise
2.1.3	(b) 100.00°C
2.1.4	(a) 226.49 kPa 309.88 kPa (b) 100.58°C
2.1.5	4.76 kPa
2.1.6	6 mg
2.1.7	(a) 37 700 K (b) 5.3×10^9 Pa
2.3.1	245 ms^{-1}
2.3.2	(a) 1.35 km s^{-1}, 11.2 km s^{-1}
2.3.3	(a) 300 K (b) 33.3 kPa
2.3.4	336.0 K, 1.17 MPa
2.3.6	141 600 K

Chapter 3

3.2.4	450 kJ, -600 kJ, 557 kJ, 100 kJ, -400 kJ, 5.36 kJ
3.2.5	(a) $W = 17.7$ J, 0, -9.9 J
	(b) $\Delta U = 0$, -1.8 J, 1.8 J. No
3.5.1	(a) Yes (b) No (c) No
3.5.3	-1125 kJ, 1875 kJ, -1157 kJ, 50 kJ, 400 kJ, -3.36 kJ
3.5.6	(a) 1500 K, 476.2 K, 1.76 kJ, 10.24 kJ
	(b) 1.76 kJ (c) $c \rightarrow b$ is non-quasistatic
3.5.7	(a) 55.9 kPa, 0.179 m^2, 17.21 kJ, -17.21 kJ
	(b) 12.00 kJ
3.5.9	A: 3.74 kJ, 0, -3.74 kJ
	B: -14.97 kJ, -9.98 kJ, 4.99 kJ
	C: 9.98 kJ, 9.98 kJ, 0
3.5.10	(a) 1347 K (b) 729 K (c) T would increase
3.5.11	(a) 113.1 litre (b) 12.48 kJ (c) 0
3.5.12	0.6 m^3, -20 kJ
3.5.13	1.4
3.5.14	(a) 257 kJ (c) 323 kJ
3.6.1	(a) 0.714 m, 84.12 J, -84.12 J
	(b) 0.714 m, 100 J, -100 J

	(c)	0.817 m,	54.0 J,	0
	(d)	0.829 m,	60.0 J,	0
3.6.2	(a)	non-quasistatic	(b) 7.5 Mpa	(c) 0
	(d)	0	(e) No	
3.6.3	(a)	0.6684 m	(b) −22.34 kJ	(c) 22.34 kJ
	(d)	non-quasistatic		
3.6.4	(a)	0.6684 m		

3.6.4 (b) 22.34 kJ (on spring), −48.38 kJ (on gas),
26.04 kJ (against external force)

(c) 48.38 kJ

3.6.5 (a) 271.67 K, 0.6129 m

(b) 18.78 kJ (on spring), −42.50 kJ (on gas),
23.72 kJ (against external force)

3.6.6 (a) 286.28 K 0.6417 m

(b) 20.59 kJ (on spring), −20.59 kJ (on gas)

3.6.7 (a) 2.5 kPa, 5 kPa, 0.3 m

(b) −30 J (c) 30 J

(d) The process is non-quasistatic. The work cannot be obtained
from the end states alone.

3.6.8 (a) 2.5 kPa, 5 kPa, 0.3 m

(b) −48.5 J (c) 48.5 J

(d) −27.7 J, −20.8 J

3.6.9 (a) 167.97 K, 232.86 K, 2.35 kPa, 5.31 kPa,
0.139 m

(b) 13.20 J, 10.07 J

(c) The process is non-quasistatic. The work done on the
individual cylinders of gas cannot be expressed in terms
of the initial and final elevation.

Chapter 4

4.1.2 (a) $T_b = 1500$ K, $T_c = 476.2$ K, $T_d = 1500$ K,
$T_e = 753.6$ K, $V_d = 0.179 \text{m}^3$

(c), (d), (e) 11.47 J K^{-1} (f) 17.21 kJ, 11.47 J K^{-1}

(g) $b \rightarrow c$ is a non-quasistatic process

4.1.3 Q: −11 kJ, −10 kJ, 6 kJ, −6 kJ, 785 J, 6 kJ
W: 17.2 kJ, 16.2 kJ, −24.7 kJ, 6 kJ, −785 J,
6.5 kJ

4.1.4 (a) 1.68 kJ K^{-1} (b) 2.17 kJ K^{-1} (c) 1.323, 548.0 K

(d) 539.7 K

4.2.1 (a) 300 K

(b) −12.74 J K^{-1}, 20.23 J K^{-1}, 7.49 J K^{-1}

4.2.2 (a) 300 K (b) 998 kPa, 5 litre, 10 litre

(c) 19.0 J K^{-1}

4.3.2 (b) External work done: 1.15 kJ

Chapter 5

5.5.2	0.03 m^3						
5.5.3	(a)	150°C, no	(b)	0.022 m^3, 8 m^3, 4 m^3		(c)	-4 MJ

5.5.3 (d) 38 MJ (e) 42 MJ (f) 75°C, 18 kg (g) 38 MJ
5.5.4 (a) 0.02, 0.996 (b) 250°C, 4 MPa, 4.8 MJ, 0
5.5.5 (a) -380 kJ, 0 (b) -410 kJ, 29 kJ
5.5.6 (b) 1700 kg
5.5.7 (a) 33°C (b) -2.4 MJ
5.5.8 200°C
5.5.9 (a) 15 MJ (b) 20 kg (c) 470 kJ (d) 78 kW, 0.03
5.7.1 (b) 100°C (c) 160°C
5.7.2 (a) 70 MW (b) 15 MW
5.7.3 (a) 2.4 MW (b) 0.46 MW (c) 130 m^3 s^{-1}, 1.3 m
5.7.4 (a) 6.9 MW (b) 0.47 MW, 270 W (c) 0.35 MW
5.7.4 (d) Heat input is approximately 13 times the compressor work input.
5.7.5 (a) High-pressure side 0.497 kg s^{-1}; low-pressure side 0.503 kg s^{-1}
5.7.5 (b) 1.14 MW, 2.26 MW (c) 100 W
5.8.1 (b) 362 ms^{-1}
5.8.2 (a) 126 kW (b) -62 kW
5.8.5 (a) 8.25 mole (b) 485 K

Chapter 6

6.5.1 (a) 417 MW, 500 MW, 82.6 MW
6.5.1 (b) 454 MW, 500 MW, 46.4 MW
6.5.1 (c) (i) irreversible, (ii), (iii), (v) unphysical, (iv) reversible
6.5.2 (c) (i), (ii) unphysical, (iii), (iv), (v) irreversible
6.5.3 (c) (i), (iv) reversible, (ii), (iii) unphysical, (v) irreversible
6.5.5 (a) 98 kJ (b) 2.23 MJ (c) 24.5 MJ (d) 5.60 MJ
6.7.2 (a) 250°C, 36 MJ, -26.5 MJ (b) 16.5 MJ, -7 MJ
6.7.3 (a) 430 K (b) 4.8 MPa
6.7.5 (a) *ab*: 1.42 MJ, -0.15 MJ *bc*: 0, -0.59 MJ
6.7.5 *cd*: -0.97 MJ, 0.06 MJ *da*: 0, 0.23 MJ
6.7.5 (b) 0.317
6.7.7 0.03 m^3, 500°C, 1.2 MJ
6.7.8 (a) 9 m, 36 kJ (b) 27 ms^{-1}
6.8.1 3.8 kJ K^{-1}
6.8.2 5.3 kJ K^{-1}
6.8.3 (a) 337.9 K (b) -168.7 J K^{-1}, 194.8 J K^{-1}, 26.1 J K^{-1}
6.8.6 6.8 kJ K^{-1} s^{-1}
6.8.7 98%

Chapter 7

7.2.1   (a) -3.46 kJ   (b) -2.88 kJ

7.2.2   (a) -1.25 kJ, -1.74 kJ, -2.99 kJ

        (b) -0.11 kJ, -2.88 kJ, -2.99 kJ

7.2.3   388 kJ

7.2.5   (a) 2.56 kJ K^{-1} s^{-1} (b) 721 kW   (c) 86 m

7.2.6   (a) $\left\{ \dfrac{aT_1^2 + bT_2^2}{a+b} \right\}^{0.5}$ (b) $\dfrac{aT_1 + bT_2}{a+b}$, $\dfrac{-ab}{2(a+b)}(T_1 - T_2)^2$

        (c) $T_0\{aT_1 + bT_2 - \sqrt{(a+b)(aT_1^2 + bT_2^2)}\}$

7.2.7   (a) $\dfrac{T_1^2}{T_2}$

7.2.8   (a) $\dfrac{v^2}{2c}$, $mc\ln\left\{1 + \dfrac{v^2}{2cT_0}\right\}$ (b) $mcT_0\ln\left\{1 + \dfrac{v^2}{2cT_0}\right\}$

7.2.9   176 K, $\dfrac{\delta u}{\delta s} = 176$ K for every step of the process

7.3.2   (a) 63.04 MW   (b) 71.18 MW   (c) 8.14 MW

7.3.3   (b) 481.4 MW   (c) 30.7 MW   (d) 0.068, 0.064

7.3.4   (a) 68.31 MW   (b) 82.56 MW   (c) 14.25 MW

        (d) 75.00 MW   (e) 7.55 MW   (f) 75.00 MW

Chapter 9

9.3.1   (8) $P = \dfrac{2}{3}A\left(\dfrac{3}{5}T\right)^{5/2}$, $U = AV\left(\dfrac{3}{5}T\right)^{5/2}$

        where $A = \left(\dfrac{7.46\pi\, mk}{h^3}\right)^{3/2}$

       (10) $PV = NRT$, $U = \dfrac{r}{2}NRT$

       (12) $PV = \dfrac{NA}{2}\left\{1 + \left(\dfrac{\pi RT}{A}\right)^2\right\}$, $U = \dfrac{3NA}{2}\left\{1 + \left(\dfrac{\pi RT}{A}\right)^2\right\}$

        where $A = \left(\dfrac{\sigma N}{V}\right)^{1/3}$

9.3.5   (a) Not homogeneous of first order; S diverges when $T \to 0$.

        (b) Not homogeneous; S diverges when $T \to 0$.

        (c) U not a monotonic increasing function of S.

        (d) S not a monotonic increasing function of U.

9.7.1   (a) $\dfrac{3m_1 + m_2}{4m_2}$; $\dfrac{3m_1 + m_2}{4m_1}$

(b) $\left(\dfrac{m_1}{m_2}\right)^{3/4}$; $\left(\dfrac{m_2}{m_1}\right)^{1/4}$

Chapter 10

10.3.2 An ideal gas in two dimensions (example 10, p. 194).

Chapter 11

11.4.7 (a) 1.45×10^{11} Pa, 14.8 MJ, 388 kJ (b) 522 K (c) 0.019
11.5.5 (d) 0.11

Chapter 12

12.1.2 35 Pa
12.1.3 Yes; above $-4.7\,°C$ the equilibrium state is the liquid phase at the pressure under the skate.
12.1.6 6.67×10^{-4}
12.3.1 c_v, c_p, κ_T, κ_s, must all be positive.

References

Abramowitz, M. and Stegun, I. A. (ed) (1970). *Handbook of mathematical functions with formulas, graphs, and mathematical tables*. National Bureau of Standards Applied Mathematics Series, 55. US Department of Commerce.

Adkins, C. J. (1983). *Equilibrium thermodynamics*, (3rd edn). Cambridge University Press.

Anderson, O. L. (1966). J. Phys. Chem. Solids, **27**, 547–65.

Bleaney, B. I. and Bleaney, B. (1976). *Electricity and magnetism*. Oxford University Press.

Callen, H. B. (1985). *Thermodynamics and an introduction to thermostatistics* (2nd edn). Wiley, New York.

Carathéodory, C. (1909). Math. Ann., **67**, 355.

Carnot, S. (1824). *Reflections on the motive power of heat and on machines fitted to develop this power*, (trans. R. H. Thurston, 1890). New York.

Clausius, R. (1850). Annalen der Physik und Chemie, **79**, 368–97, 500–24.

Clausius, R. (1865). Annalen der Physik und Chemie, **125**, 351–400.

Clausius, R. (1867). *Mechanical theory of heat with its applications to the steam engine and to the physical properties of bodies*. Collected papers and notes in English translation, (ed. T. A. Hirst, trans. J. Tyndall). J. van Voorst, London.

Denbigh, K. G. (1968). *The principles of chemical equilibrium*. Cambridge University Press.

Dugdale, J. S. and MacDonald, D. K. C. (1953). Phys. Rev., **89**, 832–4.

Einstein, A. (1979). *Autobiographical notes*, (A centennial edition, trans. and ed. P. A. Schilpp). Open Court, La Salle & Chicago, Illinois.

Endem, R. (1938), Nature, **141**, 908.

Fowler, R. H. and Guggenheim, E. A. (1939). *Statistical thermodynamics*. Cambridge University Press.

Galgani, L. and Scotti, A. (1968), Physica, **40**, 150–2.

Gibbs, J. W. (1873). *Graphical methods in the thermodynamics of fluids*, Transactions Connecticut Acad. II, 309. (Reprinted (1961) in *The scientific papers of J. Willard Gibbs, PhD, LLD*. Vol I. Dover Publications, New York.)

Guggenheim, E. A. (1967). *Thermodynamics*, (5th edn). North-Holland, Amsterdam.

Haar, L., Gallagher, J. S., and Kell, G. S. (1984). *NBS/NRC steam tables*. Hemisphere, Washington.

Haywood, R. W. (1980). *Equilibrium thermodynamics for engineers and scientists*. Wiley, Chichester.

Heller, P. (1967). Reports on Progress in Physics, **XXX**, 731–826.

Jayaraman, A. (1986). Review of Scientific Instruments, **57**, 1013–31.

Jones, R. V. (1970). Notes and Records of the Royal Society of London, **24**, No 2, 194–220.

Kaye, G. W. C. and Laby, T. H. (1959). *Tables of physical and chemical constants*. Longmans, Green and Co., London.

Korn, G. A. and Korn, T. M. (1968). *Mathematical handbook for scientists and engineers*, (2nd edn). McGraw-Hill, New York.

Kubo, R. (1976). *Thermodynamics. An advanced course with problems and solutions*. North-Holland, Amsterdam.

Landau, L. D. and Lifshitz, E. M. (1980). *Statistical physics*, (3rd edn, revised and enlarged by E. M. Lifshitz and L. P. Pitaevskii, trans. J. B. Sykes and M. J. Kearsley). Pergamon Press, Oxford.

Landsberg, P. T. (1961). *Thermodynamics with quantum statistical illustrations*. Interscience, New York.

Landsberg, P. T. (1978). *Thermodynamics and statistical mechanics*. Oxford University Press.

Margenau, H. and Murphy, G. M. (1956). *The mathematics of physics and chemistry*. Van Nostrand, Princeton, New Jersey.

Massieu, F. (1869). C. R. Acad. Sci., Paris, **69**, 858–62.

Maxwell, J. C. (1872). *Theory of heat*. Longmans Green, London.

McLellan, A. G. (1980). *The classical theory of deformable solids*. Cambridge University Press.

Michels, A., Levelt, J. M., and de Graaff, W. (1958). Physica, **XXIV**, 659–71.

Perry, R. H. and Chilton, C. H. (1973). *Chemical engineers handbook*, (5th edn). McGraw-Hill Kogakusha, Tokyo.

Pippard, A. B. (1964). *The elements of classical thermodynamics*. Cambridge University Press.

Planck, M. (1897). *Treatise on thermodynamics*, (trans. A. Ogg). Longmans Green, London.

Rankine, W. J. M. (1859). *A manual of the steam engine and other prime movers*. Charles Griffin, London.

Redlich, O. (1968). Rev. Mod. Phys., **40**, 556–63.

Reichl, L. E. (1980). *A modern course in statistical physics*. University of Texas Press and Edward Arnold.

Reid, R. C., Prausnitz, J. M., and Poling, B. E. (1986). *Properties of gases and liquids*, (4th edn). McGraw-Hill, New York.

Reif, F. (1965). *Fundamentals of statistical and thermal physics*. McGraw-Hill International, Singapore.

Riedi, P. C. (1988). *Thermal physics. An introduction to thermodynamics, statistical mechanics and kinetic theory*, (2nd edn). Oxford University Press.

Roberts, J. K. (1940). *Heat and thermodynamics*, (3rd edn). Blackie and Son, London and Glasgow.

Robinson, F. N. H. (1973). *Electromagnetism*. Oxford University Press.

Rosenberg, H. M. (1963). *Low temperature solid state physics*. Oxford University Press.

Rosenberg, H. M. (1988). *The solid state*, (3rd edn). Oxford University Press.

Rowlinson, J. S. (1988). *J. D. van der Waals: On the continuity of the gaseous and liquid states*. Studies in Statistical Mechanics, Vol XIV. North Holland, Amsterdam.

Schmidt, E. (1982). *Properties of water and steam in SI-units*, (ed. U. Grigull). Springer-Verlag, Berlin and R. Oldenbourg, Munich.

Shapiro, S. L. and Teukolsky, S. A. (1983). *Black holes, white dwarfs, and neutron stars. The physics of compact objects*. Wiley Interscience, New York.

Stanley, H. E. (1971). *Introduction to phase transitions and critical phenomena*. Oxford University Press.

Thomson, W. (Lord Kelvin) (1848). Proceedings Cambridge Philosophical Soc., **1**, 66–71.

Thomson, W. (Lord Kelvin) (1851). Transactions of the Royal Society of Edinburgh, **20**, 261–8, 289–98.

Tisza, L. (1966). *Generalized thermodynamics*. M1T Press, Cambridge, Mass. (Also, (1961). Annals of Physics, **13**, 1–92.)

Treloar, L. R. G. (1973). Reports on Progress in Physics, **36**, 755–826.

Waldram, J. R. (1985). *The theory of thermodynamics*. Cambridge University Press.

Wang, Z. X. and Guo, D. R. (1989). *Special functions*. World Scientific Publishing, Singapore.

Wilks, J. (1961). *The third law of thermodynamics*. Oxford University Press.

Zemansky, M. W. and Dittman, R. H. (1981). *Heat and thermodynamics. An intermediate textbook*, (6th edn). McGraw-Hill International, Tokyo.

Index